Mixed Messages

Mixed Messages

*Cultural and Genetic Inheritance in the
Constitution of Human Society*

ROBERT A. PAUL

The University of Chicago Press   Chicago and London

ROBERT A. PAUL is the Charles Howard Candler Professor of Anthropology and Interdisciplinary Studies at Emory University. He is the author of *Moses and Civilization* and *The Tibetan Symbolic World*, the latter published by the University of Chicago Press.

The University of Chicago Press, Chicago 60637
The University of Chicago Press, Ltd., London
© 2015 by The University of Chicago
All rights reserved. Published 2015.
Printed in the United States of America

24 23 22 21 20 19 18 17 16 15   1 2 3 4 5

ISBN-13: 978-0-226-24072-5 (cloth)
ISBN-13: 978-0-226-24086-2 (paper)
ISBN-13: 978-0-226-24105-0 (e-book)
DOI: 10.7208/chicago/9780226241050.001.0001

Library of Congress Cataloging-in-Publication Data

Paul, Robert A., author.
 Mixed messages : cultural and genetic inheritance in the constitution of human society / Robert A. Paul.
   pages ; cm
 Includes bibliographical references and index.
 ISBN 978-0-226-24072-5) cloth : alk. paper) — ISBN 978-0-226-24086-2 (pbk. : alk. paper) — ISBN 978-0-226-24105-0 (e-book)
 1. Nature and nurture. 2. Heredity. 3. Behavior genetics.
4. Sociobiology. I. Title.
 BF341.P38 2015
 155.7—dc23
                                            2014043939

♾ This paper meets the requirements of ANSI/NISO Z39.48-1992 (Permanence of Paper).

To the memory of Lois and Ben Paul

Contents

Acknowledgments ix

Introduction: The Social Consequences of Dual Inheritance 1

1. A Brief History and Outline of Dual Inheritance Theory 14
2. When Genetic and Cultural Reproduction Diverge 40
3. The Cultural Channel 61
4. Two Kinds of Sociality 87
5. Society beyond the Genetic Program 116
6. The Asymmetry of Cultural versus Genetic Reproduction 152
7. The Society of Men 176
8. Symbolic Reproduction and Reproductive Symbolism 202
9. Beyond Gender Asymmetry and Male Privilege 230
10. Sources of Human Sociality 255

Conclusion: The Giant Yams of Pohnpei 283

Notes 311   References 323   Index 341

Acknowledgments

Kenneth Burke, the great theorist of symbolic action, spent some of his last winters escaping from the northern chill as a scholar in residence at the Graduate Institute of Liberal Arts—the ILA—at Emory University in Atlanta, where I have spent my own academic career since 1977. On one occasion, Burke, who had originally included a much longer "Definition of Man" in his book *Language as Symbolic Action*, remarked that he had now gotten it down to four words: "animals that learn language." That, in a nutshell, is the premise of this book, which takes seriously the fact that we humans are animals and that we have added to our animal state the ability to form ourselves through systems of symbols (of which language is a prototypical exemplar) without ceasing to be animals.

The ILA, in addition to allowing me to rub shoulders with giants such as Burke, has more modestly in my case been a nurturing and liberal intellectual environment in which I have been able to pursue an interdisciplinary path that has included both anthropology and psychoanalysis. I want to express my gratitude to the ILA and to its current director, Kevin Corrigan, for support through the years. I submit this book as a final tribute to a unique academic institution as it closes its doors after sixty years of scholarly ferment and creativity.

It was Ralph Roughton, then director of the Emory University Psychoanalytic Institute, who first encouraged me to undertake training as an analyst, and this set me midcareer on a new path that has been infinitely rewarding. It

ACKNOWLEDGMENTS

will be obvious to many readers that while the present book is focused almost entirely on anthropological concerns, with little direct mention of psychoanalysis, the entire conception is infused with a deeply Freudian perspective. I want to express my gratitude here to Ralph and to all the others in the Institute, as well as to my analysands, from whom I have learned so much.

I was fortunate to be trained in anthropology at the University of Chicago by some of the best thinkers in the field; I want to single out in particular the late Clifford Geertz and the late David Schneider. The late Mel Spiro, my main anthropological mentor, especially encouraged me on the path that my subsequent intellectual journey was to take, and he read and made valuable comments on portions of this manuscript. The influence of all three—different from each other as their fundamental orientations are—is evident throughout this book.

At Emory, I have benefited greatly from ongoing conversations and dialogues with wonderful anthropological colleagues, including Mark Auslander, the late Ivan Karp, Bruce Knauft, Ellen Schattschneider, Bradd Shore, and Carol Worthman. Mel Konner has long encouraged me to write a book presenting a synthesis of the traditions of Darwin, Durkheim, and Freud, and this is that book. Of particular value to me was my encounter, during the years he taught at Emory, with Rob Boyd, who introduced me to the theory of dual inheritance and to whom I owe a great debt, even where I have differed with him.

The people at the University of Chicago Press have made it a real pleasure to work with them in producing this book, and I want to give thanks to Richard Allen, David Brent, and Priya Nelson in particular. The reviewers for the press made unusually extensive and constructive comments.

As always, my most helpful interlocutor and critic has been my wife, Leslee Nadelson. I have dedicated this book to the memory of my mother and father, who conceived me while doing anthropological fieldwork in Guatemala and who were my first and in many ways my most influential teachers.

Stone Mountain
August 2014

Introduction: The Social Consequences of Dual Inheritance

Humans live in societies: there is no such thing as a human who is not a member of a social group. Furthermore, humans come into being by the incorporation of information that is shared within the social group, and this information in its totality is what we may call the culture of that society. Any humans who have not incorporated a vast array of the cultural information circulating in the society into which they have been born, including but certainly not limited to language, are not complete human beings.

A human society is only made possible because of the collective reality of shared cultural information, but at the same time, a human culture must be embedded in a society; the two kinds of collective phenomena are inseparable, and together they form entities we can call socio-cultural systems. Just as humans, like other living organisms, are collections of cells but have properties at their own level including the ability to reproduce as a unit, just so human societies are collective entities comprised of human beings having their own internal structure and organization and also capable of reproducing as units. One evolutionary thinker puts it this way:

> The major transitions in evolutionary units are from individual genes to networks of genes, from gene networks to bacteria-like cells, from

bacteria-like cells to eukaryotic cells with organelles, from cells to multicellular organisms, and from solitary organisms to society. (Michod 1999, 60)

At each of these higher levels, the entities in the lower levels are arranged into new systems with emergent sets of rules and modes of interaction that do not exist at the lower levels. Human socio-cultural systems are such new emergent systems and can be described and analyzed in their own terms: this is a fundamental premise of much of socio-cultural anthropology as originally set forth in the seminal formulations of Durkheim, Boas, and their followers.

At the same time, however, the humans who comprise any socio-cultural system are organisms, and as such they have been constructed according to information contained in the genetic material, encoded in the DNA of the zygotes from which they developed, like other living organisms; obviously no human can come into being without the guidance of a vast amount of information transmitted to it in the reproductive DNA of her or his parents.

We thus have, in the human case, organisms that can only come into fully realized being on the basis of the operation of two different kinds of information: that contained in the DNA, and that communicated by the society in which they are raised. Cultural information without human organisms to learn and enact it is without effect, like an indecipherable document in an unknown language lying buried the sands of a desert; but a human organism without cultural information is, as Geertz memorably put it, "not an intrinsically talented but unfulfilled ape, but a wholly mindless and consequently unworkable monstrosity" (1973a, 68).

Any human society is composed, then, of organisms that, unlike other organisms, are the product of the effects of two separate channels of information transmission, one cultural and one genetic, each necessary for the creation of the finished product, an adequately functioning human being. And while the organization of a socio-cultural system is by virtue of being an emergent unit not the same as the organization of an individual human, any more than a human organism has the same system of operation as a cell in its body, nonetheless any human society will reflect the material out of which it is constructed, and that material is a collection of beings requiring two very different kinds of instructions to create them. Therefore we may ask what sort of relationship exists between the two kinds of information, and we may further ask what impact this relationship might have on the way human socio-cultural systems are organized. That is what this book is about.

To answer these questions we must turn to two different intellectual traditions: on the one hand, the study of socio-cultural systems that historically has been the domain of socio-cultural anthropology, and on the other, the study of evolutionary theory and of genetic information and its mode of operation and transmission, which have been the province of the field of biology.

Here we run into an initial stumbling block: the fundamental premises and paradigms of these two fields have diverged radically, so that investigating the relations between socio-cultural systems of information and genetic systems of information in a unified way would require overcoming a deep disciplinary and theoretical divide. While much of socio-cultural anthropology has recognized the existence of socio-cultural systems as realities that can be understood at their own level, just as can genes, cells, and organisms, with their own internal structure and organization, evolutionary theory in biology became, with the creation of the so-called neo-Darwinian synthesis, more and more committed to a premise of theoretical and methodological individualism: when evolutionary thinkers began seriously attempting to apply their principles to society and culture in the 1970s, the collective phenomena of culture and society were typically regarded as the sum of the behaviors of the individual human actors who comprised them rather than as entities with their own forms of structure and organization. It followed therefore that, from this perspective, society and culture could be understood to be statistical patterns of collective behavior, and could be understood to serve the genetic interests of individuals comprising them just as could the behavior of any organism.

When the formulations of such thinkers as E. O. Wilson, Robert Trivers, Richard Dawkins, Leda Cosmides and John Tooby, and others began to be known outside biological circles, socio-cultural anthropologists were by and large repelled by what they perceived to be not only the reductionism of the approaches but the negation of the very existence, much less the importance, of the socio-cultural systems that were their own disciplinary bread and butter. At the same time, many evolutionist thinkers in biology who turned their hands to the study of socio-cultural systems grew exasperated with socio-cultural scholars who, as it appeared to the evolutionists, refused to recognize that human life, including its socio-cultural dimension, is an aspect of life in general, and as such something that can and should be understood within the principles of the theory of evolution—without which, as the great biologist Theodosius Dobzhansky famously put it, nothing in biology makes sense. There was thus an estrangement between socio-cultural anthropologists, on the one hand, and biological anthropologists and their evolutionist fellows in other disciplines on the other. Indeed, so-called "four-field" anthropology departments, encompassing socio-cultural, linguistic, archeological, and biological anthropology, began to split along the divide created by the commitment to, or rejection of, evolutionary theory and the scientific research methods appropriate to it versus the more interpretive, humanistic, and critical methods favored by many socio-cultural anthropologists.

The 1970s represented, in retrospect, a moment of lost opportunity: if the two sides had been able to agree that what each was studying was a kind of information, and that the interesting question was about how the two different kinds of information interacted with each other, each would have found in the other rich material for productive cross-fertilization. It was a period of creative ferment in the study of cultural information systems, in various sub-schools of socio-cultural anthropology: symbolic anthropology, ethnoscience, cognitive anthropology, structuralism, psychoanalytic anthropology, sociolinguistics, semiotics, and others, each of them focusing in different ways on some aspect of what cultures are made of, what meanings and values they construct, and how they organize information and convey it to and disseminate it among humans in society. At the same time, the explosion of understanding in the biological sciences of how biological structure and function are dependent on the transmission of information at every level, from the molecular to the physiological to the reproductive, led to an equivalent efflorescence that, if paired with the understandings of cultural information emerging in anthropology and cognate fields, could have led to a unified paradigm that might have sought to integrate biological and socio-cultural thinking under the unifying concept to "information." To his credit, Lévi-Strauss saw this quite clearly, but his quarter of an hour of influence came and went. Geertz also understood this, as I will discuss at greater length in chapter 3 of this book, though his main influence on subsequent socio-cultural anthropology, it is true, went in other directions.

But such a synthesis, though it was theoretically possible, did not occur, and instead a process along the lines of "schismogenesis" took place, in which each side reacted to the other by distancing itself further and further from the other: most evolutionary thinkers in the fields of sociobiology, evolutionary psychology, and behavioral ecology insisted that culture was, in the end, just another means by which individual humans sought reproductive advantage, and any appearance it had of entailing altruistic or prosocial motives was illusory; while socio-cultural anthropologists developed an allergic reaction to any hint of the possible relevance of biology, genetics, or evolution to their field of study.

Yet I cannot imagine that any reasonable person today would seriously dispute the fact that humans really are a hybrid creature whose self-construction from embryo to adult is completely dependent on two different channels of information, one genetic and one cultural; and that therefore any full understanding of human life, and of the socio-cultural systems that they comprise and that in turn make possible their existence, will be bound to reflect the peculiar nature of the two different channels and of the interaction between them. It is toward furthering the project of forging such a theoretical synthesis that this book is dedicated. This synthesizing project has been called "Dual Inheritance Theory."

Dual Inheritance, Fact and Theory

Just as it may be said that "evolution" is an empirical fact, while "evolution by natural selection" is a theory to explain that fact, so too, dual inheritance, by which I mean the self-creation of human beings guided by information flowing along the cultural as well as the genetic channel, is an empirical fact, about which there also ought to be a theory. And, indeed, there is such a theory. While I have painted in broad strokes a discouraging picture of the theoretical landscape in the integrative study of socio-cultural systems and biological evolution of the last four decades or so, there has at the same time emerged a countertrend, and I am by no means the first to attempt to take both the cultural and the genetic side of human life in society seriously and weave them into a research paradigm. In 1978, Peter Richerson and Robert Boyd coined the term "dual inheritance" and began formulating a theoretical and methodological approach to studying it; these two prolific scholars have, with their colleagues and students, emerged as the leading figures in a school of thought that has grown more and more influential ever since.[1]

Working within an evolutionist paradigm, Richerson and Boyd, unlike many of their evolutionist colleagues, recognized the importance and relative independence of culture and cultural variation in the determination of the course of human life in society. They saw that information transmitted through social learning is a necessary track parallel to the genetic one, and developed a set of concepts—such as

INTRODUCTION

the kind of biases that would affect the likelihood of elements of culture being transmitted from one person to another—as well as sophisticated mathematical models to explore and describe the evolutionary processes such biases in social learning might produce. Anyone, such as myself, who is interested in thinking about the dual inheritance process must begin from and build on their work. Since I come from a background in socio-cultural anthropology, and my approach arises from and remains within that tradition, the direction in which I will take the exploration of the social consequences of dual inheritance will inevitably diverge from theirs in certain respects; but my aim is to advance the dual inheritance project and try to help bridge the gap that divides socio-cultural and evolutionary approaches to the study of human life in society.

The basic premises of the dual inheritance paradigm, which I certainly agree with, and which ought to seem almost entirely uncontroversial to any reasonable person in the present intellectual climate, include the following:

- The theory of evolution is the best available set of ideas for understanding and explaining the non-human realm of living things.
- Human beings—the species *Homo sapiens*—have evolved, like other life forms, by natural selection.
- Unlike other forms of life, human life, which takes place in social groups, requires a massive quantity of additional instructions beyond what is contained in the DNA; this information is collectively the culture of the social group, and is composed to a highly significant degree of systems of symbols. "Symbols," the constituent units of much (though by no means all) of that which is transmitted via social learning, that is, culture, are bits of information just as are the codons, genes, etc. that compose the instructions inscribed in the DNA.
- Both cultural symbols and genetic information are aspects of the material world whose importance lies in the fact that they communicate messages that can influence a living organism to act—including by constructing itself—in a particular way.
- Cultures built up from symbols arranged into symbol systems differ from society to society, as do the forms of social organization that are enabled by the symbol systems.
- Genetic variation, though it very probably plays some role in the variation among human groups, is not significant enough to account for the observed variation in human socio-cultural systems. *Homo sapiens* is a single species, but one whose socio-cultural forms differ widely across time and place.

- These differences between societies and their cultural systems are thus the result of differences in the cultural symbol systems themselves, not of genetic information. To take an obvious example, all human groups have language, the capacity for which is no doubt an evolved genetic trait; but languages, which are systems of symbols, vary widely and are mutually unintelligible; this variation is in the symbols, not in the genes.
- Both genetic instructions encoded in DNA and cultural instructions encoded in symbols are required for the construction of a complete person. DNA does not give a human organism enough information for it to survive and flourish, while cultural symbols systems can only come to life through the agency of humans who have been produced as organisms through genetic reproduction and development controlled in large measure by instructions inscribed in the DNA.

I add to these premises the rather more controversial one that human socio-cultural groups are units that have their own internal structure and organization; and that these forms are encoded in and transmitted via symbol systems (along with other non-symbolic forms of social learning), and can, like other such biological units, in particular individual organisms, reproduce themselves. While this seems completely self-evident to me, I am aware that it is not a universally accepted idea; I will explain how this works and defend it further in chapters 2 and 3.

Accepting these premises, my strategy in this book is to explore the similarities and differences between cultural information and genetic information, and how each is transmitted across generational lines so that socio-cultural systems maintain themselves in form and content despite the fact that the individuals who comprise them come into and go out of existence. I will show that because of inherent differences between the two channels of information transmission across generations, the relations between them are characterized by a certain degree of opposition, conflict, or tension: their agendas are, in fact, to a significant degree at cross-purposes. Having shown why this is so in principle, I argue that this tension must be and is reflected in the forms that different socio-cultural systems take. To do this, I must undertake a project of comparative ethnography.

Evans-Pritchard observed that the only real goal of socio-cultural anthropology should be comparison, but that this is impossible. One reason it seems impossible is that there is just too much of it; over a hundred years of ethnography has produced a vast library that cannot be mastered by any one individual. Despite this obstacle, or rather by turning that to a virtue, one successful comparative approach has been

INTRODUCTION

that pioneered by G. P. Murdock in his creation of the Human Relations Area Files. Using these coded compilations of ethnographic data, scholars can create random samples of socio-cultural systems and test hypotheses, for example by examining the occurrence of two (or more) cultural traits as variables in the selected sample of cultures to see if they are correlated. The totality of world ethnography, conveniently sorted, coded, and catalogued, can then serve as a broad data base that can be mined by random sampling for the testing of comparative proposals about human culture and society.

My reason for not using this method, however, is that from my viewpoint, ethnography is an inherently discursive and interpretive practice that is best when it stays close to the complexity of the subject under study. Socio-cultural forms are vast assemblages of institutions, practices, symbol systems, and so on, and require large and detailed books to even begin to describe them adequately. The kind of investigation in which I engage in this book does not allow for the winnowing out of a small number of "variables" that might be statistically tested, but rather requires a minimally useful picture of at least some key aspects of the socio-cultural systems under study. Here is where Evans-Pritchard's objection arises: a truly comparative survey of the world ethnographic record would be as long as all the books and articles of ethnography strung together, and no scholar could possibly master it all. Anyone attempting a comparative ethnographic study therefore has to limit the number of societies to be examined, the questions that will be asked of them, and the degree of detail in which they will be presented. How can one legitimately do this without resorting to the HRAF method that inevitably filters out much too much ethnographic detail to be useful in an explanatory comparative project such as the present one?

My solution has been to use ethnographic examples that illustrate with particular clarity the theoretical points I want to make. This is not the same as testing a hypothesis: I have no hypothesis, in the sense usually understood, to test. Rather, I have proposed that societies will in every case have to reveal in their own highly idiosyncratic way their solution to the paradoxes and dilemmas posed by the undeniable fact that in recreating themselves over time they must rely on two fundamentally incompatible modes of information transmission. My choice of examples is intended not as a random sample but rather as a selection, based on my knowledge of what is admittedly only a small fraction of the world ethnographic literature, intended to illustrate the kinds of features of human society that I argue must arise in consequence of dual inheritance. Skeptics who wish to test the validity of my premises and arguments thus ought to be able to substitute a different set of ethnographic examples and see for themselves whether they, too, exhibit the traces of the same tension as the ones I have used.

My justification for my approach is that the premises I have begun with, and the necessary implications to which they lead, have specifiable consequences for the particular form that human socio-cultural systems take. It is the fruitfulness of my approach in accounting for a great deal of what might otherwise seem puzzling or inexplicable in the ethnographic record that provides for its validation. The dynamic I am proposing—the conflict between the two modes of information transmission—leads, as I hope to show, to an account of human society that explains its distinctive peculiarities better than other ways of looking at them. I say "human society" and not "human societies" because while I accept the fact of the great variation in the ethnographic record, I want to argue that any given socio-cultural system has to be a more or less effective solution to the problems posed by the fact of dual inheritance, whatever else it is. The solutions are as varied as the human creative imagination allows; but the problem is always the same, which gives a certain distinctive flavor to any human social system—a flavor it is my aim to discern and present. Thus my model aims to respect both what is unique and what is general in ethnography; in other words, to address the fundamental anthropological problem that people around the world are the same, only different.

This method has its analogue in biology: any living organism has to be shown to have solved the fundamental problems of energy capture, self-organization, metabolism, and reproduction; it is a question not of whether but rather of how it has done so. But there is almost no limit on the ways organisms have succeeded in doing this in the course of evolution. In the same spirit, my effort here is to identify one of the fundamental problems that any socio-cultural system must solve, and to illustrate some of the ways in which existing socio-cultural systems have done so.

As for how I limit the presentation of ethnographic examples, my main focus will be on the institutions, and the rules that govern them, that are implicated in the reproduction of society over time. Historically, the anthropological study of social organization has focused on systems of kinship, marriage, descent, inheritance, alliance, and so on. Ostensibly, this was because the more elementary social systems that were the initial focus of ethnographic attention were organized around these matters. In more recent years, after some withering critiques of the study of kinship, of which the most important was that of David Schneider (1984) and the most recent that of Marshall Sahlins (2011, 2013), "kinship" has fallen from its privileged position in anthropological thought (but see McKinnon and Cannell 2013).

But from a dual inheritance perspective, there is a different reason to focus on these areas: these are the institutions that are implicated in the problem of reproduction over time. A human society must re-people itself with new human beings, and must also recreate in the new generation replicas of the institutions and symbols that comprise the social system. (Needless to say, the fact that society replicates itself does not mean that it doesn't also change—evolve—over time.) Therefore, in the main I will be looking at aspects of socio-cultural systems that are involved with reproduction, both biological and cultural. In terms of the selection of material, I have tried to bring out the most salient features of the ethnography without betraying it, while keeping within the scope of a book of manageable length. It is my hope that the cumulative effect of the examples I present will convince the reader that the ethnographic record really does support the expectations arising from the theoretical views of dual inheritance theory as I put them forward here.

The form my argument in this book takes is as follows. In chapter 1, I present a necessarily condensed overview of how, despite the theoretical individualism of much evolutionary thought, and its tendency to explain social and cultural facts as if they were only individual adaptations to enhance reproductive fitness, a number of thinkers have offered innovative ideas about socio-cultural systems in evolutionary perspective that have created much more congenial conditions under which a productive rapprochement and collaboration with socio-cultural anthropology is possible—one in which society and culture are not negated or reduced to overly simple dependent variables but are understood in their own right and with their own importance.

In chapter 2, I give some ethnographic illustrations of cases in which practices or beliefs that seem to be maladaptive from the point of view of biological fitness can nonetheless be seen to be adaptive from the socio-cultural point of view. Practices that actually reduce the reproductive capacity of individuals, and of groups of individuals, can at the same time be offset by others that enhance the ability of the socio-cultural system to reproduce itself. I thus demonstrate how the two channels of cultural and genetic reproduction can operate at least partly independently and in opposite directions at once.

In chapter 3, I present my view of how to understand "culture" as consisting in large part of systems of symbols. Symbols have an existence both in the minds of individuals and also as real objects of perception in the external world. This fact of their inscription in some material medium available to the senses of human actors is critical to understanding how they can act, in the same manner as DNA, as packets of information that can reproduce themselves in human phenotypes and thus replicate over time across generations. But whereas culture is transmitted by way of sensory perception, DNA is transmitted by means of copulation. This basic difference turns out to have enormous consequences for how the two channels influence the organization of human socio-cultural systems.

In chapter 4, I take up one of the most valuable suggestions made by Richerson and Boyd, to the effect that there must be two different kinds of evolved social behaviors among humans, one focused on kinship and governed by the evolutionary principles of kin selection and inclusive fitness, and the other on what they call "tribal" society, referring to social relations that go beyond the elementary units produced by biological relatedness. The former obeys the agenda of the genetic program, the latter the agenda of the cultural program. I provide ethnographic examples to show some of the ways real social institutions corresponding to these behavioral tendencies manage the mismatch between these two agendas.

In chapter 5, I address the question of how and why it is that humans are able to form harmonious social groups beyond the kin group. This requires that they solve the very real problems of antisocial, self-interested competition for genetic success, especially among unrelated males, which makes true prosocial behavior, of the kind and to the extent found in human societies, quite rare in non-human sexually reproducing species. I place human social organization in the context of the social life of other non-human primates and focus particularly on the hamadryas baboon. This species appears to have solved the problem in a way that holds the key to understanding how humans, too, have accomplished this, as other authors have noticed. But whereas the hamadryas baboons have solved the problem by what appears to be a genetic novelty, in humans it is the cultural symbolic system itself that, by virtue of its distinctive characteristics, enables the formation of cooperative social groupings consisting of nongenetic kin. Human society encompasses both heterosexually reproducing bonded mixed-sex pairs and a wider arena of nonsexual affiliation made possible by shared participation in the symbolic system.

In chapter 6, I present a number of ethnographic examples to illustrate how the tension between genetic reproduction, which requires copulation, and cultural reproduction, which does not, is resolved in a variety of creative ways, producing a range of social formations that, while very different from one another, all can be understood by seeing them as responses to the necessity of the larger tribal society to organize, contain, and isolate the potentially disruptive necessity of sexual reproduction.

Having demonstrated the fundamental incompatibility between social organization based on genetic inheritance and that based on cultural inheritance, and shown how this tension is often resolved to the detriment of the status of reproductive sex within the cultural symbolic system, I then show in chapter 7 how other forms of reproduction besides the sexual one can create analogues to sexual reproduction in the realm of symbols, and how and why it is often—but not necessarily or universally—the case that all-male groups within a larger society claim to monopolize the cultural channel of inheritance, thus creating the ideal of a society that could reproduce itself without recourse to sexual reproduction (which is consequently devalued). Male/female power and status asymmetries, so relentlessly evident in the ethnographic literature, are not essential in human society, as I argue, but are rather the result of only one—admittedly widespread—kind of resolution of the conflict between sexual and cultural reproduction.

In chapter 8, I continue to examine ethnographic examples of symbolic reproduction—meaning both the reproduction of the symbol system and also the translation of the imagery of sexual reproduction from biological into symbolic forms.

So as not to give the impression that male supremacy is always the way in which the problems of dual inheritance are resolved, but rather that it is the cultural channel, however understood, that is granted higher status than sexual reproduction, in chapter 9 I give a number of examples of social systems in which the typical male/female asymmetry is absent, and in which women have rights to aspects of cultural reproduction in such areas as religion and ritual, cosmology, warfare, economic affairs, and social power and prestige.

In chapter 10, I raise and address the question of how human prosociality is produced in a new generation. While granting that there must be a strong evolved basis to human tendencies to act in prosocial ways, as I discussed in chapter 4, I also show how most societies arrange matters so that these tendencies are encouraged, nourished, and realized in individuals as they develop and mature. Given that, because of dual inheritance, the prosocial dimensions of human life are offset by powerful motives of competition and pursuit of self-interest, including reproductive self-interest, social groups and their symbolic systems must take steps to insure that it is the prosocial tendencies that maintain the upper hand.

Finally, in a Conclusion, I turn my attention to one particular ethnographic case, that of the Micronesian island of Pohnpei (formerly rendered in the literature as Ponape). I chose this case because Boyd and Richerson (1985) address it specifically in their consideration of the role of symbols in dual inheritance theory. I offer some additional dimensions to Boyd and Richerson's analysis, and illustrate the difference between our uses, suited to different theoretical purposes, of the premises of dual inheritance theory. Since I knew nothing at all about Pohnpei before looking into the ethnographic materials in order to write this chapter, I also intend it to serve as a test case to see how the principles I adduced throughout the book might present themselves in an ethnographic situation with which I had previously been unfamiliar. The examination of this case also allows me to rehearse, in this final ethnographic example, what my fundamental conclusions are about the nature of human society and how they are applicable in this—and as I propose potentially in any—case one might care to examine.

Finally, I should make very plain what it is that I am not attempting to do in this book. I do not inquire into the possible evolutionary history of the socio-cultural systems I consider, nor do I try to chart, either through modeling or through historical research, how and why they have taken on their present form (and of course it will be understood that I refer to the ethnographic present). I do not attempt to plot trajectories of evolutionary or historical change, past or future, nor is this a study of demography. I am, furthermore, not interested in passing any form of judgment on the societies I describe, including whether they might be adaptive or maladaptive, although that issue comes up from time to time since it is quite salient in the work of many dual inheritance theorists. The fact that these societies exist and have persisted over time maintaining a distinctive structure and organization is good enough for me. I also do not address the manifestations of dual inheritance at the level of individual psychology and behavior, though it should go without saying that they are plentiful. I consider it sufficient in one book to have used ethnography to make a case for a certain hitherto largely unnoticed dynamic in the construction of human society that I believe sheds new light on what is actually to be observed in the ethnographic record.

ONE

A Brief History and Outline of Dual Inheritance Theory

In this chapter I offer a survey of the most important ideas that have led to the emergence of the contemporary theory of dual inheritance. Among the authors to be considered are Luca Cavalli-Sforza and Marcus Feldman, Richard Dawkins, Donald T. Campbell, E. O. Wilson, David Sloan Wilson, William Durham, and Robert Boyd and Peter Richerson. I make no claim to completeness, but intend rather to touch on those key contributions that allow me to trace and explain the basic assumptions of the theory. As I hope to show, taken together, many of these assumptions are far enough away from the most dogmatic and reductionist versions of evolutionism (in relation to the study of culture), and compatible enough with ideas in socio-cultural anthropology, to make a rapprochement between the two streams of thought possible. It is my project in this book to contribute to that rapprochement.

It is helpful to state right at the outset that the phrase "dual inheritance" is in some ways a misleading one. On the one hand, there is the genetic code, but on the other side of the duality there are several other ways in which instructions and information are transmitted among people and across successive generations. So, for example, Evelyn Fox Keller (2010) writes at the conclusion of her critique of the binarism of "nature and nurture": "Let us acknowledge that . . . almost all human traits *are* transmitted from one generation to the next; but at the same time, let us also accept the fact that the mechanisms of transmission are very varied. They may be genetic, epigenetic, cultural, or even linguistic" (80). Or to take another example, Eva Jablonka and Marion Lamb (2005) encapsulate the whole argument of their valuable book in its title and subtitle: *Evolution in Four Dimensions: Genetic, Epigenetic, Behavioral, and Symbolic Variation in the History of Life.* Here genetic evolution is contrasted with not one but three other modes of information transmission across generations. In short, the "duality" in the dual inheritance model is based on the binary opposition between the genetic program and everything else that is not genetic. This latter category includes a congeries of phenomena which may loosely be categorized as "culture," in that they are acquired by a new organism from its predecessors by some form of "learning" or "experience" undergone by the phenotype itself rather than given by means of sexually transmitted genetic information.

The emerging field of epigenesis studies how at the molecular level the genetic code itself can be manipulated by the organism, and indeed how some "acquired characteristics" may be transmitted to a future generation in apparent defiance of evolution's "central dogma" that rules out such so-called "Lamarckian" inheritance. This phenomenon is extremely interesting and has potentially important implications for the understanding of genetics and of the transmission of information across generations more generally; however, it is beyond the scope of my present project, as well as of my competence, and so I will not consider it further here.

Behavioral learning, or social learning, whereby information is passed from a phenotypic "teacher" to a "learner," is found among many species as well as in humans, and involves the imitation of observed or shown behavior. In humans, this form of information transmission can be quite intentional; to what extent that is so among animals such as some primates—one thinks here of chimpanzee tool use, for example—is open to debate. Thus if a child picks up its parents' accent, gait, or bodily comportment, or learns from an elder how to string a bow or make a blow-pipe by watching and imitating, these are examples of social learning in humans that need not involve much or any explicit encoding in symbolic form.

It is only among humans that there also exists an extrinsic system of encoded information comparable in complexity and in its mode of operation to the genetic code. That is the cultural code, which in my view should be understood to include language (in contrast to Keller who, in the passage quoted above, seems to view language as if it were something other than culture). Language, like DNA, depends on the creation of significant differences in a material vehicle—molecules forming proteins in the one case, sound waves shaped by the human larynx in the other—in the form of primary binary oppositions, on the basis of which complex systems of representation can be built.[1] Language is not, however, the only form of symbolic information by any means, and human culture is rich in all sorts of vehicles for the conveyance of meaning from one person to others in what Susanne Langer (1957) called "significant form." Langer's concept includes not only discursive forms, such as language, but also presentational symbols, such as are found in art styles, ritual objects, spatial arrangements, and the like. Human "cultures" as actual existing entities are best understood as socially constructed symbolic worlds within which (reasonably) coherent human lives are led by individuals in social groups acting in meaningful patterned relationship with each other. What both social learning and symbolic systems have in common, in opposition to genetic information, is that they are both transmitted via sensory perception in a social arena.

There is also in the determination of human action the external factor referred to as "the environment": many aspects of the wider non-human world impinge on the organism and affect the organism's form and behavior. While the environment does not usually "intend" to teach, since non-human nature does not (so we Westerners assume) have the capacity to act on such "intentions," nonetheless environmental factors instruct learners in the sense that the latter draw lessons from their experiences with the environment.[2]

Of these nongenetic factors in the inheritance of information and instructions by humans from one another, the one on which I will concentrate is the symbolic dimension, both because it so closely parallels the genetic channel of information transmission (albeit with highly significant differences that I will specify more fully later) and because it is so salient in informing actual human existence, in individuals and collectively in society. Since the form in which much extrinsic information flows between and among people, including from parent to child, is along the symbolic channel, I will for convenience often refer to the symbolic code as if it were the same as "culture," but it will be understood that other nonsymbolic factors of the sort I have just described are in play as well. What all these forms of cultural information share, however, is that they are made available and transmitted in the perceptual sensorium of phenotypic organisms, either fully formed or in the process of development, rather than via the gametes. It is this feature that plays a key role in my theoretical formulations, so that the question of whether a particular piece of behavior is acquired by the transmission of symbolic form or by extra-symbolic, imitative social learning, while interesting, is not of crucial importance in this context. I will expand on how I understand cultural symbolism in chapter 3.

Cavalli-Sforza and Feldman

The origins of dual inheritance theory can be charted beginning in the 1970s with the work of Luca Cavalli-Sforza and Marcus Feldman. While various earlier authors had seen a parallel of some sort between genetic and cultural evolution in an often vague or intuitive sense, it was the development of sophisticated mathematical operations that could be used to model processes in population genetics and their application to cultural ideas by Cavalli-Sforza and Feldman (1973) that paved the way for the future development of a "dual inheritance theory" that was scientifically sound enough to provide the basis for a substantial research paradigm along well-established Darwinian lines. Evolutionary theory had made a great stride forward when it was established as a principle, in the years during which the neo-Darwinian synthesis was being forged, that what Darwin had called "species" (as in "The Origin" thereof) can usefully be thought of as collections or "populations" of individual organisms capable of interbreeding and thus transmitting their genetic material to offspring.[3] Evolution in this view proceeds by the differential rates of reproduction and mortality among these individuals, affecting the changing form of the gene pool of the population over time. These differential rates, in turn, are a result of the fact that individuals within a population exhibit variations, and that there is inherent competition when, as is typically the case, there are resource limitations. Some traits allow the phenotypes exhibiting them to be more successful in reproducing offspring in future generations than others, and this process leads to evolution and, in some cases under particular conditions, speciation.

Traits, including behavioral traits, that are more or less adaptive in the sense of enhancing reproductive "fitness," are in turn the expression in the organic form and behavior of the individual phenotypic organism of instructions encoded in the genome about how to develop ontogenetically. For the most part, genetic instructions are stable, but occasional random or blind changes, or "mutations," in the genes introduce a new trait, which, when expressed in the phenotype, is usually deleterious but in a few instances proves to be advantageous to the individuals instructed by them. This advantage, understood in the specific sense of enhancing reproductive success, would then spread in the population since the phenotype exhibiting the new adaptive trait (and the genetic material it passes on to its offspring) provides an advantage over other phenotypes in the population. One can construct sophisticated and complex mathematical models of how percentages of genes might wax or wane in a population and thus make predictions that can be tested against empirical observation; this is the basis of the science of population genetics.

Cavalli-Sforza and Feldman realized that the same method could be used to study the ebb and flow over time of "ideas" or "beliefs" within a population of humans if these ideas or beliefs were treated as heritable "traits" that could be passed from one generation to the next; and they were able to model the way new inventions or ideas might spread in a given population using the concepts already in use in population genetics. At the same time, they were well aware of the limitations of the analogy of socially transmitted ideas to genes in a gene pool and noted some obvious differences. One of these is that the reproduction of ideas or innovations in a human population does not entail differences in shifting patterns of the birth, survival, reproduction, and death of numbers of individual organisms with or without a certain trait. Such "cultural" information can be transmitted directly in much less than the lifetime of a human generation, which is the key unit required for the blind "trial and error" method of natural selection. The information is acquired directly, sometimes in a matter of minutes or seconds, and its spread has nothing necessarily to do with the genetic reproductive rates of the individuals who have learned the new ideas (though of course it may). In this sense, as others before Cavalli-Sforza and Feldman had already noted, cultural evolution is "Lamarckian" rather than strictly "Darwinian," in the sense that innovations acquired during the lifetime of the phenotype can be inherited by future generations. In the sexual transmission of DNA, by contrast, it is assumed that the DNA in the gametes changes only by random mutation; it does not alter itself in response to lived experience that the phenotype bearing the DNA has undergone.[4]

Cultural inheritance is likewise not Darwinian in that it could be said to involve "artificial" rather than "natural" selection.[5] The latter—Darwin's great contribution to science—operates blindly and automatically as phenotypes carrying genetic instructions or "alleles" for various traits are simply observed to die out or spread in a population without any intent to do so. Indeed, the value of Darwin's formulation is that it eliminates the need for anything resembling foresight, intention, will, purpose, choice, value, or any related phenomena from every stage of evolutionary process, from "random" mutation to the shifts in the gene pool of a population that result from the accumulation of more adaptive traits as new environmental pressures shape differential survival and mortality within the population. This elimination of "anthropomorphic" factors enabled evolutionary biology to become congruent with the natural sciences of physics and chemistry, which had long since dispensed with anything but physical matter whose properties and changes over time could be described in mathematical terms. The elimination of intention from evolutionary process did not, to be sure, negate the observable fact that individual organisms produced by natural selection can and do exhibit purposive behavior, but the process itself is understood to be teleologically blind.

But "artificial" selection, with a discussion of which Darwin opens his master work, *does* involve choice, specifically choice on the part of humans, such as pigeon fanciers who can, through selective breeding, produce all sorts of phenotypic variation in their pigeon subpopulations. If breeders want pigeons with fluffier feathers on the crest or tail, they can mate those among their flocks who already have something of those features over several generations. Darwinian evolution involves substituting involuntary environmental pressure for the human breeder who acts with a conscious goal in mind, thereby eliminating the factors of choice, intention, or foresight from the process.

But it would seem to follow, if we grant that humans are capable of making choices guided by preferences, intentions, purposes, and so on when it comes to selecting the kind of pigeon they want to breed, that the same would go for any human invention. People are self-evidently capable of envisioning a desired outcome, figuring out what needs to be done to get there, and doing it; so that when a cultural "innovation," whether transmitted through symbolic means such as via language or imitated in pure "social learning," is adopted by individuals, it is with the hope, wish, or expectation that it will in fact provide some benefit to themselves. Thus, not only does cultural evolution appear to be "Lamarckian" rather than "Darwinian," it also seems capable of being the result of "artificial" (that is human, and hence intentional) selection rather than of "natural" selection, as Christopher Boehm (1978) argued many years ago. It must be added that if one understands human thought, including the forming of intentions, as something that can occur unconsciously, as contemporary cognitive science recognizes, then one need not suppose that all such artificial selection is made by conscious deliberation.

Finally, Cavalli-Sforza and Feldman's analogy of cultural to genetic evolution in populations implies that whereas genetic material can only be transmitted from biological parents to their own offspring, that is "vertically," cultural information can also be passed indirectly or horizontally just as well. Thus, for example, American children might absorb English from their biological parents, be taught Spanish by a genetically unrelated school teacher, and learn how to play games in the playground from other (unrelated) children. The effect of this would be perhaps the most dramatic of all in differentiating a cultural evolutionary process from the genetic one, because it would mean that cultural evolution could proceed to some degree independently of considerations of genetic fitness. If cultural instructions can be transmitted between or among individuals with no genetic relation to each other, but nonetheless are to the benefit of those individuals, then there is no inherent reason to suppose that the cultural traits selected for will necessarily confer any benefit to those individuals in terms of genetic as opposed to cultural fitness.

Before this, and in many if not most evolutionist circles to this day, including many who attempt to understand human behavior in evolutionary terms, such as sociobiologists, evolutionary psychologists, and human behavioral biologists, it has been or is assumed that any cultural trait would have to be adaptive not only in a social sense but also in the sense that it would enhance the inclusive reproductive genetic fitness of those whose behavior included it. Culture, which the ancestral evolutionary anthropologist Leslie White had defined as humans' "extra-somatic means of adaptation," would therefore be understood in the final analysis as a trait acquired by natural selection in the course of human evolution that, like flippers for seals or sonar for bats, was selected for because it enhanced "adaptation," that is, genetic reproductive success in individuals.

Richard Dawkins

The theoretical delinking of cultural evolution from service to the interests of genetic fitness received a major boost from the work of Richard Dawkins (1976, 1982) with his introduction of the concept of the "meme." Although he was a champion of what is in some respects an extreme form of methodological individualism and genetic reductionism, arguing that phenotypes, including humans, are basically nothing but "lumbering robots" produced by the genes for the sole purpose of reproducing themselves, he also recognized that "Darwinian" evolution was a process that in principle could occur in any things in any domain in nature that could replicate themselves with variation, so that selection could operate on them. While biology took it for granted that genes were replicators of this sort, there was, Dawkins argued, no reason to suppose that they were of necessity the only kinds of replicators that could or did exist. And indeed, he proposed, just this sort of thing happens in the non-organic realm of human artifacts. Among these are some that are capable of being copied, and to these he gave the name "memes." Anything extragenetic that can be replicated and so transmitted from one individual to the next, such as a catchy tune, a recipe, a Xeroxed piece of paper, or a religious dogma can in principle be mapped over time as it replicates itself, or fails to do so, in human populations.[6]

The term "meme" was well-chosen. It sounds like "gene," and its penumbra of associations suggests imitation (as in *mime*), identity (the French word *même*), memory, memoranda, "mean," as in "meaning," self-interest (me! me!), and the goddess Mnemosyne, and is thus altogether perfectly suited to catching on. At that time, the search was on among scholars interested in expanding neo-Darwinian concepts and methods into the study of culture from an evolutionary perspective for a unit of cultural replication comparable to the gene in the field of biological reproduction. Cavalli-Sforza and Feldman had used the relatively straightforward term "idea" to refer to a trait that was reproduced, Lumsden and Wilson (1981) had floated "culturegen," and Cloak (1975) had suggested "instruction" (which I find particularly useful, since it actually conveys what a unit of information does). Many people were and remain satisfied with "information" as an over-all descriptor of what it was that was being transmitted. But it was "meme" that had reproductive success, and besides lending its name to a field of study, "memetics," entered popular discourse to a degree few scientific neologisms manage to achieve.

As I have mentioned, the history of this kind of thought might have been very different if, instead of making up a new word, evolutionists had realized that what they were searching for was already well theorized within anthropology, linguistics, semiotics, philosophy, literary studies, and related fields, and had taken a concept such as "sign," "symbol," "representation," or "signifier" as the unit of culture they sought. This might have led to a mutual cross-fertilization of the evolutionary and socio-cultural thought-worlds, but this did not happen. Instead, once Dawkins had shown in principle that the propagation of memes could be independent of the genes, memeticists tended toward the view that culture, insofar as it is composed of memes, is as likely as not to be inimical to genetic interest, and thus to operate like a malignant parasite on the organism rather than as a necessary component in its self-construction and functioning in the world.

A leading example of this tendency can be observed in the treatment by some evolutionists of the proclivity of humans to believe in religious ideas. Religion, these theorists propose, based as it is on nonempirical beliefs, confers no adaptive advantage at best, and at worst is a dangerous folly; accordingly, it should have been eliminated long ago by any effective process of selection. For scholars such as Sperber (1985a, 1985b, 1996), Boyer (2001), Atran (2002), and other thinkers grounded in evolutionary thinking who theorize about religion and other such "irrational" beliefs, therefore, the explanation for the persistence of religious ideas must lie in the evolved cognitive capacities of the brain.[7] Boyer, for example, emphasizes the fact that, as he shows in cognitive laboratory experiments, those ideas that are most easily remembered are ones in which most features are relatively normal or prototypical, but with one anomalous feature. Too many anomalies and the idea is incoherent, too few (i.e., none) and it is routine, dull, and forgettable. But an idea such as belief in a God who is like a human being in every respect except that he is invisible is just the sort of idea that is attractive to the human brain. Others argue that it is the evolved human capacity to attribute intentions to other humans that allows it to be fooled into extending this belief not only to an animistic universe but also to a belief in unseen divine or spirit beings who possess intentionality toward humans. In this theoretical formulation, the evolution of human behavior by natural selection is preserved, but with the proviso that the cultural system can lead people astray, producing superfluous maladaptations that endure over generational time. There is thus a tendency among many evolutionists to view culture, when they grant it any place at all in their thinking, as a mistake or pathology; a good example of this is Sperber's (1985a) comparison of culture to a virus.

Despite this move away from a conceptualization of culture that might lead to an alliance with socio-cultural anthropologists, Dawkins's declaration of the independence of "memetic" replication from genetic reproduction helped pave the way for a dual inheritance view in which cultural phenomena do not exist only in the service of the genetic fitness of individuals competing within a population but rather have their own evolutionary trajectories.

Donald T. Campbell

The next important contribution to the development of contemporary dual inheritance thinking that requires our consideration is an influential article by Donald T. Campbell (1975). This key document was a presidential address Campbell delivered to the American Psychological Association, and it marked a turning point in that field away from the strict behaviorism that had previously been hegemonic in the discipline. Campbell's paper injected evolutionary concepts into psychology in what was to prove to be a major shift in orientation.

Campbell's stated purpose in his address, which is entitled "On the Conflicts Between Biological and Social Evolution and Between Psychology and Moral Traditions," is to argue against the opposition most psychologists then exhibited to traditional religious and moral values, which (so he assumed) they generally viewed only as repressive and restrictive. He instead proposes to view them as adaptations at the group level: the aspect of Campbell's position most relevant for our current discussion of the development of dual inheritance thinking is thus his espousal of the possibility that the Darwinian principles of variation, selection, and retention operate not only on organisms but also on social systems.

Innovation at the social level in the form of new ideas or inventions may be thought of as either blind or intentional—it does not matter to a Darwinian process, Campbell argued, as long as variation is produced somehow. Selection then operates not on individuals bearing the ideas or innovations but rather on the social systems into which these have been introduced: "it is necessary to make a plausible case for selection at the social system level" (1106). This process enshrines in heritable social traditions attitudes, beliefs, and instructions for action that have survived a rigorous test equivalent to natural selection at the genetic/organismic level. This means that we must view any cultural tradition, including religious beliefs, not as a set of bizarre or misguided superstitions and errors that has hijacked the human brain, but rather with the respect due to anything with demonstrated adaptive value. After all, we do not scoff at even the apparently oddest behavior encountered in the study of the animal kingdom, assuming rather that upon analysis it may well turn out to have conferred some sort of adaptive advantage. How then can we reject as anti-human or irrational the values or practices that, by the evidence of their persistence, have helped enable a group to survive through the course of its history?

In arguing thus Campbell is setting himself against a foundational belief in contemporary neo-Darwinian theory according to which selection at the social level, or "group selection," is if not impossible than highly unlikely. He is also opposing views, such as those later to be espoused by thinkers such as Sperber, Boyer, and Atran, that religions are "maladaptive" and have been produced by the tendency of particular cognitive capacities of the human brain, originally evolved for adaptive reasons, to misfire. Campbell thus introduces into the nascent dual inheritance discourse the idea that cultural evolution at the group or social level can actually be both independent of the genetic program and also adaptive and advantageous at the level of the group taken as a whole. I find this formulation very congenial, as subsequent chapters will make evident.

Campbell is well aware that contemporary evolutionary theory takes a dim view of group selection, but he counters with the observations that humans are empirically highly cooperative and altruistic, even though at a genetic level it must be assumed that individual organisms compete with each other for reproductive advantage. From this it may be concluded, he argues, that cooperation and altruism among humans must not be genetic. And if these traits are not transmitted genetically, then they must be transmitted at the level of the social group itself. He goes on to draw the conclusions that human social complexity has been made possible by social rather than biological evolution, and that social evolution has had to counteract individual selfish tendencies selected for in the biological evolution of the organism as a result of competition at the genetic level. If religious creeds of the great early civilizations of human history uniformly inveighed against "selfishness, stinginess, greed, gluttony, envy, theft, lust, and promiscuity, all close to biological organization for self and children" (1119), it is because the social system, operating on the basis of interests independent of those of the individual organisms that compose it, must impose on a self-interested, hedonistic, and backsliding populace a morality that preaches against the very human selfishness promoted by the genes.

Campbell thus proposes a genuine dual inheritance model of human evolution according to which "on the one hand, there is biological evolution optimizing an individual person and gene frequency system. On the other hand, there is a social-organizational-level social evolution optimizing social systems functioning. For many behavioral dispositions, the two systems redundantly support each other. For others, the two are in conflict, and curb each other" (1116). In this view, sociocultural evolution is not only to some degree independent of genetic evolution; it often operates in opposition to it. This is far from detrimental, however, because in so doing it imposes on genetically evolved and hence inherently competitive individuals a set of moral principles that enable them to rise above their own petty interests and act cooperatively for the benefit of each other and for the group as a whole.

Sensible and persuasive as these ideas seem to me (although I would maintain that humans do have evolved genetic propensities for prosocial and well as for competitive, self-interested behavior), in that they appear to describe the human world as it actually is rather than how it would have to be if genetic fitness alone determined the forms of individual and social life, they are fighting words to many orthodox evolutionist thinkers. And they bring us to a consideration of the whole problem of "group selection."

Group Selection: The Case of E. O. Wilson

Group selection is the idea that some individuals in a group, such as a flock of birds, by altruistically curtailing their own reproductive advantage, may enhance the overall fitness of the group. Thus, for example, if birds, by flying in flocking formation, observe themselves and conclude that their numbers have grown too great to be supported by available resources, then it may be advantageous for some group members to constrain their own fitness potential and inhibit themselves or refrain from reproducing. The theory of group selection was put forward most forcefully by V. C. Wynne-Edwards (1962), but was quickly rebutted by a number of prominent evolutionary theorists, for whom the only operative factor in evolution can be competing individuals pursuing maximal reproductive success. Nothing in nature, so it is argued, could be understood to dissuade organisms programmed by genes designed to enhance fitness from doing so, even if it would lead to the disadvantage of the group. The genes, it is assumed, having been selected to advance their own reproductive advantage, have no way of foreseeing or doing anything about such looming disasters for the population as a whole as overpopulation and resource depletion. Yet despite its apparent incompatibility with neo-Darwinian theory, the problem of empirically observed altruism, sociality, and cooperation in animal species as well as their exponentially increased representation in humans remained to be explained.

The much-heralded solution to the matter was provided by the work of William Hamilton (1964) two years after Wynne-Edward's book appeared. Hamilton argued that since the only thing that mattered in the evolving genome of a population was the reproduction of genes in the future generation, it made no difference how these genes managed to get themselves reproduced. And since closely related individuals, such as parents and children, or siblings, share many genes, it is not necessary that individuals should pursue only their own personal fitness. As long as their behavior enhances the reproduction of genes, many of which they share with close kin, genetic instructions for it will remain in the gene pool. If, therefore, individuals act altruistically—that is, in such a way as to compromise their own reproductive fitness, but in doing so enhance the reproductive fitness of their close kin bearing genes identical to their own—then the effect will be the same as if they had enhanced their own individual fitness. The genes shared by kin would be among those replicated in the next generation of offspring. The remarkable cases of some social insects, which had puzzled evolutionists since Darwin, among which reproduction is handled only by a queen while many members who contribute altruistically to the life of the colony are infertile, is resolved in Hamilton's theory by the observation that peculiarities of the reproductive systems of these organisms make them much more closely related genetically than is the case in which offspring and siblings share only about a half of an individual's genes. After all, a society of clones—and indeed a multi-celled organism could be seen as just such a "society" of cells all bearing identical DNA—would have no trouble acting for the collective good, since that would simultaneously correspond to the genetic good of each. Ants, being much closer because of a peculiarity in their genetic make-up to clones than are most sexually reproducing organisms, are therefore capable of a level of group altruism that is beyond the reach of most other species.

Hamilton's theory overcame some initial resistance to become widely accepted under the name of "inclusive fitness"—"inclusive" in the sense that altruistic behavior was after all possible insofar as it enhanced the fitness of the genes borne by the altruistic individual by enhancing the fitness of his or her close kin, who were thus "included" in the reproductive self-interest of the individual organism and its genome.

Meanwhile, Robert Trivers (1971) provided yet another solution to the problem posed by prosocial behaviors like cooperation and altruism, one that remained consistent with the idea of the commitment of the genes to their own continued reproduction, with his introduction of the concept of "reciprocal altruism." Individuals, he argued, might indeed help out other nonrelated individuals at some cost to themselves but only on the condition that the recipients of their largesse were relatively likely to do the same for them some day, evening out the score so that in the end no one's genetic fitness was actually compromised. Thus armed, evolutionary theorists were now emboldened to attempt to explain human social life in terms that rejected group selection (*contra* Campbell's proposal) and relied on neo-Darwinian theory augmented with inclusive fitness theory and the theory of reciprocal altruism as explanatory principles underlying humans' (apparent) cooperative and prosocial behavior. Among those championing the application of such an approach to the study of human social life was Edward O. Wilson, whose influential book *Sociobiology* (1975) caused a veritable firestorm in the intellectual world, stirring up virulent opposition while at the same time inaugurating a strong tradition of research, both of which—the opposition and the research program—survive to this day.

As a specialist in the study of ants, it seems understandable that Wilson was happy to take up Hamilton's idea and expand on it. In his books *On Human Nature* (1979) and (co-authored with Charles Lumsden) *Genes, Mind and Culture: The Coevolutionary Process* (Lumsden and Wilson 1981), E. O. Wilson (and his colleague) developed a "coevolutionary" model that sought to integrate the study of culture and society within a thoroughgoing neo-Darwinian evolutionism. According to this view, while culture can be thought of as distinct from biology, it coevolves with it and is held on a "short leash" by the genetic program with its single determining factor of inclusive reproductive fitness, to which culture must conform if it is to survive the forces of selection.

In taking on "culture" and "society," concepts traditionally associated with the disciplines of anthropology and sociology, Wilson was not shy about the fact that he did not feel obliged to engage, in this "new synthesis" of his, with the already existing study of culture and society by those whose work was actually devoted to such study. This was because, like many other strict neo-Darwinian theorists, Wilson did not actually believe that there were any such things as society or culture to be studied. Hence, he was perfectly willing to dismiss any scholarly research traditions that built their theorizing on the recognition of groups as entities with their own systems of organization that could be analyzed on their own terms. Wilson wrote: "Despite the imposing holistic traditions of Durkheim in sociology and Radcliffe-Brown in anthropology, cultures are not superorganisms that evolve by their own dynamics. Rather, cultural change is the statistical product of the separate behavioral responses of large numbers of human beings who cope as best they can with social existence" (1978, 78).

I confess that I have never entirely understood why people who claim to have a "scientific" worldview seem perfectly willing to grant to blind Nature the ability to produce prodigies of organizational complexity and emergent ontological novelty and autonomy at the level of atoms, molecules, cells, and organisms, but balk in principle at going any further up the ladder of biological systems, refusing to allow for the possibility that organisms too can be grouped into more comprehensive wholes that plainly exhibit their own internal organization in the form of social structure and organization.

Be that as it may, however, there are several good reasoned arguments for why, in a natural world ruled by natural selection acting on genes, group selection should be disallowed. One is that groups are relatively fluid, with individuals moving back and forth between them, so that they are not the stable, bounded entities of the sort that, like genes, could compete with one another. Furthermore, the prevalence of genetic competition among the individuals composing them would further preclude their having the unitary status necessary to compete with other groups. Except in very rare cases, individual competition would overrule the possibility of coordinated groups competing as units, as was famously argued in Hardin's "Tragedy of the Commons" (1968).

Perhaps the strongest argument against the possibility of groups acting as units that compete and undergo selection among themselves is that if a group were to achieve the situation in which individuals constituting them agreed somehow not to compete for reproductive advantage among themselves, in a spirit of altruism and for the benefit of the group as a whole, they would be vulnerable to "cheating" on the part of deviant individuals who would take advantage of the good will of their fellow citizens, benefiting from their sacrifice while pursuing their own antisocial aim of individual inclusive reproductive advantage. The result would be that the genes programming the cheaters to cheat would spread through the population, and the social agreement to cooperate would soon collapse as the individuals composing the group would have morphed over a few generations from altruists to (kin-inclusive) egotists.[8]

It was, then, a momentous occasion when, decades later, E. O. Wilson himself, who had done so much to discredit the idea of group selection and advance the cause of "inclusive fitness" in the first place, announced that he was recanting his earlier beliefs (D. S. Wilson and E. O. Wilson 2007; E. O.Wilson 2012). Working with some colleagues sophisticated in mathematics, he claimed to show that Hamilton's models did not in fact work out as he originally thought they did or ought to, and he also reminded himself that there were species that did not have haploid-diploid genetic systems that nonetheless seemed to be capable of altruistic, that is self-limiting, reproductive behavior in the interests of group cooperation. He thus ignited a second firestorm by declaring in his book *The Social Conquest of Earth* (2012) that while group selection was indeed rare, it had in fact occurred in evolution, and when it did occur it provided great selective advantages to its constituents. Among the beneficiaries of this process, he now held, were humans, whose forms of life seem self-evidently to be inherently social and cooperative (except of course when they aren't). Ant colonies, termite societies, bee hives, and human societies could now also be seen for what they had appeared to be all along, that is, emergent superorganisms that act as units and have an organized existence over and above the organisms that populated them.

Among humans, E. O. Wilson went on to say, the genetic self-interest of the individuals under the regime of group selection would now became a source of potential disruption of the group, while behaviors that had evolved to serve group interests struggled, through group norms and values, to keep actions based on such self-interest in check:

> The dilemma of good and evil was created by multi-level selection, in which individual selection and group selection act together on the same individual but largely in opposition to each other. Individual selection is the result of competition for survival and reproduction among members of the same group. . . . In contrast, group selection consists of competition between societies. . . . Individual selection is responsible for much of what we call sin, while group selection is responsible for the greater part of virtue. Together they have created the conflict between the poorer and the better angels of our nature. (2012, 241)[9]

This perspective, which has brought him a great deal of opprobrium from many evolutionary scientists who see it as a betrayal of fundamental neo-Darwinian principles, brings E. O. Wilson's ideas into line not only with those of Donald Campbell, as described earlier, but also, ironically enough, with that ancestor of sociological thought, Durkheim, whose "holistic" ideas Wilson had so airily dismissed several decades earlier. For Durkheim,

> It is not without reason . . . that man feels himself to be double: he actually is double. There are in him two classes of states of consciousness that differ from

each other in origin and nature, and in the ends toward which they aim. One class expresses our organisms and the objects to which they are most directly related. Strictly individual, the states of consciousness of this class connect us only with ourselves. . . . The states of consciousness of the other class, on the contrary, come to us from society; they transfer society into us and connect us with something that surpasses us. Being collective, they are impersonal; they turn us toward ends that we hold in common with other men. (Durkheim 1973 [1914], 161–62)

Still more ironically, perhaps, we might compare Wilson's recent about-face with this statement:

Finally, the social instincts, which no doubt were acquired by man as by the lower animals for the good of the community, will from the first have given him some feeling of sympathy, and have compelled him to regard their approbation and disapprobation. Such impulses will have served him at a very early period as a rude rule of right and wrong. . . .

As a struggle may sometimes be seen going on between the various instincts of the lower animals, it is not surprising that there should be a struggle in man between his social instincts, with their derived virtues, and his lower, though momentarily stronger impulses and desires. (Darwin 2007 [1871], 191–92)

Group Selection: David Sloan Wilson

Before E. O. Wilson had his change of heart, David Sloan Wilson had already emerged as one of the leading proponents of the validity of the concept of group selection, despite the regnant evolutionary view that it was rare or impossible. He argued this on theoretical grounds in a work co-authored with the philosopher Elliott Sober (Sober and Wilson 1998) and continued his argument in a sympathetic consideration of the role of religion in human evolution (2002)—quite at odds with the generally hostile or reductionistic approaches to religion that were emanating from many other evolutionary quarters, as I have noted. D. S. Wilson made the most extensive case for the idea that religious ideas serve to bind individuals through shared belief into a higher order group, which he was not afraid to compare to an organism: "Natural selection is a multilevel process that operates among groups in addition to among individuals within groups. Any unit becomes endowed with the properties inherent in the word organism to the degree that it is a unit of selection" (2007, 43).

Wilson turns approvingly to Durkheim's overall theory of religion as "an adaptation that enables human groups to function as harmonious and coordinated units," adding that a social group is such an abstract entity that, so Durkheim showed, "needs to be represented by a set of symbols to be comprehended by the human mind" (1973 [1914], 54); and he quotes Durkheim: "In all its aspects and at every moment of history, social life is only possible thanks to a vast symbolism" (cited in D. S. Wilson 2007, 43). These ideas are certainly very compatible with the ones I will be developing in this book.

In his analysis of Calvinism as an example of a religious system, D. S. Wilson draws on Calvin's catechism, and in doing so makes an argument that is thoroughly compatible with dual inheritance theory: "Catechisms are a potential gold mine of information for the evolutionary study of religion. They may truly qualify as 'cultural genomes,' constituting in easily replicated form the information required to develop an adaptive community" (93). Plainly, D. S. Wilson not only understands the importance of cultural symbolism in forming an organized human society but also grasps that a symbolic entity such as a catechism is quite literally parallel to a genome in containing "easily replicated information"—information that constructs Calvinists as Calvinists just as the Calvinists' genes constructed them as humans requiring a symbol system to complete them as components of a human community.

Thus one must rank D. S. Wilson as one of the most important formulators of dual inheritance theory and credit him with creating an opening to which socio-cultural anthropologists might easily respond with sympathy. Indeed, he is one of the few evolutionary theorists who, far from thinking that the social sciences need to be subsumed within the biological paradigm, is aware that any real dialogue must be a two-way street: "It often seems as if the integration of biology and the social sciences is a one-way street, more a conquest by biology than a fertile interchange. Here [learning from Durkheim's theory of religion and of the importance of symbolism in constituting human groups] is a case where the influence needs to flow the other way" (226).

William Durham

A landmark in the development of contemporary dual inheritance theory was the book *Coevolution: Genes, Culture, and Human Diversity* by William Durham (1991). Among the major contributors to this field, Durham is the rare anthropologist by disciplinary training, who spent important formative years in the Stanford anthropology department.

There he was able to engage in dialogue with cultural anthropologists, thus becoming familiar with key ideas in that field. This makes his approach, like D. S. Wilson's, more congenial to students of culture than that of some other evolutionary or even coevolutionary thinkers, such as E. O. Wilson and Lumsden. For example, he posits five assertions about culture that represent the mainstream consensus in the field of symbolic anthropology and with which he seems to agree:

1) that culture has conceptual reality (as opposed to being a mere generalization or abstraction from real individual behavior), and as such that as such it is a "pool" of information that is both public and prescriptive;
2) that it is socially transmitted;
3) that it is dependent on symbolic encoding able to create and convey "meaning";
4) that it is systematically organized and tends to form an integrated whole; and
5) that it has a social history that has molded it over many generations.

These ideas, in contrast to many other conceptualizations of culture by evolutionists, still seem very compatible with how culture is seen by many cultural anthropologists to this day, myself included, as I will argue in chapter 3.

Rare among his "coevolutionary" colleagues, Durham also recognized that Clifford Geertz, far from being the touchy-feely postmodernist flake some evolutionists caricature, was actually among the first on the socio-cultural side to anticipate and lay the groundwork for a dual inheritance model that recognizes symbolic information as parallel and equal to genetic information in human life. Durham cites this passage from Geertz's paper "Religion as a Cultural System," one of the key texts in modern cultural anthropology:

As the order of bases in a strand of DNA forms a coded program, a set of instructions, or a recipe for the synthesis of the structurally complex proteins which shape organic functioning, so culture patterns provide such programs for the institution of the social and psychological processes which shape public behavior. Though the sort of information and the mode of transmission are vastly different in the two cases, this comparison of gene and symbol is more than a strained analogy. . . . It is actually a substantial relationship. (1973b, 92)

This perspective, which I find wholly compatible and remarkably prescient, and which I will develop in chapter 3, led Durham to view culture as a second system of inheritance in human life that may variously complement, oppose, or vary indifferently with respect to genetic change. This means that the key empirical questions for him include these: "How do the processes of genetic and cultural evolution relate? What is the range of possible relationships between genetic and cultural evolutionary change, and which of these relationships are most important to understanding of the evolution of human diversity?" (37). In answer to these questions, Durham proposes three hypotheses.

The first of these is premised on the observation that unlike the mechanism of genetic evolution, selection is performed on cultural symbols (he calls them "memes" rather than "symbols") by human actors themselves.[10] Humans have the capacity to exercise choice and indeed to experiment with possible choices by making symbolic models of real-life situations to be used as guides or templates, much as Geertz proposes when he compares a cultural actor interacting with cultural symbols to a driver running a finger over a map to plan the trip s/he is about to undertake. Following the lead of Christopher Boehm (1978), Durham differs from many other evolutionary thinkers in ascribing intention and foresight to human actors when they decide on the choice of an "allomeme" (his term for the parallel in cultural heredity of an allele in genetic inheritance). Therefore, both their own personal experience, as well as the sum of the accumulated wisdom gained from the experience of others and of predecessors encoded in cultural symbols (which he calls the "secondary values," genetic values being the "primary values") direct them to act in ways they themselves judge to be likely to advance their interests and lead them toward their own personal and culturally determined goals.

The contrast between this view and that of many mainstream neo-Darwinists could not be starker: the latter view has as one of its key premises the idea that the variation upon which selection operates is generated completely blindly with regard to the benefit of the phenotype, so that, to take an absurd example, the DNA of a proto-giraffe on the African savannah, which would gain advantage by being able to nibble the leaves of the acacia trees, might just as well produce mutations leading in the direction of enabling it to hop like a kangaroo or quack like a duck as to generate instructions for longer necks. Only happy chance would result in a mutation for longer necks. For many evolutionists it is a founding premise of the field that biology must conform to the presuppositions of the fields of physics and chemistry, from which purpose, meaning, value, and intention are excluded on principle. From that point of view, then, Durham's first hypothesis has given away the store in allowing factors such as these to participate in a cultural selection process.[11]

Durham's second hypothesis is that there are five possible modes of relationship between genes and culture, two "interactive ones" and three "comparative" ones. By interactive modes, Durham means that

1) culture can adapt to the constraints and opportunities presented by the gene pool at any given moment in evolutionary history; or
2) genes can adapt to the "secondary values" encoded in the cultural symbol system at any given time.

Unlike some more rigorous neo-Darwinians, such as Alexander (1979), for whom "culture" is no more than a particular part of the "environment" in which genes always express themselves, Durham distinguishes between the encoded information systems, including both genes and culture, on the one hand, and the environment external to the phenotypes that exerts selective pressure on them, on the other. The latter—the "environment"—is not encoded information, in contrast to both DNA and cultural symbol systems. In the instance of the interactive modes, however, Durham recognizes that genes and culture can each function as part of the selective environment for the other. In the "comparative modes" of interaction, by contrast, genes and culture are subject to selection pressure independently of each other. When this happens, the result can either be

3) that culture enhances genetic reproductive fitness; or
4) that it is neutral with regard to genetic fitness; or
5) that it opposes or diminishes genetic fitness.

Using these five relational modes between genes and environment, Durham then examines a number of historical and ethnographic cases in detail to illustrate each one. I will comment further on one of these, Mundurucu head-hunting, in the next chapter.

One might have thought, given the independence of each other that Durham grants to the genetic and symbolic channels of inheritance, that he would allow for the possibility that culture and genes might not infrequently act at cross purposes, but he does not go there. Rather, he argues that secondary values (that is cultural values) "must have acted as true 'surrogate values,' helping the system to make decisions . . . in general agreement with what was effectively the decision criterion of genetic selection" (208). That is, "as a general rule, cultural values that promoted decisions with maladaptive . . . consequences for their makers must themselves have been weeded out." This leads to his third hypothesis: that the main, though not exclusive, effect of human decision making, which after all is executed by a brain itself shaped by natural selection, is to promote a pattern of positive covariance between the cultural fitness of allomemes and their inclusive genetic fitness values for the individual phenotypes who embody and perform them.

Here it seems to me Durham gives back to the neo-Darwinian fold with one hand what he had just taken away with the other. Having admitted that humans could make intelligent choices that effected the direction of their own evolution, based on their own and their culturally inherited observations of the results of past or trial actions—something explicitly ruled out in the strict understanding of the mechanism of natural selection—and seeming to open the door to the possibility of cultural development heading off any which way by showing the two systems of inheritance to differ greatly and to be capable of operating independently of each other, he nonetheless supposes that when humans do decide and choose, they will do so in a way that actually does enhance, and therefore serves, their genetic fitness after all.

Durham in my view makes a theoretical wrong turn here. As I will argue in the next chapter, it is not necessary for someone making a decision to act in a way that is maladaptive from the genetic point of view to be "weeded out" by selection if the action is adaptive from the cultural point of view. Precisely because genes and culture are to a great extent independent of each other, a cultural trait passes along cultural, not genetic, lines, and therefore is not subject to natural selection acting on the genes via the bearer's comparative genetic reproductive success or failure. Cultural selection derives its adaptive advantage for humans precisely because it does not rely on the differential reproduction and mortality of individual phenotypes in a population in order to evolve. Human evolution, as Campbell and D. S. Wilson argue, and as I will argue at greater length in the course of this book, has had to create this second channel of information transmission precisely in order to overcome its own propensity to disallow, undo, and undermine extensive sociality, so as to obtain the benefits of the cooperation such sociality enables. To be able to do this, culture necessarily had to serve interests other than those of individual genetic fitness, if it was to do what it was evolved to do. This is a fundamental premise of my own view of dual inheritance, and it informs my ethnographic analyses throughout this book. It is also the view held by Richerson and Boyd, to whose work I now turn.

CHAPTER ONE

Boyd and Richerson (or Richerson and Boyd)

Beginning in the fertile intellectual milieu of the late 1970s, Peter Richerson and Robert Boyd began developing a research agenda explicitly asserting the reality and at least partial independence of both a genetic and a cultural track along which human inheritance proceeds, and indeed it was they who introduced the term "dual inheritance" (Richerson and Boyd 1978) to describe theories that start from this assumption as well as models derived from such theories. Bringing together the population thinking and associated mathematical modeling of Cavalli-Sforza and Feldman, the idea of cultural evolution as not only varying from genetic evolution but in fact restraining its imperatives in order to make social life possible as suggested by Campbell, and various other strands of co-evolutionary thought, they, and students and colleagues of theirs influenced by their thinking, established and have maintained a lively research enterprise that in many respects best represents the current state of the art of dual inheritance theory.

For Richerson and Boyd, "population thinking is the key to building a causal account of cultural evolution" (2006, 6), while "culture is an evolving product of populations of human brains" (7), in which ideas are stored and manipulated. As for what culture is, it is *"information capable of affecting individuals' phenotypes which they acquire from conspecifics by teaching or imitation"* (33, emphasis in original). The essential feature of culture, for these writers, is social learning, "the nongenetic transfer of patterns of skill, thought, and feeling from individual to individual in a population or society" (34). Starting from these premises, they are able to construct complex mathematical models of the flows of differing "cultural variants"—Durham's "allomemes"—among populations and subpopulations of individuals (and their brains).

Key to their theory is the idea that in the process of social learning, the naive individual, the "learner," has certain biases toward choosing some variants rather than others. These biases in turn have been evolved in the human brain by the forces of natural selection insofar as on average they result in behaviors that serve the ultimate evolutionary goal of inclusive genetic fitness for the individual actor. These biases include direct bias, that is, trial and error by the learner of the different variant models offered by cultural "teachers" to see which one works best; indirect bias, which influences the learner to choose the cultural variant employed by the prospective teacher who seems to be the most successful at the task at hand; and frequency-dependent bias, which involves choosing the variant that is most frequent in the local social environment, that is, conformity bias. These biases will, so it is assumed, usually lead to the learning of cultural variants that enhance genetic fitness.

The wild card among the biases is that which produces a "runaway effect." Runaway processes within the domain of genetic evolution are those that emerge and flourish because of the one exception in Darwin's model to natural selection, namely, sexual selection. In sexual selection, an individual member of a certain species chooses among prospective mates on the basis of features these latter possess that indicate qualities that will result in offspring with a high probability of genetic fitness for the one doing the choosing. But this process can misfire when the trait advertising fitness becomes divorced from real adaptive advantage. Thus, in a classic example, the huge antlers of the males of an extinct variety of elk are said to have evolved because while at first large antlers were indicators of fighting prowess in combat over mating opportunities, and thus qualities that it would be in the genetic interest of females to bequeath to their own male offspring by allowing such males to mate with them, gradually female brains evolved to choose larger antlers in and for themselves. The result was a "runaway" process in which antlers surpassed the optimum size for effective combat and became maladaptive because they were chosen for "aesthetic" and "sexual" reasons rather than because they actually did enhance an individual's reproductive fitness.

Among humans, the principle cause of "runaway" effects is the analogous selection of traits by one form of indirect bias, that in which learners model their behavior on that of individuals who are accorded high prestige by their neighbors, rather than those who have demonstrable success in some arena that actually enhances genetic fitness. Boyd and Richerson offer as an example the fact that on the Micronesian island of Pohnpei, men devote energy that might instead have been spent in productive agricultural pursuits to the growing of yams of enormous size whose value is principally ceremonial rather than nutritional. Men who grow such yams thereby gain great prestige in the community (Bascom 1948). So, according to the admittedly speculative hypotheses of Boyd and Richerson, what might have happened was this: originally, naive learners observed successful yam farmers who were able to bring the largest yams to the feasts, these being good indicators of the prospective "teacher's" farming ability. But once the size of yams became seen as indicators of farming success, then "beliefs and practices that lead to larger yams would increase. Individuals with a stronger tendency to admire large yams will be more likely to acquire these beliefs. This will cause the two traits to be correlated—and therefore, when the practices that lead to larger yams increase, so too will the admiration for the ability to grow large yams" (Boyd and Richerson 1985, 269–70). Such runaway processes are able to arise, these authors argue, because of the independence that cultural evolution can exhibit in relation to genetic adaptation.

Boyd and Richerson are dubious of the claims some cultural anthropologists such as Marshall Sahlins (1976a, 1976b) make for the importance of symbols in the constitution of culture, insofar as they understand Sahlins to be saying that symbol systems are not functional. They argue that many symbols really do serve important evolved functions. But for symbols that appear to be purely arbitrary, or not "sensible," to use their term, Boyd and Richerson argue that such symbols arise by beginning as indicators of some genuine value (in terms of real fitness) and evolve by a runaway process that makes that indicator acquire apparent significance as an index of prestige. Such significance may seem internally coherent, lending the symbolic system an air of "meaning," but this internal logic will have more in common with aesthetic than with functional design. "Much as peacock tails and bowerbird houses are thought to result from runaway sexual selection, the indirect bias runaway process will generate traits with an exaggerated, interrelated, aesthetically pleasing but afunctional form" (278). Many aspects of the cultural symbolic system are not, then, for these authors, of any actual value, seen from the point of view of fitness, but may be understood as epiphenomenal and perhaps even maladaptive. After all, peacock's tail feathers are indeed beautiful to human eyes, but (according to the theory) they do not do anything to actually enhance the adaptedness of the birds that grow them.

There is no point in denying that I, like most anthropologists interested in cultural symbolism, find the interpretation of the aesthetic coherence and meaning-producing dimensions of symbolism fascinating—something of little interest to most evolutionists, to judge by their writings. But it is also true that I, unlike some of my culturalist colleagues, like to find underlying explanations for why things are as they are, and that includes elaborate symbolic constructions. My response, however, is not a reduction of symbolic systems to some supposed genetic fitness-enhancing value. After all, I agree with Richerson and Boyd when they write: "Culture is interesting and important because its evolutionary behavior is distinctly different from that of the genes. . . . Culture would never have evolved unless it could do things that genes can't!" (2006, 7). In my discussion of their analysis of the symbolism of giant yams, in my Conclusion, I will in fact show that Richerson and Boyd's theories, when understood in the ways I will propose in this book, can be shown to make much more sense of the Pohnpei data than they themselves allow.

I have shown in this chapter that dual inheritance theorists have already made a number of adjustments and emendations to standard neo-Darwinian evolutionary theory as it is understood in the world of biology, some of them quite radical, in order to allow it to encompass or accommodate culture. In granting that cultural evolution is "Lamarckian," capable of transmission along lines of descent independent of genetic ones, subject to human choice and intention, composed of systems of symbols that are public and communicable, amenable to group selection, often at work in ways contrary to the best interests of the genetic program, and essential to the establishment of society by making possible selection for altruism and values that mitigate the (inclusive) selfishness of the genes, many dual inheritance and coevolutionary theorists have really moved a long way from strict neo-Darwinism in highly significant aspects of theory. Most important, perhaps, is D. S. Wilson's recognition of the right of socio-cultural systems to be seen as "organisms" that have their own integrity as expressed in their ability to reproduce themselves by means of the replication of symbols. In making these innovative emendations to the standard orthodox evolutionary doctrines, these thinkers have, so it seems to me, given to socio-cultural anthropology an opening for a productive encounter despite the gaps and differing assumptions that still separate them. It is to this challenge, and opportunity, that I respond in the pages that follow.[12]

TWO

When Genetic and Cultural Reproduction Diverge

In this chapter I examine some ethnographic examples that illustrate how in some cases practices that are enjoined in the cultural symbolic system can have deleterious effects on the fertility of the individuals in the society, without leading to overall harmful consequences to the socio-cultural system itself. This is because while the genetic program operates by aiming toward the optimal biological reproductive success of individuals in the society, the socio-cultural system itself pursues an independent reproductive goal, which is to replicate itself in the future generation of individuals who will populate the society in coming years. As I will show, these two reproductive aims can lead to very divergent outcomes; in the cases I consider, cultural practices that manifestly lead to reduced fertility on the part of individuals are offset by other cultural practices that replenish the population of the society with means other than biological reproduction. These examples thus demonstrate the relative independence from each other of the genetic and cultural channels of inheritance, but they also demonstrate that while the two may go off in opposite directions in some cases, the end result may well be the overall reproductive viability of the socio-cultural system. In looking at any particular case, one must take into account both the cultural and the genetic forms of reproduction, and their combined final result. Yet while the cultural channel, serving the reproductive interest of the symbolic system itself, does not, as these examples will show, directly serve the reproductive success of the genetic channel, their relationship is not one of lethal opposition but rather what might be understood as a necessary tension that gives rise to an array of creative socio-cultural attempts at resolution.

Just to be clear: the genetic system has been shaped by evolution to lead to behaviors and motives in individuals that will reproduce the gene pool itself in subsequent generations. By the same token, once it is granted that the symbolic systems constituting the social institutions of the socio-cultural system are themselves quasi-autonomous entities, these symbol systems can likewise be seen in a parallel light to produce motives and behaviors in individuals that will lead to the perpetuation of the socio-cultural system, though not necessarily of the people who comprise it. Both the genome and the symbol system "program" individual phenotypic people to whom they attempt to endow traits that would further each of their own reproductive interests. These may or may not coincide, but most often there is a degree of tension between them leading to inherent conflict both in individuals and in the social systems they constitute. The cases I present here illustrate, precisely by being quite extreme examples, how it is possible for the two to differ radically without leading the whole socio-cultural system to ruin.

Before turning to these examples, I want to introduce my discussion by addressing the relation between culture and genes that William Durham in his book on *Coevolution* (1991) calls "opposition," which might be confused with what I am arguing.

Durham on Opposition: The Mundurucu Case

As I discussed in the previous chapter, Durham proposes "opposition" as one of the five possible relationships between cultural and genetic evolution. In principle he does not believe that there can be very many genuine cases of this, because real opposition by culture to genetic fitness would lead to serious maladaptation and thus to serious decline, or, in the worst case, extinction. He rejects some cases that have been put forward as possible examples of opposition, such as Chinese footbinding, Pharaonic infibulation, the use of drugs, alcohol, and tobacco, and such still more apparently anti-fitness practices as abortion, infanticide, fosterage, adoption, and spouse exchange, arguing that we lack sufficient evidence that these allomemes do actually lower viable fitness values more than do other available options (1991, 374–75). And

CHAPTER TWO

it is certainly true that degrees of adaptation or maladaptation are very difficult to quantify, measure, and test.

But having dismissed these suggested cases, Durham considers at length two other candidates for being examples of real cultural "opposition" to genetic fitness. One of these, that of the spread of a deadly disease called *kuru* among the Fore people of New Guinea, is quite literally a no-brainer. The introduction of cannibalism in the mortuary practices of this society led to the unintended consequence that a scrapie-like replicating agent was transmitted by eating human flesh, including brains, causing an illness. Obviously, if people persistently eat anything, including the brains of deceased people, that routinely transmits a serious disease, this is going to be maladaptive in the strict sense of reducing fitness for them and for their society. And not surprisingly the Fore and their society collectively suffered very deleterious consequences.

Since Fore cannibalism is such a straightforward case of true opposition, and therefore merits no further commentary from me, I will turn to Durham's other, more interesting and debatable case of purported opposition, that of Mundurucu headhunting. The Mundurucu Indians, who inhabit the area south of the Tapajos River that feeds the Amazon in Brazil, live by fishing, hunting, and the cultivation of manioc. Before pacification in the early twentieth century, they were relentless warriors who carried out frequent headhunting raids on their neighbors near and far (Murphy 1960). They did not thereby acquire new territory, nor were their attacks directed at groups who had harmed them, or out of any motive of retaliation. A main reason they gave for headhunting was that the heads taken, processed and turned into trophies, not only raised the prestige of the men who took them but also played an important role in the Mundurucu belief system and ceremonial life: the trophy heads were thought to enhance the fertility and hence abundance of game animals.

So far, one quite frankly fails to see anything in this account of their headhunting beliefs and practices that would qualify them as illustrating the opposition of cultural beliefs and practices to genetic success. The only aspect of Mundurucu headhunting that seems maladaptive in the strict sense of lessening reproductive success (as opposed to being distasteful or reprehensible to present Western sensibilities) is the fact that the successful head-taker is awarded the status of *Dajeboiši*, which literally means "mother of the peccary" (white-lipped peccary being the chief game animal pursued in collective hunts). This title refers to the fact that the trophy head is thought to lead to the reproduction of the main game animal, and such fecundity is considered to be a "maternal" function.

The man who achieved such a distinction was surrounded by strict precautions: "Both the taker of the head and his wife were prohibited from sexual intercourse and were . . . supposed to avoid looking upon any person who had recently indulged in the act. . . . The Dajeboiši spent most of his time lying in his hammock and his wife was assigned special assistants to perform her household and garden chores" (Murphy 1957, 1025, cited in Durham 1991, 181). Since he did not hunt, other men brought him food during the years of his ritual isolation. If the Dajeboiši violated the terms of his elevated status, a matter that was publicly tested from time to time by ritual means, his trophy lost its power to produce abundant game.

Durham writes: "I should think one would be hard pressed to find a better candidate for opposition through cultural reason" (181). After all, a prohibition on all sexual intercourse would seem clearly to lead to a sacrifice of competitive reproductive fitness for the head-taker in relation to his fellows. Durham then sets about making the argument that Mundurucu headhunting is not, despite appearances, a case of opposition after all. He spends most of his effort arguing that headhunting does really increase game, which is a scarce resource also sought by enemy groups. It achieves this result, on his analysis, not by magical means but by scaring off would-be competitors from accessible hunting grounds. I have no objection to this argument, except that it addresses a non-problem, namely the supposed "opposition" of headhunting to genetic fitness which, as I have said, was never apparent in the first place.

The part of the Mundurucu headhunting complex that might represent actual "opposition," on the other hand, namely the restriction on the successful headhunter's sex life, and hence on his reproductive fitness, is dealt with by Durham in a passage in which he argues that among the egalitarian Mundurucu, headhunting, which poses equally distributed risks, would only be acceptable to the men of the community if it also provided equally distributed benefits to all participants in the raid, whether they took a head or not. This might not seem to jibe with the fact that the Dajeboiši was wined and dined by his non-trophy-taking peers, and thus raised to a higher status than them, but in any case the crux of Durham's argument is this rather obscure suggestion:

Consider the Dajeboiši in this light, then: lounging around the village without having to hunt for more than two years after the successful raid [elsewhere the length of Dajeboiši status is given as three years], his sexual activity would surely have threatened to absorb a disproportionate share of the group's potential gains. . . . Any more immediate self-indulgence after the head-hunt would have converted the participation of other, less fortunate members of the raiding party into costly acts of self-sacrifice. (1991, 390)

If I interpret Durham correctly in this opaque passage, he seems to be saying either (or both) (a) that the Dajeboiši needed to make some sacrifice to offset his enhanced prestige resulting from his success in the raid so as to re-equilibrate the expected egalitarian status of all the men in the group, or (b) that if he weren't restricted from sexual activity while he was alone in the village and the other men were off hunting, the Dajeboiši would be able not only to reproduce more efficiently with his wife but also to seduce the wives of the absent hunters, and thus gain a reproductive advantage over them. If he isn't saying this, these nonetheless seem like good arguments one could make if one wanted to take Durham's side of the argument.

All this might or might not be true, and certainly the maintenance of strict egalitarian ideals is a high priority in this as in many comparable societies. But it offers nothing by way of an argument against the supposition that, for the successful headhunter elevated to Dajeboiši status, there is a net reduction in his competitive fitness compared with his fellows who each now have two (or three) years of reproductive opportunity available to them that the head-taker lacks. I therefore can agree with Durham that headhunting is very likely a potential net gain for the Mundurucu collectively in their competition with other groups, for the reasons he adduces, namely, that it increases their access to game and denies it to competitive groups. (I may also observe that the successful providing of meat by men is the preferred ticket to sexual success with wives and/or lovers—and hence to enhanced reproductive fitness—in much of lowland South America.) But the restriction of the successful head-taker's sex life on the individual level, Durham's explanatory efforts to the contrary notwithstanding, is a net loss in his viable fitness values any way you slice it, and thus on the face of it a genuine case of cultural "opposition" to genetic fitness for the successful headhunter. Durham seems here to be motivated by a desire to show that any cultural practice really is at bottom genetically adaptive, for the sake of preserving the integrity of his view of evolutionary theory, rather in order to understand what is really happening.

I will offer another interpretation of this data, but before I can propose a better explanation of what may well be actually going on here,

I need to prepare the ground by examining some far more dramatic cases of apparent cultural opposition, that is, situations in which cultural beliefs, practices, and values really do lead to dramatic reductions in genetic fitness—but without the result being deleterious to the overall health of the society.

The Marind-Anim

In this book I put forward the argument that culture is regularly and routinely, by its very nature, always to some degree in "opposition" to genetic fitness. But this opposition is not of the sort that leads to the decline or extinction of the group as a viable socio-cultural system, but is rather an inevitable tension that results from the difference between the ways in which cultural inheritance by means of symbols and sexual reproduction by means of genes operate. Before beginning to make that case on a theoretical basis (in chapter 3), however, I will develop the discussion of headhunting raids more generally, through some interesting comparative ethnography. This will illustrate how a dual inheritance model that does not choose sides between culture and genetics, as Durham's does by denying the possibility of widespread opposition that is nonetheless not destructive to society, puts the matter in a quite different perspective. I begin this excursion by discussing another ethnographic case that clearly illustrates quite radical cultural opposition to genetic success, and that also complicates the whole picture of how to think about "opposition," "maladaptation," and the co-evolution of culture and genes. I then add a couple of other cases, and finally return to give an explanation of what I think might really be going on with the Mundurucu.

The Marind-Anim of the southern coast of Irian Jaya, the western half of the island of New Guinea now forming a part of the nation of Indonesia, were, like the Mundurucu, avid headhunters before their pacification under the colonial Dutch in the first part of the twentieth century (van Baal 1966, 1984; Knauft 1993). Their complex ceremonial life revolved around the enactment of myths that brought back to life the founding supernatural beings of their culture, the *Dema*. The taking of heads as part of their ritual cycles was necessary both because all children had to receive a name from a person whose head had been taken, and because heads were associated with the fertility of coconuts, which in turn were symbolically linked both to human heads and to human and non-human natural fertility. It is not hard to see that coconuts resemble human heads not only in their shape and size, and with their "eyes," but also in that both contain a milky inner pulp. This in turn links both coconuts and heads by association to the single most important substance in Marind ritual life, namely semen. Semen, thought to be the sole source of vitality and fertility, is used not only in all major rituals and in magic, but is also used as a body and hair ointment, as an ingredient in food, and so on. The widely encountered belief in the value of heads as sources of fertility often rests on the assumption that brains, spinal fluid, and semen are the same or closely related substances (La Barre 1984; Onians 1951).

Moreover, the semen that was put to these various growth-promoting uses could not, except in a few specified circumstances, be obtained through masturbation, but had to be mixed with vaginal fluid extruded from a woman's vagina after ritual copulation undertaken expressly for this purpose. In addition to such ritual occasions for intercourse, when a woman married or returned to active village life after childbirth, it was mandatory that she participate in a practice called *otiv bombari*, in which all the men of her husband's subclan, often as many as ten or twelve, have intercourse with her in succession, usually over the course of a single night. *Otiv bombari* was also performed on other ritual occasions as well. All these practices, which demanded frequent sexual intercourse, were intended to promote both human and natural fertility for the community.

In the early twentieth century there was an epidemic of the venereal disease cervical granuloma among the Marind (the word "anim" in their ethnonym simply means "people" and can therefore be omitted), ultimately affecting up to 25 percent of the women. A Dutch medical team was authorized to attack the problem, which came on the heels of the great influenza epidemic of 1917–19, and was accompanied by a serious population decline in Marind country. The Depopulation Team found, to its surprise, first that infertility had been an endemic problem among the Marind before the introduced diseases took their toll; and second, that the incidence of female infertility, which was abnormally high, did not actually seem to correlate with those women who had contracted granuloma.

The Depopulation Team concluded, on the basis of strong circumstantial evidence, that depopulation among the Marind, while certainly exacerbated by the granuloma and influenza epidemics, was "probably due to chronic inflammation of the *cervix uteri* of the female genital organs in consequence of excessive copulation" (cited in van Baal 1966, 27). In other words, it seems highly likely that the Marind practices of *otiv bombari* and the collection of semen for ceremonial purposes requiring ritual copulation—practices intended by the Marind themselves to increase fertility—actually had the opposite effect. Of women interviewed in 1953 who had been sexually mature during the last years during which the traditional Marind culture was still intact and their ritual and sexual practices still prevalent, a very high 33 percent had never been pregnant. In those generations of women coming of age after the Dutch Administration put a stop to Marind "sexual promiscuity," by contrast, only 6 percent had never been pregnant.

Here, certainly, is a case in which cultural beliefs and practices—ironically the very ones undertaken for the purpose of enhancing the fertility of the group—clearly worked in opposition to the genetic fitness both of the group and of the people who composed it. Calculating the percentage of young people in the population to gauge the trajectory of the local demography, the ethnographer writes: "40 [percent] should be adopted as the normal percentage of youngsters in the pre-contact situation [out of the total population]. As . . . the age at marriage was . . . almost certainly higher than 18, 40 [percent] in reality represents the highest possible percentage for the age-group of youngsters under pre-contact conditions. That implies that the rate of reproduction was insufficient to keep numbers constant . . ." (van Baal 1966, 31). But though the rate of fertility was thus on a clear downhill trajectory, the Marind did not suffer cultural decline or die out, as would be the inferred ultimate outcome of their "maladaptive" practices; in fact, they were an expanding society. This paradox is explained by the second half of the last sentence of the previous quotation: ". . . and that consequently, the kidnapping of children and their adoption as full-fledged members of the tribe was not a matter of indulgence, but necessity." In other words, in addition to the taking of heads for ceremonial and prestige purposes, a prime goal of Marind raiding expeditions was the capture of children who would replenish their otherwise diminishing numbers due to their self-induced infertility as a group.

Van Baal cites demographic data on the Marind collected by the missionary van de Kolk in the early years of the twentieth century. Of 211 couples, van Baal reports, van de Kolk found that they had 389 children among them:

Of these there were probably some 30 or 40 who had been kidnapped. In Okaba alone he counted 22 individuals between aged between 10 and 35 who had been abducted on head-hunting raids and subsequently adopted. . . . It was bad form among the Marind to allude to someone's being of foreign extraction. . . . Kidnapped children were raised in precisely the same way as own children and their real descent kept a secret from them as long and as completely as possible. It follows that the number of own children was in reality low. (van Baal 1966, 29)

So the high number of kidnapped and adopted children recorded by van de Kolk is probably even higher given that people either did not know, or preferred not to mention, the foreign origin of some of the children he counted.

There would then be some logic in arguing that the Marind belief in the fertility-enhancing power of headhunting had some validity, since the net effect was to offset the population decline among their own genetic offspring through the capture of young children of other groups, who would then be adopted and raised exactly as if they had been children produced by the sexual procreation of the Marind themselves. But, of course, from the evolutionary point of view, this would have been an illusion: overall the Marind beliefs and practices described here would have to be counted as "maladaptive" insofar as the gene pool came over time to be composed less and less of Marind genes and more and more of introduced genes from foreign competitors.

So the key question is this: do we want to say that the Marind have successfully reproduced themselves, or not? Clearly, from the point of view of genetic fitness, they have not. Their genes dwindle in the face of competition from foreign genes over time, while those of their headhunting victims continue to exist in the persons of their kidnapped children and whatever further genetic offspring they may engender. As individuals, and collectively as a population, Marind people are doing something genetically highly maladaptive.

From the point of view of an actual Marind individual, however, the situation is much more ambiguous. We have seen that captured children are taken very young and fully adopted by the tribe and by Marind individuals, who raise them in every respect as their own children, so much so that others, and even the children themselves, may have no knowledge of their origins as war captives. People who reproduce via adoption as opposed to sexual procreation may not be enhancing their own genetic fitness, but from a social point of view they are certainly reproducing themselves by producing successors, heirs, and bearers of the cultural symbolic system passed on to them by their adoptive parents. From the Marind cultural perspective, these adopted children are real descendents—just as real as if they had been biological offspring.

This allows us to say, further, that from the point of view of the

Marind socio-cultural system and its constituent beliefs and practices, communicated to offspring who are genetically related or adoptive indifferently, the symbolic code that encodes and transmits those beliefs and practices has, in a manner analogous to the genetic code that reproduces itself in each new phenotype, also created a new cultural "phenotype," independently of whether that new enculturated person is born of genetically Marind parents or not. Even the practice of *otiv bombari*, deleterious though it may have been to the fertility of the married women who engaged in the "excessive" copulation it required of them, would be reproduced in subsequent generations of Marind without any harmful effects on the overall Marind society. On the contrary, as a result of their cultural beliefs and practices, the Marind maintained viable if perhaps diminished population levels, transmitted their incredibly complex mythic and ceremonial cycle intact across generations, expanded their territory, and, into the bargain, protected themselves from attack by foreign groups, all of whom lived in terror of their legendary ferocity and would never have dreamed of encroaching on Marind territory. To call such a result "maladaptive," in any broader sense, would therefore seem entirely inappropriate, even though it is certainly "maladaptive" in the strict evolutionary sense. Indeed, for all we know, at some point in their history, there may have been very few or even no "Marind" people who were actually genetically descended from genetically Marind ancestors. Whether or not this ever occurred, it is certainly a theoretical possibility.

This example is instructive in a number of ways. Suffice it to say here that it demonstrates that the cultural channel of information that informs the rising generation in a social group may operate from its own point of view, successfully and therefore "adaptively," not just independently of the genetic channel but even when it actively interferes with the latter's success in demonstrable ways.

And, to take the argument a step further, if there are indeed, as this example clearly shows, two separate channels of inheritance at work in human life, by what right do we give the genetic one preferential treatment in judging the whole system just described as "maladaptive" when that is only the case with regard to the genetic channel? I think we here encounter the bias in Western culture itself, from which evolutionary theory itself springs, to the effect that, as Schneider (1984) argued, "blood is thicker than water." We tend to assume, in our cultural symbolic system, that genetic relatedness is "real" relatedness. However, the direction of the whole field of cultural anthropology has, for the last several decades, been toward arguing that there is no warrant

CHAPTER TWO

for assigning greater "reality" to the DNA than to the symbolic code in the study of kinship.[1] In a true "dual inheritance" model, they are parallel methods by which humans reproduce, and we will be better off theoretically if we refrain from judging what is adaptive or maladaptive solely on the basis of one mode of inheritance and instead examine the totality of genetics and culture working simultaneously, which is always entailed in the continuation of human life in society over time.

This bias in favor of genetic over cultural inheritance is clearly evident even in the thinking of so open-minded a coevolutionary theorist as Durham: thus for example, he refers to symbolic effects as "secondary functions," compared with the "primary" genetic ones. But there is nothing intrinsically primary about genes over symbols, except that we can safely assume that genetic inheritance preceded cultural inheritance in evolutionary history. Given that human life is now as fully informed by cultural instructions as it is by genetic ones, however, that history is of no real relevance for a good understanding of the current situation, in which genes and culture both play determinative roles in shaping human lives and societies.

Caduveo/Mbaya

One may well counter, at this point, that the situation I have just described among the Marind is a unique anomaly in the ethnographic record.[2] To respond to this argument, let me turn next to the case of the Caduveo Indians, who live today as a remnant of the once mighty Mbaya people in the Gran Chaco region of Brazil. At their height, the Mbaya were divided into powerful chiefdoms who dominated and exploited other neighboring Indian groups. Today, the Caduveo, Pilaga, and Toba are the surviving groups speaking languages derived from the language family currently designated "Guaicuru" (or "Guaycuru"), which was the name by which the Spaniards and Portuguese referred to these peoples' ancestors. (So "Mbaya," "Caduveo," and "Guaicuru" or "Guaycuru" all refer to the same group.) The heyday of the Mbaya ascendency, which began shortly after Spanish contact, and culminated in the late eighteenth century, is well documented in the work of the Jesuit missionary José Sanchez Labrador, who lived in close contact with the Mbaya from 1760 to 1767, as well as in the works of other Spanish observers.

Originally a hunting and gathering group, the Mbaya were transformed and launched into a new and successful cultural adaptation with the coming of the Spaniards to South America and more particularly by their adoption from the Spaniards of horseback riding as a regular mode of existence. Thus outfitted, they were able to dominate, raid, and enslave neighboring horticultural peoples, and style themselves as nobility. Sanchez Labrador writes:

> They [the Mbaya] are masters over a great extent of land and inspire terror in all other nations. Nowadays there are quite a few chiefdoms each having its independent chiefs. . . . The principal chiefs of these groups are Napidigi and Apagamega. Their sons and some of their relatives are already shown honors due to chiefs, being acknowledged by the people as legitimate masters. (Cited in Oberg 1949, 52)

Lévi-Strauss, who visited the contemporary Caduveo during his travels in Brazil, described in his classic book *Tristes Tropiques*, and who greatly admired their distinctive style of face painting, amplifies the picture of the historical Mbaya:

> These Indians had kings and queens, and like Alice's queen, the latter liked nothing better than to play with the severed heads brought back by their warriors. . . . The Guana [a neighboring tribe, ancestors of the contemporary Tereno] cultivated the land and handed over part of their produce to the Mbaya lords in return for their protection against the pillaging and depredation practised by bands of armed horsemen. A sixteenth-century German who ventured into these regions compares the relationship between the two populations with those existing in his day in Central Europe between feudal lords and their serfs. (1973, 179–80)

Unlike the Marind, who valued natural fertility and sought to increase it by ritual means, the Mbaya were, quite to the contrary, dramatically hostile toward genetic reproduction, which they saw as beneath their dignity as feudal masters. Lévi-Strauss writes:

> It was a society remarkably adverse to feelings that we consider as being natural. For instance, there was a strong dislike of procreation. Abortion and infanticide were almost the normal practice, so much so that perpetuation of the group was ensured by adoption rather than by breeding, and one of the chief aims of the warriors' expeditions was the obtaining of children. It was estimated at the beginning of the nineteenth century that barely 10 per cent of the members of a certain Guaycuru group belonged to the original stock. (1973, 181–82)

In addition, when children were born and survived, they were not brought up by their parents but rather were given to another family and visited only rarely.

CHAPTER TWO

Sanchez Labrador, our main source on the historical Mbaya, writes in a similar vein:

> Guaycuru cruelty is exhibited in abortion and infanticide. . . . Spinsters practice this cruelty in a hidden manner, as if they were committing a sin. As soon as they feel the burden of their imprudence they seek to provoke an abortion by whatever means their inhumanity dictates. . . . The married people do not beat around the bush, but openly try to kill their children in their entrails, or they redouble their cruelty by killing the little creatures at birth. (Cited in Oberg 1949, 55)

Even today, Oberg reports, the Caduveo say that they had to capture children from other tribes in order to keep up their numbers. He cites Sanchez Labrador, who asserts: "due to these cruelties there are not many children in Guaycuru settlements. Among all I know but four couples, who being exceptions, had two children. All the rest had either one or none" (55).

What accounts for such a negative attitude toward procreation? One answer is that the Mbaya lived as stockbreeders, keepers and traders of slaves, and raiders for booty, who viewed sexual reproduction as something fit only for lower orders of being than themselves. As long as they could replenish their numbers through slaves and captive children, they saw no need to engage in sexual relations for any reason other than the pursuit of pleasure. Felix de Azara, another early observer of the Mbaya at the height of their power, gives a lengthy account of the preference of women for either one or no children (cited in Boggiani n.d., 22–23). He reports that he himself reproached women and their husbands for sacrificing their children and "thus exterminating their own nation." They replied, smiling (he tells us), that men should not meddle in the affairs of women. Finally a woman confided to him that if they continued to bear children, it would cause their bodies to lose their form and to grow old, and their husbands would not be attracted to them. Therefore they induced abortion as soon as they sensed themselves pregnant in order to remain beautiful and sexually active.

Another reason for this anti-procreation attitude, cited by Boggiani as well as by Oberg (1955), is that many South American groups whose mode of subsistence required them to be constantly on the move, as was the case with the horseback-riding Mbaya, fared best when their social units were composed of fit individuals capable of keeping up with the group in its nomadic expeditions. In practice, this meant that people between the ages of about fifteen and forty were preferred as group members, and if these could be best attained by capturing them, adopting them, and enslaving them, as the Mbaya were able to do, then so be it.

Lévi-Strauss, meanwhile, offers yet another explanation: the Mbaya were divided into three strictly segregated castes, with differing levels of prestige, so that a leading anxiety of any members of a given caste was the fear of losing face by marrying below their station. The result was that each caste shut itself off from interaction with the others, thus imperiling the cohesion of the society as a whole:

> In particular, the inbreeding practised by the castes and the ever-increasing gradations in the hierarchy must have made it more difficult to have marriages that corresponded to the concrete necessities of collective life. This alone can explain the paradox of a society being opposed to procreation and, in order to safeguard itself against the dangers of improper alliances within the group, having recourse to a form of inverted racialism through the systematic adoption of enemies or foreigners. (1973, 194)

Whichever of these explanations one prefers, singly or in combination, the situation among the Mbaya is similar to that of the Marind in that their cultural beliefs and practices led them to drastically limit their own procreative fertility, and to offset the resulting depopulation with nongenetic offspring captured from enemy peoples on whom they made war. As in the case of the Marind, the Mbaya genetic line of inheritance was severely compromised, but practices encouraged in the cultural line prevented the society from dwindling toward nonviability by means of cultural instead of sexual reproduction. The big difference, of course, is that while the Marind were trying to enhance procreative fertility, but went about it in a way that proved to be deleterious, the Mbaya were openly hostile to the whole enterprise of producing genetic offspring. This example introduces a new element into the equation: not only do some cultural ideas and practices actively oppose genetic fitness, but some cultural ideologies actively oppose the very idea of sexual procreation essential to genetic fitness; yet they nonetheless do not thereby undermine the long-term replication over generations of the society as a society, as a strict evolutionist model would lead us to expect. They do so by finding other ways to recruit new immature members to the group by nonprocreative means—in the two cases just considered, by stealing them.

Comanche

An intriguing parallel to the transformation undergone by the Mbaya is the case of the Comanche Indians of the North American plains. They too had been a relatively small-scale, socially and technologically uncomplicated group of hunters and gatherers until the latter half of the eighteenth century, at which point, after they became acquainted with horses through contact with the Mexicans who were expanding northwards, they migrated from their original home in the high plateau to the southern plains. In short order, the Comanches, as they then became known, learned to rustle, breed, and ride horses with great expertise; most sources credit them with being the finest horsemen among the various Plains Indian groups who adapted to the new way of life made possible by this great technological innovation. The horse not only enabled a new way of exploiting the bison herds of the plains on a massive scale, but also led to new and much more efficient techniques of raiding and warfare.

Once mounted, the Comanches were able to travel great distances and to impose their will by force not only on other Indian groups but also on the encroaching Mexicans and Texans, ruling over what is often referred to as an "empire" called "Comancheria" that covered a vast territory in the southern part of the North American West. By raiding, as well as by trading and dealing in horses that they bred or stole, as well as in buffalo hides and guns, they became major players in the history of the United States west of the Mississippi until their defeat by the U.S. Cavalry in the 1870s. At the height of their domination of the southern plains, they were as much a traveling business enterprise and mobile gang of terrestrial pirates as they were an "Indian tribe" in the usual sense.[3]

When living as horseless hunter-gatherers, they had not practiced the capture and adoption of women and children, since they were not predominantly a warring people. When they became the Comanche Empire, by contrast, these practices became both profitable and necessary. The capture of women was advantageous because the new Comanche enterprise required a large workforce to process buffalo hides, work traditionally done by women. The capture of children was necessary because the Comanches suffered from low fertility rates that created the demographic need for recruits to the group obtained as children from outside who could be socialized as Comanches and added to the workforce as well as to the fighting force. The best-known of all Comanche captives was Cynthia Ann Parker, kidnapped at the age of nine in a Comanche raid on her family homestead in Texas south of what is now Dallas–Fort Worth. She became assimilated into the Comanche social and cultural world and married a Comanche chief. Her son, Quanah Parker, emerged as one of the great legendary military leaders on what was eventually the losing side of the wars of conquest waged by the United States against the Indians of the North American West.

The cause of Comanche infertility was neither ritual sex that misfired, as among the Marind, nor outright hostility toward procreation as among the Mbaya. Some of it no doubt resulted from introduced Western diseases, but Fehrenbach intriguingly though perhaps somewhat speculatively (1974) suggests that some of it may have been the result of relentless and continuous horseback riding itself, which can produce both lowered male fertility as well as miscarriages in pregnant women. Whatever the cause, lowered fertility was offset by the recruitment of new members by nonprocreative means, specifically by kidnapping them. If we follow Fehrenbach's suggestion, then we have another situation in which a cultural practice—horseback riding—leads to a net reduction in genetic fitness, while also enabling the military prowess by which that deficit can be overcome by the replacement of the group with members captured in successful war.

Mundurucu Headhunting Revisited

I now return to the Mundurucu case left hanging earlier in this chapter. In his article on Mundurucu warfare, on which Durham bases his analysis of that practice that I considered earlier, Robert Murphy (1957) addresses the question of why such warfare occurs at all. Quoting an article on war among the Mohave by George Fathauer (1954), he asserts that their "conception of war was largely non-instrumental: it was an end in itself." Among the Mundurucu, war was "considered an essential and unquestioned part of their way of life, and foreign tribes were attacked because they were enemies by definition" (Murphy 1957, 1025). Informants did not themselves identify as reasons for war the defense of home territory or any provocation by the other groups. Nonetheless, through direct questioning, Murphy was able to identify a number of reasons why the Mundurucu went on headhunting raids when there was no apparent provocation to do so. One of these, in the wake of contact with white people, was for pay: the Mundurucu were highly regarded by whites as loyal and effective mercenaries. But material gain was only secondary, since the Mundurucu mercenaries took no interest in looting defeated enemies for their own benefit.

Other motivations for the Mundurucu man to join a headhunting expedition, ones which unlike the previous one would have predated contact, included the prestige that went with success in war and the spiritual power that was thought to inhere in the trophy head, as I have mentioned above. Another motivation Murphy adduces is relief from boredom: fighting was a source of sport and excitement. After rejecting these various motivators as insufficient to explain the enthusiasm for war, Murphy turns, in the most important part of his article, to uncovering the functions of war for Mundurucu society, in contrast to motivations that might spur individual warriors. His argument is that the peculiar arrangement whereby matrilocal residence was paired with patrilineal descent created multiple cross-cutting ties of affinity and descent that facilitated intervillage cooperation among the men, which in turn created a military advantage over competing groups. Within the village, there was a sharp contrast between male and female behavior. Whereas the women, who were mostly close relatives, often quarreled among themselves, it was considered absolutely essential that there be no breach of harmony among the in-marrying men, who were neither biologically nor culturally related to each other. Since friction among humans living together is always present, suppressing overt conflict among the men by redirecting hostility to outsiders was the only alternative to serious disruptions of social life.

Adding to the complicated situation, chiefly families, unlike the rest of the community, practiced patrilocal residence, and the result was that in any given community there was a conflict of interests between the authority of the chiefly lineage and the in-marrying men who had no kinship ties to bind them to that lineage and who might resent the headman and his kin. "Under these conditions of potentially divisive institutions of leadership, descent, and, residence, warfare functioned to preserve, or at least to prolong, the cohesiveness of Mundurucu society" (1030) by focusing ingroup aggression outward against "enemies." This all seems reasonable enough, but as Murphy himself points out, the Mundurucu probably only became matrilocal after contact. This complication leads him to conclude that, since the Mundurucu were vigorous warriors before contact, when they were organized along patrilineal, patrilocal lines, "warfare acquired new functions subsequent to that change and consequent upon it" (1031).

Therefore, as Durham notes, Murphy's "social cohesion" argument rests on a rather weak foundation. This leads Durham to propose his own hypothesis about the social and adaptive function of Mundurucu warfare in resource acquisition and defense, which I have discussed earlier in this chapter.

In the present context, however, I want to focus on one other dimension of Mundurucu headhunting that both Murphy and Durham mention in passing, but without pausing to examine its implications. Writing of the spoils of raiding, Murphy notes:

> No objects of significant economic value were taken, and the raiders maintained their mobility by traveling light. Captive children were a far more important class of booty than were material items. Captive-taking was a means of strengthening the group through the addition of new members, but it is not certain whether the increment compensated for the loss of mature warriors and their female followers. If, however, one accepts intense warfare as a given factor, the capture of children was important to its successful continuance. Moreover, the Mundurucu valued and desired children, and the captives were treated as the captor's own. (1027)

At this point in my argument, after a tour through the cases of the Marind, Mbaya, and Comanche, I need hardly point out the relevance of this factor in understanding headhunting raids among the Mundurucu: they were a way of reproducing the group without procreating sexually. What was the source of the problem in this particular case? The Mundurucu were not repulsed by procreation, as were the Mbaya, nor did they indulge in ritual practices harmful to female fertility, as the Marind did, or reduce their own fertility by horseback riding as the Comanches may have done.

Murphy states that it is not certain whether the taking of captive children was a way of making good the Mundurucu individuals killed in the course of the raids. If that were the case, warfare would still seem rather pointless insofar as there would have been no lost group members to replace with kidnapped children if they had just stayed home and not put their lives at risk. The vicious circle inherent in this explanation leads us to inquire whether other demographic considerations might have been at work.

In their ethnographic study of women in contemporary Mundurucu society, Robert and Yolanda Murphy (1974) state that the population level of the Mundurucu is relatively stable. However, it appears that the situation is a bit more complicated. Not only are disease and infant mortality constant threats to successful procreation, but both abortion and infanticide are practiced as well. "Given the conditions of public health, then, it requires a high conception rate for the Mundurucu to just hold their own" (161). Moreover, there is a conflict between men and women regarding the place and value of childbearing. Men, though they in fact have little to do with young children, very much want progeny. Women, however,

> are resentful of the continued cycle of pregnancy and birth, regarding it as an encumbrance and physical handicap and the source of their principal preoccupations and labors. Most of them are quite affectionate mothers after the child is born, and they usually want children during the early years of marriage. After three or four live births, however, their desire for children declines markedly. Men want children more than do women, they say, and men also favor boys over girls, a sentiment which is not shared by the mothers. (161)

Noting that women employed several birth control methods, made from root concoctions, and that these same roots can also induce abortions, the Murphys add that "the attitude of the women toward male disapproval of contraception and abortion was quite simply that it was not their business" (161).

In short, the demographic situation among the Mundurucu is that population stability was precarious, even more so since the introduction of diseases spread by white contact, but always on the margin given the ever-present threats of stillbirth, infant mortality, and the birth of unwanted children such as twins and deformed infants, who were put to death. But adding to this situation was the mismatch between the goals of men and women when it came to having children by means of sexual procreation. Men wanted many children, and they wanted male children—probably so as to give them an advantage in warfare once the children were grown. Women also wanted a few children, but only up to a point, and they had the power to prevent or end pregnancy at will. This was resented by the men who did not like having control over this vital resource taken out of their hands and managed by their wives, who operated independently in this regard.

What we can surmise about men's motivation for headhunting, then, is that, along with other considerations, it was a means of redressing the balance of power between men and women over reproduction of the group. If the women would not produce an endless stream of (mostly male) children by means of biological birth, as the men desired, a cultural remedy was close at hand. One could obtain children by the expedient of killing their parents and stealing their children, as we have already seen in the other cases I have examined. As with the Marind, headhunting would indeed increase "fertility," insofar as it really did add to the progeny of the successful warrior. That the Dajeboiši—the successful head-taker—was enjoined from engaging in sex for a few years would thus have been offset by the fact that he had very likely already acquired a child (or children) of his own through his success in the raid that gave him his status, but by means of cultural rather than sexual reproduction. That is to say, the raid itself would have made him a father by virtue of the fact that he would seize the child or children of his victim(s). The high prestige and value placed on war would indeed have had a direct reproductive payoff for Mundurucu men, as long as one accepts the proposition that adopted kidnapped children, though not contributing to genetic fitness, do nonetheless enhance cultural fitness in a very real way in ensuring new viable generations of culturally Mundurucu people.

Is this, then, a case of "maladaptation" or, in Durham's terms, "opposition"? Child capture and adoption are not adaptive in the Darwinian sense of enhancing genetic reproductive fitness in individuals; and certainly in the cases I have discussed, the cultural and genetic channels of inheritance seem to be working at cross purposes, and sometimes indeed in very plain opposition to the goal of genetic fitness. If, however, once accepts the reality of true dual inheritance along separate genetic and cultural channels, then it is misleading to discuss the cases in terms of something going wrong. In each of the cases I have discussed, the societies in question successfully reproduced themselves—even if their genes did not.

In this chapter I have given some examples from ethnography that demonstrate the reality of dual inheritance. Cultural practices may work against what would be best for genetic fitness, but they also can offset the damage by means of reproduction that can properly be called cultural rather than biological. It is entirely possible to think about human social life as a composite that reproduces itself over time both by means of sexual reproduction and by means of the reproduction of the vast array of symbolic information transmitted externally to new human beings once they have been reproduced. In the various cases I have given, there are factors that inhibit the rate of sexual reproduction, but of course genetic reproduction does after all have to occur somewhere in order for there to be children to enculturate and socialize in accord with the group's symbolic system. As we have seen, however, the biological reproducing does not have to be accomplished by the same people who are doing the enculturating and the socializing.

Although the cases I have given are, admittedly, not the typical ways social reproduction is accomplished in human groups, they serve to illustrate the starting point of my argument, which is that there really are two different channels of inheritance, that neither one can be considered more real or more basic than the other, and that they need not work in harmony for social reproduction to occur. On the contrary, I will show that they almost always work in some degree of disharmony, and that this is a necessary outcome of their different intrinsic characteristics. Starting in the next chapter, I will lay out this argument in detail, beginning with a discussion of how best to conceptualize the cultural dimension of ongoing social life.

THREE

The Cultural Channel

This book is not a contribution to the study of how culture may have evolved in our prehistory; rather, it takes it as a given that human beings, and the societies in which they come into being, live, and reproduce themselves are formed by the combined operation of instructions flowing along two different channels, encoded in different media, and operating in differing ways. How this arrangement may have emerged in actual evolutionary time is not my concern here.[1] My aim is to think about, and explore ethnographically, the implications and consequences for human society of humans being the product of this unique configuration of information transmission over the generations.

    The ethnographic examples in the last chapter served as illustrations of the ways in which a social group, the cultural system that sustains it and which it in turn sustains, and the descent lines of the individuals within it can have continuity over the span of mortal generations despite culturally prescribed and transmitted practices that are maladaptive in genetic terms. This phenomenon supports the view that the two channels of the dual inheritance model are, in principle, to a large degree independent and capable of operating on the basis of different agendas and values. It further shows that the cultural channel, rather than being a parasite on the genetic channel (as is stated or implied in many evolutionary approaches to the study of human life, according to which it is at best supposed to serve the aims of genetic fitness), is itself as real a source of essential information capable of engaging in real social reproduction as is the genetic channel.

In the last chapter, we saw that cultural practices that work to reduce genetic fertility, such as the *otiv bombari* among the Marind, can reproduce themselves over time through alternate means of recruitment of new generations into Marind society, that is, by the kidnapping, adoption, and re-socialization of non-Marind. We also saw that a society such as that of the Mbaya can have cultural beliefs that are openly hostile to biological reproduction and can nonetheless continue over generations thanks to nonsexual reproduction via child capture. As I will show in later chapters, such hostility to sexual reproduction, though perhaps not as extreme, is common in the ethnographic record. In the present chapter, though, I focus on some key differences between the cultural channel and the genetic channel and spell out how these differences may be expected to lead to certain interesting consequences. I will then, in subsequent chapters, show that ethnographic data from around the world supports the theoretical picture I will have drawn of how dual inheritance works, and what effects it has and what traces it leaves on the characteristic shape of any human society.

Why Did Culture Evolve?

While as I have said it is not my intention to address the evolutionary history of the human capacity for culture, it is helpful to begin by asking the question of what adaptive advantages it offers that must have led to its being selected for in the course of hominin evolution. It goes without saying that unless culture did in fact enhance genetic reproductive fitness for the hominin line and ultimately for *Homo sapiens*, including at its origin as well as over evolutionary time, it would not be here today. Of the many advantages it represented, I will focus on two. The first is the fact that culture enabled humans to live together in highly cooperative societies. Social cooperation enhances the fitness of the individuals who participate in it by enabling work projects that are more efficient and productive than could be achieved by individuals alone, making possible defense and offensive warfare that benefits the members of the group in competition with other groups, and making available to them resources, in the form of human helpers and allies, as well as in the form of material resources and distributed information that in their totality far surpass their own capacity or that of any single individual. Although it may be assumed that in the course of evolution humans also adapted to the cultural environment by genetic changes encouraging prosociality, a main point of the present chapter is to argue that culture, being a collective phenomenon, by its very nature requires and entails enhanced cooperation.

The second adaptive advantage is that once culture gets off the ground, it enables adaptation to new niches, situations, climates, and ecologies in a vastly more efficient way than can be achieved by ordinary natural selection. Societies with culture, and thus the individuals constituting them, can adapt quickly to changed circumstances of any kind, taking advantage of new opportunities and avoiding threats to their way of life, without waiting for the cumbersome process of natural selection to do its work.[2] That process, it must be recalled, depends on the differential reproduction and mortality of phenotypes with varying traits that result from different alleles in the genome. The genetic variation, in turn, is thought to be the product of mutations that, if not strictly speaking "random" in the mathematical sense, are in any event "blind," in that they are not based on feedback from the phenotype about what would be useful in adapting to an altered environment. Evolution by natural selection achieves fitness only through a very inefficient process that does not involve foresight, an ability to plan, an ability to remember and thereby learn from past experience, an ability to survey the environment and decide what changes are likely to prove beneficial and then put them into effect, or an ability to utilize feedback from the environment via conscious trial and error or other experimental methods—in short, an ability to set goals, recognize or invent what needs to be done to achieve them, and do it.

Put in this way, it seems obvious that an organism that was in fact able to do all these things in the course of its phenotypic life would be able to run evolutionary rings around almost any competitor that lacked such abilities. While I do not speculate about how or when in evolutionary time these capacities emerged, it is self-evident that the genome of *Homo sapiens* did develop such capacities, just as it developed the human ability to cooperate effectively in groups. So the next question is, how does culture make these feats possible?

Richerson and Boyd on Culture

An answer offered by dual inheritance theorists is that humans' enormous capacity for social learning enables adaptive information to be passed from one generation to the next, and that the human brain has evolved under natural selection to allow learners of culture to model their own actions on those of already enculturated "teachers" (though without the implication that they necessarily "teach" in any self-conscious or explicit way). Learners may conform to the most prevalent observed behaviors, or they may model their own behavior after that of particularly successful, or particularly prestigious, individuals. But what is culture?

For Richerson and Boyd, culture is the information that is transmitted in the course of social learning: *"Culture is information capable of affecting individuals' behavior that they acquire from other members of their species through teaching, imitation, and other forms of social transmission"* (2006, 5, emphasis in original). What is the nature of this "information," and how does it get transmitted—along what channel, or in what medium, does it move in the course of its spread or decline in the population? Here, Richerson and Boyd's theory rests on the following assumption: "Our definition is rooted in the conviction that most cultural variation is caused by information stored in human brains—information that got into those brains by learning from others" (ibid., 2).

In locating cultural information in the brains of individual actors, Richerson and Boyd are able to apply the kinds of mathematical modeling already so successfully used in population genetics to the study of the spread or decline of some trait, programmed by cultural information, in a population over time. The success or lack thereof in terms of the fitness of any given behavioral trait, in this view, may be measured by modeling, and/or observing, the demographic rise or fall of numbers of people in whose brains are lodged the various "alleles," or variants of the information in question. Note that in this model, it is not the genetic fitness of the person whose brain contains the information that is being counted but rather the relative preponderance of brains containing one or another competing cultural allele in a population. Since information does not have to replicate among genetic kin, nor does it require the differential biological reproduction and mortality of numbers of individuals to determine the outcome, this model grants real autonomy to the cultural channel. The success of this approach, demonstrated in numerous research works in the dual inheritance tradition, confirms that this is a useful and robust set of assumptions. One might characterize it as a population genetics, *mutatis mutandis*, of competing alleles of ideas located in individual people's heads rather than of competing alleles in their genes.

Richerson and Boyd further argue: "The vast store of information that exists in human culture must be encoded in some material object. In societies without widespread literacy, the main objects in the environment capable of storing this information are human brains and human genes" (2006, 2). Since brains, and the neurons that compose them, are material objects, and since the brain has doubtless been shaped by the forces of natural selection over the course of evolution, one can comfortably make the analogy with the genes: cultural information, in the form of ideas, beliefs, values etc., is encoded in neurons just as genetic information is encoded in genes in the DNA. And just as natural selection in the genetic realm preserves some genetic instructions and weeds out others, so too a Darwinian selection process favors neuronal information that leads to behavior that gets copied by others in the process of social learning. Some information inscribed in neurons may be capable of being inherited genetically, while other information is learned by the brain—or rather the person who owns it—in the course of life through cultural inheritance.

Richerson and Boyd acknowledge that a vast amount of information must be stored in brains for culture to operate. A linguistic lexicon, for example, so they assert, requires somewhere around sixty thousand associations between words and their meanings, while grammar entails highly complex sets of rules about how to combine lexical items. Subsistence techniques even of hunter-gatherers in small and simply organized societies require more sheer information about the environment than Western scientific observers have been able to identify and record; and these authors go on to ask us to "imagine the instruction manual for constructing a seaworthy kayak from materials available on the north slope of Alaska" (2006, 61). But since we know that human brains are indeed capable of making unimaginably large numbers of neuronal connections, it certainly seems plausible at first blush that brains are the material medium best suited to storing cultural information.

Nonetheless, Richerson and Boyd also recognize that there is another potential candidate for the material medium in which this huge sum of information might be stored:

Undoubtedly some cultural information is stored in artifacts. The designs that are used to decorate pots are stored on the pots themselves, and when potters learn how to make pots they use old pots, not old potters, as models. . . . Without writing, however, artifacts can't store much information. . . . Without written language, how can an artifact store the notion that Kalahari porcupines are monogamous, or the rules that govern bridewealth transactions? With the advent of literacy, some important cultural information could be encoded in the pages of books. Even now, however, the most important aspects of culture still tend to be those stored in our heads. (2006, 61–62)

I yield to none in my recognition of the importance of the invention of literacy as a great advance in the capacity to store information in a medium external to the human brain, as Goody (1968, 1977, 1987, 2000) has extensively shown. But the passage just quoted raises a crucial question: in the absence of literacy—the condition that prevailed throughout almost all of human evolution and history in the majority of societies in the ethnographic record—how *does* all that information about, say, the Kalahari desert enter the brain of a young hunter just undergoing enculturation? The answer of course is perfectly obvious: some of it he discovers for himself through his own experience; but for much of it, he learns it because someone who already knows either tells him, or shows him by his actions, or some combination of both. That is indeed what is entailed in the "social learning" model so central to the dual inheritance discourse about culture.

Let us examine this model more closely by comparing someone reading an instruction manual for how to construct some cultural item—let's say a kayak, to follow Richerson and Boyd—with someone who is constructing it under the guidance of an illiterate teacher who is telling and/or showing him how to do it. In both cases, we would certainly not be wrong to say that the source of the instructions is information that is stored in the kayak expert's brain. But, as Richerson and Boyd also realize, this information is of no value to the learner as long as it merely stays stored away in someone else's brain. To enter into a "social learning" situation, the knowledge of how to build a kayak has to get out of the expert's brain and be translated into some external form in the public social—that is, interpersonal—arena so that the learner can perceive it and then go on to internalize it. That is, it must leave the neurons and realize itself in external, material form in the public world—in the learner's sensorium, that external, observable "public" space which is the object of our sense organs.

In the case of a written instruction booklet, the expert has transferred his knowledge into written form, perhaps accompanied by pictures or diagrams, which the learner can see, interpret, and try to match with his own actions. In the case of the illiterate expert, he translates his "neuronal" knowledge into spoken words, while also "showing" with gestures or while saying "watch me," or "do it like this" or "no, not like that." In this case, too, the knowledge has been translated into a public, observable form in the external world available to perception by the learner. The situation is identical to that of the written instruction manual, except that the medium in which the expert has materialized his knowledge is sound waves, if the information is aural, or light waves conveying images of the teacher's gestures and actions. I believe this is where some people sometimes get confused, since they fail to remember that sound waves shaped into language by the human larynx are just as "material" a medium as is a printed page, or a neuron—or DNA. All information, to do its work and get itself transmitted, has to take the "significant form" of meaningful signals in the perceptual sensorium of the learner (to allude to the influential work of Susanne Langer [1957]). There is really no reason to privilege the knowledge contained in books over that which can be transmitted orally as far as its reality, or material existence, is concerned. The only difference is that the former has a greater range of dissemination and its transmission can be delayed in time.

This is not a mere quibble over the role of books in cultural processes. It is an assertion that while it is true that cultural knowledge is stored in brains, there is also always another medium, either external to the body, enacted by it, or inscribed upon it, in which cultural knowledge is recorded. Knowledge is of no use to anyone until it gets out of the brain and puts itself in action in the observable world. Even when no explicit teaching is involved—when, for example, the learner simply observes the expert doing something and imitates it, or examines the expert's product itself and uses it as a model—the knowledge stored in the neurons is embodied in actions that make or have made physical traces in the world and can be seen, heard, or otherwise sensed. Those traces in the external, material world represent the actual way in which the knowledge gets into the brain of the learner, and thence into his own actions in building his kayak. To get out of one brain and into another, cultural knowledge has to be translated into another material medium, namely some aspect of the external, perceptible world that, by its pattern, shape, or system of differences, is able to convey highly complex information.

To see the importance of the point I have been making, consider the analogy with the genetic transmission of instructions in sexually reproducing species, such as humans. The instructions for how to construct a new human organism are contained in the nuclei of the gametes, encoded in the DNA of the reproductive cells. These gametes in turn are stored in the testes and ovaries. As long as they stay there, they have no effect in the world, and certainly do not generate a new phenotype upon which selection can act. It is only when they leave their storage place, and are brought into union with a gamete of the opposite sex through the act of copulation, that the gametes can begin to actually transmit instructions to a "learner," that is, an embryo that constructs itself under the guidance of the parents' DNA that is now housed in an identical copy in each of the embryo's proliferating cells.

In just the same way, to say that cultural information is stored in brains is no more a complete picture of the whole process of information transmission than is the statement that DNA is stored in the gonads. While true, this is only a partial picture of the process. Genetic information may be stored in gonads, but it is only activated by being transferred to the new organism by means of the inscription of the DNA in the sperm and ova, leading to the union of the gametes of two fertile adults of different sexes. Cultural information, on the other hand, is only transmissible when it is put into the observable world in perceptible material forms that serve as vehicles for meaning to someone who can perceive and make sense of it. The implications of this distinction, which I will spell out in the course of this book, are vast. *Genes transmit their information to new individuals by means of copulation; cultural information is transmitted by inscription in some material medium in the external, perceptible, "public" interpersonal world.*[3] This is another of the key premises underlying the theory and ethnographic analyses I offer here.

Clifford Geertz on Culture as Systems of Symbols

I have said that a great deal of what we call culture consists of the significant forms in which knowledge—instructions, information, ideas, beliefs, values, attitudes, hopes, dreams, and the rest of it—enters into the perceptual material world and thus into the process of ongoing social life. A vast number of these significant forms that convey meaning are what I will call "symbols." I stress once again here that not all cultural knowledge is encoded in symbols: some social learning occurs by direct imitation without any or much by way of symbolic trappings, and some cultural knowledge people figure out or invent for themselves. But symbolic forms in the broadest sense, including both discursive and presentational symbols in Langer's terms, are so pervasive and predominant in human life that I will focus my attention on them, taking them, *pars pro toto*, as the distinctively human components of the cultural channel of inheritance.

A problem with the word "symbol," however, is that it conveys different meanings in different scholarly discourses and this leads to a great deal of misunderstanding. Despite this, I have chosen to use the term "symbol" in preference to the alternative term "sign," which would otherwise have worked as well, because there is no adjectival form of the word "sign" analogous to the very useful adjective "symbolic." Just to give a taste of the range of uses of the term "symbol," consider that in mathematics and symbolic logic, a symbol is something that can be assigned a precise and unambiguous meaning or referent, whereas in Jungian psychology and similar discourses a "symbol" is inherently multivocal and highly charged, carrying a penumbra of emotional as well as referential meaning that has no clear boundaries. Thus people may use the word "symbol" to refer on the one hand to something as precise and logical as "x" in "x + 2 = 4," and also to something such as the American flag or the Christian cross or things with even less clearcut referents. In the linguistic school of Saussure and the structuralism that emerged from it, "signs," rather than "symbols," are the generic term for the representations of meaning, and in the semiotics derived from the work of C. S. Peirce, signs are further subdivided into a trichotomy of trichotomies. For Cassirer and Suzanne Langer, symbols are the true human vehicles of meaning, whereas "signs" are natural phenomena with no intention to communicate, as when we say that a dark cloud is a sign that it is going to rain. And of course Freudian "symbols" imply something else entirely. But what all these different usages, and many more besides, do share is the idea that something perceptible in the external world conveys meaningful information by its distinctive shape, form, or structure, and the way this form or structure conveys meaning may be usefully analyzed or interpreted, both by actors within the socio-cultural system within which the symbols circulate and by the ethnographic observer.

Many thinkers understand the "symbolic" as if it stood in opposition to the rational, or the technological, or the practical, or the commonsensical. Thus, one sometimes reads that, for example, planting a seed is practical but saying a prayer over it is symbolic.[4] The prayer said over a plant may or may not be efficacious in our judgment, but it is certainly real and material in that it is encoded in the physical medium of sound waves, and that pattern of sound waves conveys meaning that can be understood or interpreted. The question of whether the prayer is a real thing, which it is, is completely different from the question of whether it is efficacious in the ways claimed for it. In my present usage, both the "practical" technique of planting and the "ritual" technique of praying are "symbolic": knowledge of both is conveyed across generations in encoded form, whether by linguistic signs or through meaningful gestures, and is not programmed in the genes.

The way to proceed then, which I follow here, is to recognize that all human knowledge and action, whether we think it is rational or effective or otherwise, takes place through the use of material vehicles that bear meanings that travel from person to person through the perceptual realm of the external world. All of these are what I will refer to by the generic and inclusive term "symbols," with no implication about whether they are motivated or arbitrary, precise or imprecise, presentational or discursive, fuzzy or clear and distinct. Any act of human social learning, sometimes even apparent simple imitation, may be a symbolic act, first insofar as it entails the transmission through space from one person to another of significant form, and second because the expert and the learner are both wholly or partly enculturated persons whose very existence as humans requires them to have already learned and embodied a huge array of cultural symbols, into which any new information is incorporated in a way determined by the already existing symbolic matrix. If one does not limit one's understanding of symbols to discursive symbols such as linguistic ones, but allows for the word to connote any "significant form" inscribed in a material substrate, then one may call a teacher wordlessly showing a learner how make a bow and arrow symbolic—especially since the resultant bow and arrow will have a distinctive style or form that distinguishes it from the bow and arrows constructed in other socio-cultural groups. In the end, however, as I have indicated, the important point about cultural knowledge for our present purposes is that it passes through the material world by giving it recognizable and communicable shape or structure that can be perceived and imitated or understood by a learner. Therefore I need not prolong a discussion of what exactly is or is not covered by the term "symbol" other than to say I intend it here in the broadest and most inclusive possible light, as the generic term for a unit of culture.

Importantly for my present purposes, human social organizations, even while obviously involving the natural realities of birth, death, and sexual reproduction, do so within the context of symbolic constructions such as marriage rules and prohibitions, ethnobiological beliefs, kinship terminologies, afterlife conceptions, and so on. Furthermore, they generally are nested within an overarching cultural construction about the meaning, order, and morality of the cosmos. Forms of social organization and the constructions of a cosmos that uphold them—what Rappaport (1999) usefully conceptualized as "logoi" (the plural of "logos")—are organized systems of symbols, and they give form to any ongoing socio-cultural system. And it is mainly on these domains that I focus in this book.

No theorist has surpassed Clifford Geertz in eloquently and convincingly articulating the view of "culture" as a system of symbols. He writes:

> So far as culture patterns, that is, systems or complexes of symbols, are concerned, the generic trait which is of first importance for us here is that they are extrinsic sources of information. By "extrinsic" I mean only that—unlike genes for example—they lie outside the boundaries of the individual organism as such in that intersubjective world of common understandings into which all human individuals are born, in which they pursue their separate careers, and which they leave persisting behind them when they die. By "sources of information," I mean only that—like genes—they provide a blueprint or template in terms of which processes external to themselves can be given a definite form. As the order of bases in a strand of DNA forms a coded program, a set of instructions, or a recipe, for the synthesis of the structurally complex proteins that shape organic functioning, so culture patterns provide such programs for the institution of the social and psychological processes which shape public behavior. (1973b, 92)

Since symbols exist out in the world in material form, thinking itself is not a purely intracerebral process, but is rather the interaction of the human mind with external representations and models that provide plans for action and information about how to implement them:

> A fully specified, adaptively sufficient definition of regnant neural processes in terms of intrinsic parameters being impossible, the human brain is thoroughly dependent upon cultural resources for its very operation, and those resources are, consequently, not adjuncts to, but constituents of, mental activity. In fact, thinking as an overt, public act, involving the purposeful manipulation of objective materials is probably fundamental to human beings. (1973a, 76)

The upshot of these considerations is that while it is true that human brains do house in material form the stores of knowledge that allow us to function as complete human beings, the actions of the "mind" derive from an interaction between the brain and the extrinsic array of cultural symbols and the meanings embodied in them, into the midst of which every child is born. Indeed, and this is crucial, the symbols are, from the point of view of the developing child, out there in the world *before* they are in the brain. Children are surrounded, from some time during gestation in the womb, by the sounds of the language of their community, a language that was there out in the public airwaves before they were conceived, and that once born (and even *in utero*) they learn by exposure to this external reality. So while it is true that symbols from the world can be retranscribed in the course of learning and enculturation into neuronal codes and stored in the brain, they do not originate there, nor is that their only or even their primary location.[5]

In order for me to argue that culture, and the symbols that constitute it, reproduce themselves over time, it is crucial for my argument to understand that it is in their existence as observable features of the ongoing external public social arena, not as neuronal patterns encased in mortal brains, that they are able to play this role, one analogous to the role played by the genes in genetic inheritance across generations.

It is the easy communicability of information encoded in external symbols that makes possible the rapid change and proliferation of socio-cultural systems in the face of changed environmental or other external changes, and allows humans to remain one species while creating novel strategies for adapting and living that any other organism would have to evolve genetically in much slower evolutionary time.

Consequences of Viewing Culture as an Extrinsic System of Symbols

I have come therefore to the conclusion that the crucially operative locus of culture "in itself"—if I may permitted that phrase—is not its engraving in the neurons of individual brains, though that is a necessary aspect of its operation, but rather its inscription in the external world where it can be seen, heard, touched, tasted, smelled, and experienced by the person via the senses. And if we think of it as having an existence in the material medium of the various waves and chemicals that convey information to us, then a number of consequences follow.

The first of these is that a human society, which is organized and coordinated by rules, beliefs, values, and attitudes embedded in systems of symbols, is not best understood solely as a population made up of individuals with cultural information in their brains; or more generously put, there are many things one can learn about human cultures other than what can be discovered by analyzing them in population terms. A human socio-cultural system has emergent properties, and these consist in large measure of structures made up of the rules, beliefs, and norms governing the organization of the constituent individuals among themselves; in other words, social structure and social organization. Richerson and Boyd, along with some other dual inheritance theorists, recognize the importance of "institutions" in constituting human society, thus acknowledging the existence of collective realities above the level of individuals. While it is true that people have to know these rules etc. with their brains in order to live in accord with them, the rules emerge in human social interaction in the form of externally perceptible symbols organized into systems in the context of which such social interaction takes place. It is therefore a fundamental principle of social and cultural anthropology that institutions such as "matrilateral cross-cousin marriage" or "male initiation rituals" can be understood, analyzed, and compared on their own terms and need not be thought of only as a collection of individuals with brains containing such-and-such instructions packed into them—any more than the function of an organ in the body is best described as the summed actions of the cells that constitute it, or than the structure of a cell can be derived from the structure of the various proteins of which it is made.

Of course, if the individuals who make up an institution by their collective action cease to act in ways supportive of the ongoing existence of the institution, the institution no longer exists. But this does not interfere with usefully thinking of a socio-cultural system and its constituent institutions as units at their own level that can reproduce themselves over generational time, while the individuals comprising them are born, enact them, and die, to be replaced by different individuals with different neurons that have to be programmed anew from the symbols in the sensory world around them.

Furthermore, if one takes the view that cultural information is lodged only in brains, rather than having a life in the space between or among brains, an analogy with genetic processes easily suggests itself; on this common understanding, information in the brain is like the genotype, and behavior in the world based on this information is like the phenotype. I believe this is wrong and misleading. The brain full of information is a feature of the mature phenotype; it only comes into being through having been constructed both by the genetic instructions and by the cultural input that has interacted with the brain as the person is developing. The view of cultural information as existing *first* in the sensory environment, and only *then* in the brain (or mind) thus reverses the more usual picture: cultural symbols residing in the perceptual realm are analogous to the genotype, and enculturated individuals with information now ensconced in their brains are thus like the phenotype—*not* the genotype. In the world of biology, selection operates on the phenotype, but what is preserved across generations and undergoes evolution is the genotype. To be true to this analogy, then, it is preferable to say that it is the cultural system of symbols existing in the external sensorium or public arena of a community that is preserved across generations, not anyone's brain or what was in it; these perish with the phenotype. It is thus, for example, that the Marind practice of *otiv bombari*, so maladaptive from a genetic point of view, reproduces itself nonetheless culturally, by instructing young recruits, including kidnapped and adopted ones along with those produced sexually by Marind parents, who observe the ritual and its symbols and learn to participate in its repetition, and who pass it along to their children, genetic or otherwise.

What emerges from this discussion is that the symbolic channel has its own reality independent not only of the genetic channel but also of the individual actors who constitute the cultural equivalent of phenotypes. Cultural phenomena replicate themselves the same way genes do: by instructing individuals how to construct themselves and then how to do what they need to do to survive, flourish, and reproduce. These individuals can perpetuate the genes and the cultural symbols both, by transmitting them to others in future generations. The two channels are thus actually a good deal more analogous at this level than I think they have previously been widely understood to be. Symbols and DNA both give instructions, encoded in a material medium by shaped and structured form, by means of which the complete human phenotype realizes itself. I have already referred to the main and key difference between the two channels in their method of transmission; now it is time to state it again very plainly: *the transmission of the genetic one is achieved by means of sexual intercourse, while that of the cultural one isn't.*

Culture as Cards on the Table

To attempt to give a clearer picture of what I am talking about for those to whom the view of culture as a system of symbols existing in the real material world is unfamiliar or still obscure, let me give a homely little analogy. Imagine a game of Texas Hold'em, a currently popular version of poker.[6] Several players are seated around a table and the hands, dealt from a deck of cards, consist of only two "hole" cards, dealt face down to each player, which only the player holding them can see. Five more cards are then "flopped" on the table face up by the dealer in succession, so that there are finally five cards on the table facing up. Betting takes place at each stage of the game: before the deal in an "ante," after the initial hole cards are dealt, and after each flop. Since poker hands are rank ordered on the basis of kinds of "hands," consisting of different patterns based on characteristics of the cards such as number or suit, a hand being made up of five cards in all, each player puts together the best possible poker hand using five cards from among seven in all: the five cards face up on the table and his own two hole cards. When the hands are shown, the highest ranked hand wins the game and the person holding it collects the pot—all the money that has been bet.

Let us say that the cards are cultural symbols (which of course they are; but in my analogy they can stand in for cultural symbols in general, of which they are excellent exemplars). The rules of the game—the markings on the cards that differentiate them each by number and suit, the ranking of poker hands formed from combinations of them into a hand, and the order and rules of play I have just described—organize these symbols into a system. Let us suppose I am standing around the casino and I want to bet some money, but I don't know how to play. So I either ask someone, or I read an instruction booklet of some sort, or I can watch the game as it is being played by people who already know how to play it and deduce the rules from the social interactions (moves in the game) that I observe, perhaps asking a question from time to time such as "Remind me, does a full house beat a straight?"

The information necessary to play Texas Hold'em is now in my brain, and I am able to use the external symbol system that I have internalized to put together a high hand, to bet according to the strength of my hand, and in general to develop a strategy and play the game. For example, when I ask my dumb question about whether a full house beats a straight, I may really not know, or I may want to pass as a novice to induce another player to take me for a fool and try to exploit my lack of experience with an overbold move that I can then use to my advantage. But note that the information was out there at the table before it got into my brain: I learned by some combination of symbolic instruction, social learning, and individual thought.

Now let us suppose further that all the original players at the table leave one by one—some because they have lost enough money, others because it is getting late, or because they don't want to risk their winnings, or for some other reason. As each one leaves, new players come up to the table and join the game. The players change, new dealers come on in shifts, old decks of cards are discarded and replaced with new but identically marked ones; they all enter and then exit the game—but the game itself goes on indefinitely.

Where, then, is the game? Knowledge of it must be in players' brains (or minds) for them to participate in it, certainly. But the game itself is evidently *on the table*. It is outside each individual in the symbols—the cards—that constitute the game when ordered by a set of rules that also enters my brain via some external, perceptual medium of communication: whether someone tells me, or I read about it, or I figure it out by watching. The game takes place in the public arena, in full view of anyone, which is indeed why this particular game has become a spectator sport and can be broadcast on television. The game was going on before I joined it, and it will go on with the same system of cards and rules after I exit from it. In the same way, a social system exists before any given individuals enter it, and it will exist after they die, precisely because it has an external existence that transcends the particular individuals involved at any given time.

Now suppose I am an anthropologist and I want to know what is going on here. I, too, can learn about the game in same way a novice player can. When I write my article on it, I can then describe the rules of the game, as I just did a few paragraphs ago, recording the markings on the cards and reporting on their relative rank order as well as the rank order of hands. I can also, if I want to do so, report on the various strategies employed by individual players, and I can give accounts of actual games to convey a sense of how it is played out in real time. (The one thing I can't do is look into the brains of the players, unless I have brought my fMRI machine with me, and even if I have it won't tell me much about how to actually play Texas Hold'em.)

In sum: studying cultural information by counting the number of individuals with a certain set of beliefs and rules in their heads and how these numbers wax and wane in time is indeed a good way to chart the course of an innovation in a particular group, for example, and to create a wide range of predictive or explanatory evolutionary models of changes in a population over time. Thus, in the present case, poker is a game with several different varieties: cultural alleles, if you will. That repertoire includes draw poker, stud poker, high-low poker, and many other variants. A few years ago, no one had ever heard of Texas Hold'em, and now it is widely known, played, and watched. An evolutionary model based on population thinking can plot the course of this development—no mean feat, to be sure. One could then talk about the evolutionary trajectory of poker in our time, as Texas

Hold'em becomes an allele that has succeeded. And one could then make the case that this success is due to the fact that it is the variant of poker in which the most information is visible from the start, and therefore the most suitable to spectatorship and hence to exploitation by television broadcasters who can sell it to advertisers. None of this, however, tells us how the game is actually played. Whereas seeing culture as residing in brains enables a population genetics approach to culture, then, we might say that viewing it as existing in brains in interaction with inscriptions in the external world where the symbols are publicly available to perception leads to a cultural version of comparative anatomy and physiology of socio-cultural systems with their own observable rules, structure, organization, and operation.

Symbols and Collective Consciousness

Each of the players in a game of Texas Hold'em holds a different two-card hand, and each plays the game from his own perspective and with his own self-interest in mind, and does so with his own particular strategy, style, and skill level. They are all competing against each other in the hopes of winning the pot, and everything they do can be interpreted in the light of this "jackpot" motivation. So far, so Darwinian (let's assume in evolutionist fashion that winning a jackpot enhances ones attractiveness to members of the opposite sex as the evolved basis of its being rewarding in its own right). But the game is only possible insofar as all the players also agree on the meaning of the symbols and the rules governing their organization into a system, and agree to play by these rules. In other words, the information necessary for the game to go on is the same for each participant, though their objectives and particular situations are different and, indeed, at odds.

Furthermore, most of the information necessary to play the game intelligently is available to everyone, in the form of the five cards that are face up on the table. Only each person's two hole cards are privately known to each one of them (though the others try to deduce what they might be based on the player's betting pattern, his or her playing style, statistical probabilities, facial expressions, and so on). Not only is culture thus "public," meaning registered in some shared social space, however big or small, open or restricted; the fact of its being public means that the instructions it gives to individuals can in principle be identical and shared by an indefinitely large group of people. It is "identical" in that, while people's perceptions of and thoughts about how to deploy them may vary, the actual cards that everyone sees are literally the same ones, and the rules written in the official rule book are likewise one and the same for everyone involved. The cultural symbols are units that when replicated in the minds (or brains) of several individuals may vary in terms of each individual's knowledge, experience, and wishes, but which in themselves do not because there is only one of each of them. This line of thinking is of course one that is entailed in viewing the symbols as external and analogous to the genotype, not to the phenotype.

The range of culture's ability to be shared publicly is limited only by the relative effectiveness of the media of communication available. This extends from what can be experienced together by twenty or thirty people gathered around a fire or on the dance ground of a camp of hunter-gatherers to what can be disseminated via electronic media in a text or tweet or video that has gone viral and is capable of reaching hundreds of millions of people around the globe; or, in the other direction, what is shared between only two people.

Identical genomes are shared only by identical twins. Even first degree relatives—parents and children, full siblings—share only an average of 50 percent of their genes, and the percentages fall off sharply beyond that. This means that inclusive kin selection based on shared genetic information ceases to be a powerful factor in the behavior of individual organisms once it gets beyond the immediate family. When it reaches to more distant relatives, or to members of the same clan or moiety, it is clear that the identity that is operative is cultural not genetic. I wish to stress once more my point that this does not make it any the less real. Benedict Anderson's influential work (1991) on "imagined communities" shows how media such as radio can make shared identity based, for example, on "nationality"—which has no analogue in the realm of genetic relationship but is purely constructed of cultural symbols (albeit with symbols that refer to an idea of shared biological relatedness)—and on the sharing of things like language or dialect.

Therefore, while genetic information at best can in principle achieve something approaching real unity of purpose among the very closest relatives—full parents, full children, or full siblings—and otherwise pits each individual phenotype against the others in a competition for reproductive success, cultural information can in principle unite many more people, including people completely unrelated by genetic kinship, in a shared identity. It is this fact that makes possible the adaptive advantage of culture I mentioned earlier, namely, that it enables genuine cooperation among nongenetic kin in a way that is otherwise very rare and difficult in the natural world, where genetic dynamics operate alone without culture. Individuals who are informed by identical information can regard themselves as "the same," as sharing identical substance, whether that information originates in the genes or in the cultural symbol system. And here culture definitely has the upper hand, since it can unify far more people in a cooperating group than can genetics, which normally works against such widespread cooperation.

Anyone who has ever been in a crowd at a highly charged athletic stadium during an important game, or in a political rally, knows the Durkheimian euphoria that overtakes those in attendance when something exciting happens. The crowd rises to its feet "as one man," as the saying has it. This fellow feeling lasts for a little while, and is similar to the familiar phenomenon in which people who have lived as separate individuals pursuing their own interests with little connection to their neighbors are transformed, at least temporarily, by having all been involved in some extraordinary occurrence, even a bad one such as a natural disaster. In such situations, people including total strangers often perform altruistic acts and indeed acts of selfless heroism, and treat each other with a warmth and empathy that is missing during more ordinary times of life. I would propose that an explanation for this is that if we have all just left a rally for a successful candidate, or a ball game in which the home team won, or have been caught up in the same blizzard, then there is no need to have any doubt about the ideas and intentions that are foremost in the minds of complete strangers whom we meet: we know that they are all thinking and feeling the same thing, and that it is the same thing that we are thinking and feeling. Therefore it doesn't really matter if the collective event is good, such as a political or athletic victory, or bad, like a blizzard or hurricane. What matters is that everyone has experienced it together and we can be assured that, at least for a short while, they are all feeling "kinship" with everyone else who experienced it. And this kinship is just as "real" as any other kind, since it consists precisely in the sharing of similar information. (That natural disasters and sports victories can also on some occasions lead to rioting and looting does not negate their potential to form ad hoc *communitas* as I have described.)[7]

Having mentioned Durkheim, let me quote him here, since he set the course for this line of thinking:

If left to themselves, individual consciousnesses are closed to each other; they can communicate only by means of signs [what I am here calling symbols] which express their internal states. If the communication established between them is to become a real communion . . . the signs expressing them [individual sentiments] must themselves be fused into one single and unique resultant. It is the appearance of this that informs individuals that they are in harmony and makes them conscious of their moral unity. It is by uttering the same cry, pronouncing the same word, or performing the same gesture in regard to some object that they become and feel themselves to be in unison. (1965 [1912], 262)

Durkheim was here speaking about the power of ritual, but as an elementary form of pure basic sociality in human communities it does not require any "religious" or sacred overtones. Among the smallest and least complex hunter-gatherer groups, collective singing or dancing occurs with or without any notion of sacredness, ritual, or orientation toward the supernatural. Radcliffe-Brown, in a well-known passage, was taking a page straight from Durkheim when he wrote about dancing among the Andaman Islanders, hunter-gatherers of the Indian Ocean: "The well-being, or indeed the existence, of the society depends on the unity and harmony that obtain within it, and the dance, by making that unity intensely felt, is a means of maintaining it" (1964, 250).

Dancing, like singing, is a cultural phenomenon in which a single external "significant form"—the melody and/or rhythm—is experienced simultaneously and collectively by everyone participating and draws them into a unity that is entirely real. Song and dance can both be performed by everyone in unison, or in organized or improvised "harmony," but in either case the collective whole experienced by each participant is greater than any individual. A delightful example from more recent ethnography is the description by Frank Marlowe (2010) of song among the Hadza, hunter-gatherers of northeastern Tanzania:

The Hadza sing often, and everyone can sing very well. When several Hadza get in my Land Rover to go somewhere, they almost invariably begin singing. They use a melody they all know but make up lyrics on the spot. These lyrics may go something like "Here we go riding in Frankie's car, riding here and there in the car. When Frankie comes, we go riding in the car." They take different parts in three-part harmony, never missing a beat, all seemingly receiving the improvised lyrics telepathically. (67)

Steven Mithen (2005), building on an observation of Darwin's, argues persuasively that singing and music generally preceded language in human evolution, and, given its ability to form unity among people, and thus enhance prosocial experience and feeling, it is easy to see why this would be so. Merlin Donald's theory of a stage of "mimetic culture" in human evolution is also consistent with this line of thinking. William McNeill (1995) has written what is perhaps the best description and analysis of the phenomenon of coordinated movement in the creation and maintenance of group solidarity in his book about dance and military drill, *Keeping Together in Time*.[8]

I want to stress again that singing and dancing are examples, present in human society from the simplest to the most complex, of situations in which externally encoded information, in the form of patterned sound waves producing rhythm and melody, give identical instructions simultaneously to a collectivity of individuals who feel themselves, and are, united into a larger whole by virtue of sharing the same external instruction simultaneously. The text of the song is not as important as the fact that it is sung in unison or together in harmony, and indeed repetitive nonsense syllables may form the most beloved parts of many songs—they are the easiest to sing all together. In the same way, Rappaport (1999) suggests that the most important and effective religious credos are the ones that contain the smallest amount of discursive "information" and for that very reason can command near universal participation and assent.

In an insightful discussion of songs among the Baining of northern New Britain, Jane Fajans notes that anthropologists have found Baining songs disconcerting because they have such mundane texts. She quotes Gregory Bateson: "The songs contain no reference to ancestors or other religious concepts. They are simple little lyrics referring to incidents of daily life, to love affairs, and to dreams" (cited in Fajans 1997, 260). One song Bateson collected, for example, has these lyrics: "They stared at me / They stared at me / You would have said they were frogs / You would have said they were fish" (1997, 262). The song was composed by a young man recounting his experience on being taken to court.

As Fajans points out, while the songs are composed by individuals and usually reflect on private, shameful, and otherwise socially problematic situations and events, they are learned by the group and sung collectively. In this way, they function as publicly shared "gossip" about potentially disruptive events. I would interpret this further as follows: the Baining are deeply concerned about the maintenance of social harmony. Asocial events that could disrupt social harmony, and lead to bad feelings and conflict, are contained by being expressed in a song that is then performed collectively. In singing collectively to antisocial lyrics, the group transforms the description of the events in the lyrics to something that is experienced by all, in this way turning its disruptive potential into an enactment of social unity, so that the

CHAPTER THREE

shared experience of union repairs the rift contained and described in the lyrics.

Evolutionary thinkers who study culture are often stumped by the question of how humans managed to achieve cooperation given the fact that the exigencies of competition for individual (or inclusive) genetic fitness render cooperation unsustainable and hence empirically rare in non-human living nature.[9] (I discuss some cases of social cooperation among animals in chapter 5.) The answer, I am suggesting, lies in the fact that culture, as an alternative channel of information, enables groups of people to experience themselves as equivalent to kin, in the sense of "people who share and participate in the same reality" and thus to see each other as enough alike that cooperation can trump the competition inherent in their genetic makeup.

Cultural and social anthropologists have long talked about the unifying effect of kinship not in terms of biological relatedness but in terms of "amity" (Fortes 1969) or "diffuse enduring solidarity" (Schneider 1980). The most recent contribution to this discussion is that of Sahlins (2011, 2013). After noting that "kinship fashioned sociologically may be the same in substance as kinship genealogically" (2011, 4) and giving several ethnographic illustrations of the point, Sahlins formulates a kinship system as "a manifold of intersubjective participations, founded on mutualities of being" (14). Recuperating Levy-Bruhl's idea of "participation," shorn of its association with any so-called "prelogical mentality," Sahlins understands it in terms of intersubjective being, in which people "live each other's lives and die each other's deaths . . . sharing one another's experiences even as they take responsibility for and feel the effects of each other's acts" (16).

This mutuality of being means participating in each other through jointly participating in a shared symbolically constructed social unity, as Durkheim said. It is often, though not universally, understood in terms of the sharing of substance, but it is critical to remember that the "substance" that is shared can be either genetic, or symbolic, or both. Thus, as Sahlins observes, citing Merlan and Rumsey (1991), people of the Nebilyar Valley of New Guinea conceptualize kinship as created either by sexual reproduction or by social practice, by the transmission of grease or fat, which is the essential matter of all living beings. This substance can be transmitted via the father's semen in sexual intercourse and via the mother's milk in nursing; but since sweet potatoes and pork also contain it, it can likewise lead to consubstantiality through food-sharing or eating from the produce of the same land (Sahlins 2011, 4–5). "Kin" constructed in the latter social and symbolic way are considered just as real "kinsfolk" as those who share biological parents.[10]

This creation of inter-implicated individuals is made possible by the fact that the symbol system is external and can thus unify as many people as the available media of communication will allow. Identification with others on the basis of the similarity of socially construed symbols is thus a parallel, or rival, basis for cooperation and mutuality to the one posited by inclusive fitness, and provides the basis for the sociality that enables humans to operate as a group or a society rather than as merely a collection of individuals.

Group Selection Explained (or Deconstructed)

The considerations explored in the previous section lead to another point about the consequences of the extrinsic existence of culture in its various symbol systems, namely that it solves at a single blow the problem of "group selection" that has so plagued the community of evolutionary scholars of human culture and society. As we have seen, if one views human groups as populations of individuals competing for inclusive reproductive fitness, the models from population biology show that cheaters and defectors will undermine any altruistic tendencies over time and all but eliminate the possibility of true prosocial coordination unless adequate safeguards are established. Contemporary evolutionary orthodoxy proposes that groups can only engage in selection among themselves when the constituent individuals are more similar among themselves than they are to the constituent individuals of a competing group. In that circumstance, groups that strongly cohere internally can form themselves into larger entities that compete and are selected naturally. This idea actually goes back to Darwin himself (2007 [1871]), who believed that it was self-evident that a society composed of group-spirited people loyal to each other would outcompete a rival group in which no such harmony of interest prevailed. We have seen that E. O. Wilson has recently argued in favor of this view of human societies as prosocial groups. He argued that if intergroup competition or hostility in evolutionary time was frequent and lethal enough, it would provide the climate in which group selection could take place.

Bowles and Gintis (2003, 2006, 2011) argue that the kind of strong reciprocity observed in human groups could not have evolved through individual selection, and therefore must be a product of group selec-

CHAPTER THREE

tion during hominin evolution. Like E. O. Wilson, these authors rely on a presumption that Pleistocene bands engaged in frequent violent and lethal warfare with each other; this is the condition that allows inter-group competition to trump intra-group competition and lead to the prosocial and even altruistic strong reciprocity observed in actual human societies. Sterelny (2012) has challenged this assumption on the basis of the archaeological evidence and other considerations. The debate about hunter-gather intergroup violence is itself a very heated one that I will not explore further here.[11]

In any event, the fact of dual inheritance renders unnecessary attempts to generate human altruism and prosocial behavior from selection acting either on individuals or on a population of individuals in a band confronting other hostile foreign bands. The reasons for this are twofold. The first is that if people can just as well feel themselves, and actually be, closely related to each other through the agency of the symbol system as they can be by genetic relatedness, then there is no intrinsic reason why they cannot act in concert in defense of common interests just as close biological kin can be expected to do in the case of the nepotistic altruism that is predicted by the theory of inclusive fitness. In this way, a group of individuals who have internalized the same symbolic instructions about group loyalty and cohesion could act in concert on the basis of those values inherited along the cultural channel.[12]

But actually, it is not a matter of the selection of individuals at all: what is being selected in group evolution is the cultural system that constitutes the group as a human society. This cultural system, as we have seen, is located in the world, not just in human brains, and though, like the genes, it requires the action of human phenotypes programmed by it to replicate itself, there is a strong sense in which what is being preserved and reproduced are the systems of symbols constituting the society and informing the institutions and practices that the members of the society enact or act upon. Cultural symbol systems are best viewed not as population phenomena but as individual entities in their own right and at their own emergent level.

Thus, to return to an example from the previous chapter, it would not really matter if the Dajeboiši, the successful head-taker among the Mundurucu, had more or fewer genetic offspring than other members of the group, or even if, in a particular generation, no one occupied that status. The status itself is reproduced symbolically in ongoing public discourse. And though headhunting was no longer practiced when Murphy did fieldwork among the Mundurucu, he too could reconstruct it because there were people who either remembered the practice or were told about it by predecessors who knew about it. A ritual complex, like Mundurucu headhunting or the *otiv bombari* of the Marind, exists at the social level and continues over generations while participants in it enter and exit the system through birth, enculturation, and death, just as, on a different time scale, players enter and exit in the game of Texas Hold'em while the game itself endures.

Human groups competing with other groups, like the Marind or the Mundurucu and their neighbors, consist of individuals who are more like each other, genetically and/or culturally, than they are like the people in the other groups with whom they are in contact. They recognize themselves as a distinct group, and they share certain cultural complexes that distinguish them from their neighbors. But the relative numbers of people who hold the ideas corresponding to these complexes in their heads are not what matters, because the complex is a single organized entity existing at the level of the collectivity as an emergent organization, or society. Since these complexes organize the actions of the people who participate in them, it is they that would be seen as the units upon which selection acts, not the individuals, just in the same way a living multi-celled organism can be treated as a single functional unit upon which selection can act, rather than as a population of the summed functions of its constituent cells. I do not mean to imply by this analogy that a society is just like an organism, since the latter is made up of cells sharing identical DNA, while a society is made up of individuals who may bear similar cultural instructions (or not, depending on the complexity of the society) but who also carry differing genetic instructions, whose interest and motives vary and who come into conflict regularly within the overarching social order. I do mean to suggest, however, that a society of humans and the cultural symbol systems that inform it represent an emergent higher level of organization that can and should be treated as a unified entity rather than as a mere abstraction based on the individual actions of its constituent units.

Socio-cultural systems, constructed by means of symbols, serve the fitness interests of constituent individuals by providing them with the advantages derived from cooperation in such matters as hunting, war, childcare, and the division of labor, and with a source of information that makes ongoing adaptation much more efficient and creative than ordinary genetic change would be. The symbolic systems that carry the external information also serve as the social glue that makes social cooperation possible beyond the small group of very close kin, by providing a channel of information transmission that is external, that does not itself undergo birth and (organic) mortality, that enables the use of foresight and planning, and that can create kin on the basis of shared symbolic substance.

The emergence of a symbolically encoded cultural track of information and instructions in human life has thus made it possible for humans to form cooperating groups of people related by cultural kinship as well as, or instead of, genetic kinship, it is true; but it has also introduced a new complication in human life by virtue of the differences between the genetic channel and the cultural channel. Among the many differences between the cultural and the genetic channel I have focused on two. One is that the genetic channel requires heterosexual intercourse as a *sine qua non* for its perpetuation across generations, whereas the cultural channel is transmitted in social space, in a public perceptual arena. The other is that, because of the first difference, cultural information can inform a multitude of people with identical instructions or symbols, making possible a human group as a true socio-cultural entity rather than as only a statistical construct. From this perspective, several issues that appear as problems from a strictly neo-Darwinian point of view, such as group selection, can be seen instead to present no conceptual difficulties. But the fact of dual inheritance also imposes new problems and complexities of its own, to which I now turn my attention.

FOUR

Two Kinds of Sociality

In the last chapter I presented the arguments for viewing cultural symbolic systems as potentially independent carriers of necessary information across generational boundaries. In taking on such essential status in the formation of human individuals, however, they do not thereby eliminate the need for genetic procreation to occur (somewhere) as well. But as I also demonstrated, the two channels of inheritance differ in crucial respects: first and foremost, one requires sexual intercourse in order to transmit information, while the other relies on information encoded in the external, public arena in the form of shared sensory data. This distinction in turn means that while identity of information encoded in genes shared by phenotypes can only extend to biological relatives, identity of cultural information can be shared infinitely widely, limited only by the available technology of dissemination. The consequence is that "inclusive fitness" in the genetic sense at best encompasses a small group of very close biological relatives, whereas in the cultural sense it can unite a great many more people with shared substance, in this case shared symbolic information. These differences in turn have major consequences for how a human society needs to be composed.

Two Modes of Inheritance, Two Kinds of Sociality

As Richerson and Boyd (2006) observe, while our prehuman ancestors in the distant past, before culture became a major factor, must have lived in social groups composed mainly of relatives, at some unspecified moment in evolutionary time proto-humans acquired the capacity to cooperate in larger "symbolically marked" groups—that is, in the terms I developed in the previous chapter, groups formed through symbolic relatedness rather than through genetic relatedness. How could this have happened? Richerson and Boyd propose two hypotheses about this development. The first is that, since cultural evolution vastly increases the rate of rapid, cumulative change of human groups in response to varying environmental conditions, intergroup competition produced pressure for more cooperative, more coherent groups to outcompete less cooperative, less coherent ones—much as Darwin (2007 [1871]) himself had originally supposed.

The second related hypothesis is that cooperative social conditions would have provided the social environment in which prosocial norms were favored by selection and became part of the innate, genetic endowment of constituent members of cooperating societies. In other words, the human genetic program adapted to the socially cooperative cultural environment by mutating in the direction of a greater proclivity to seek social rewards and avoid negative social sanctions.

The result of these developments, Richerson and Boyd argue, is that

> people are endowed with two sets of innate predispositions, or "social instincts." The first is a set of ancient instincts that we share with our primate ancestors. The ancient social instincts were shaped by the familiar evolutionary processes of kin selection and reciprocity, enabling humans to have a complex family life and frequently form strong bonds of friendship with others. The second is a set of "tribal" instincts that allow us to interact cooperatively with a larger, symbolically marked set of people, or tribe. The tribal instincts result from the gene-culture coevolution of tribal-scale societies by the process described above. (2006, 196–97)

Two mechanisms are adduced to explain how such group selection could be possible from an evolutionary perspective. One of these is the emergence of moral norms that are enforced by punishment; the other is the prevalence of conformity bias in social learning. Both assume that the group itself, or a majority within it, can establish and then enforce rules for behavior that minimize the disruptive effects of the defectors, cheaters, free-riders, and other deceivers whose selfishness might confound shared commitment to group norms and goals.

Richerson and Boyd's discussion of two types of instincts is reminiscent of Donald Campbell's original argument in 1975:

[The] new tribal instincts were superimposed onto human psychology without eliminating those that favor friends and kin. Thus there is an inherent conflict built into human social life. The tribal instincts that support identification and cooperation in larger groups are often at odds with selfishness, nepotism, and face-to-face reciprocity. . . . The point is that humans suffer these pangs of conflict; most other animals are spared such distress, because they are motivated only by selfishness and nepotism. (1975, 215)

I am very much in agreement with these authors about this necessary conflict in human life. And like them I attribute this evolutionary innovation to the rise of the capacity for cultural inheritance. Richerson and Boyd identify the mechanism by which culture makes possible the evolution of prosocial tendencies in the rapid change itself made possible by symbolic processes. But while that may well have been the evolutionary story, I would locate the source of conflict in the nature of the cultural system itself, and in its key differences from genetic evolution.

As Richerson and Boyd rightly argue, culture makes it possible for large numbers of people to identify with each other through shared symbolic markers, and through the shared recognition of markers of the group itself as a group. What I showed in the previous chapter, furthermore, was that the values inherent in genetic inheritance, which favor inclusive fitness and limited cooperation via reciprocity, are antithetical to the values inherent in cultural symbol systems. Insofar as the latter replicate themselves over time by instantiating themselves in phenotypic living human organisms over succeeding generations, they do so by appearing in a shared public perceptual space to the sensory fields of multiple individuals who may or may not be genetically related. And, even more important, they do so independently of the medium of sexual intercourse, upon which the genetic system is dependent.

It follows, then, that if there are two social systems combining to form the totality of human social existence, as both Campbell and Richerson and Boyd propose, and as I certainly agree is the case, one of them has as its highest value the performance of potentially fertile heterosexual genital intercourse, while the other has as its highest value the creation and communication of a set of symbols that are able to unite many disparate individuals into a higher-order, emergent social organization by making cultural kin out of them. In this latter context, sexual intercourse is not only not necessary, but a potentially disruptive factor working counter to the production of groups over and above the mating pair. In what Richerson and Boyd call the tribal system, impulses toward the maximization of genetic reproductive fitness, via the pursuit of opportunities for copulation, and its results when fertile, are at best something to be controlled, limited, and monitored, and at worst a major source of disunion and outright conflict that can threaten communal unity. It is a fundamental premise of my argument here that the relative harmony of the wider social community is dependent on containing and managing the task of procreative copulation in order to maintain itself free of the potentially destructive competition for mating opportunities inherent in sexual reproduction.

I am led to conclude, then, building on Richerson and Boyd's model, that human society and its many distinctive forms of organization can usefully be understood as so many various and more or less ingenious efforts to manage the inherent conflict between a realm of genetic reproduction and a realm of cultural reproduction. I use the word "realm" loosely, since it will be seen that many different strategies are employed to separate the two systems, to integrate them via various compromises, and to resolve conflicts between them. They may be separated conceptually, in time, in space, by ritual boundaries, or in many other ways, and it will be my purpose here to explore some of those various ways as they appear in the ethnographic record.

How Are the Two Systems Related?

There was a time in the evolutionary past when proto-human hominins did not have a system of symbolic communication. When exactly that changed, and how, is not a matter that concerns me here. The immediate question raised by the considerations just put forward is this: since it is clear that our species did at some point develop symbolic culture, in what relationship did and does that symbolic culture stand to the genetic system that preceded its emergence? Judging by the example of our close nonhuman primate relatives, the chimpanzees and bonobos, we can assume that even without symbolic culture humans were capable of some prosocial behaviors and psychological processes, that they could imitate and learn from other conspecifics, that they formed cooperative groups, and that they understood each other's behavior to a certain degree. The fact that chimpanzee societies differ widely among themselves indicates that externally transmitted as well as genetically transmitted instructions form a part of their inheritance system too. Nonetheless, we can assume that much if not most behavior was the result of genetic instruction at some point in our past. At the same time we can assume that the emergent cultural prosociality rested on a capacity already anticipated in pre-human ancestors.

When the capacity for encoding and transmitting information culturally emerged and came to its present necessary and extensive position in the formation of human beings, what happened to behaviors that had previously been directed by programs inherited via genetics? Since, as I pointed out in a previous chapter, the whole adaptive point of culture was to make possible rapid change in the face of environmental challenge, many actions formerly performed on the basis of innate genetic instructions would now require cultural instructions either in addition or instead for their optimal performance.

One can imagine several different possibilities regarding the fate of the genetic programs that were thus rendered obsolete by the advent of culture. One is that they withered away in a process of devolution until the only relevant genetically necessary human capacity was the capacity to learn the cultural system. According to this view, "human nature" as a relatively determined set of genetically inherited aptitudes and preferences either no longer exists, having been superseded by culture, or else is a minimum that can be held constant but need not be incorporated in the analysis of differing cultural systems. This view is that associated with Boasian cultural relativism and has many adherents among the ranks of cultural and social anthropologists to this day. Without denying the role of biology in human life altogether, this perspective places by far the greater emphasis on cultural capacity and the resultant variety of human socio-cultural systems. A good representative statement of this position can be found in a recent essay by Marshall Sahlins:

Symbolic capacity was a necessary condition of [human] social capacity. Not that we are or ever were "blank slates," lacking any biological imperatives; only that what was uniquely selected for in the genus *Homo* was the inscription of these imperatives in and as variable forms of meaning.... Nor am I denying the currently popular theory of co-evolution: the notion that culture and biological evolution reciprocally gave impetus to each other. But that does not mean that the effect was an equal valence of these as "factors" in human social existence. On the contrary, there had to be an inverse relation between the variety and complexity of cultural patterns and the specificity of biological dispositions. In the co-evolution, the development of culture would have to be complemented by the deprogramming of genetic imperatives or what used to be called instinctual behaviors. (2008, 106)

CHAPTER FOUR

In other words, adaptation to new and varied environments required that the cultural system be mutable, and therefore that the symbols that gave instructions for how to live in these different environments needed to be just as mutable, so there can be no genetically "hardwired" human nature; or if there is one, it can only provide at best very incomplete information needing to be completed by cultural instructions.[1]

A completely opposite perspective is that which minimizes the extent of human variation as a premise. This is typified by evolutionary psychologists for whom there is a universal evolved human psychology that arose in the Pleistocene, the so-called environment of evolutionary adaptation (EEA). In relation to this evolved psychology, observed cultural variations are mostly window dressing, surface phenomena, or local variations on universal themes. For example, Lee Cronk (1999), relying on the work of Donald Brown (1991), who provides a list of shared human universals, writes:

> Brown's "Universal People" take up ten pages of dense type, with universals appearing in everything from the details of language and grammar to social arrangements to the ubiquity of music, dance, and play. . . . If, as cultural determinist dogma would have it, culture is all-diverse and all-powerful, why are there any such universals? Why aren't human cultures more diverse than they apparently are? . . . We are in search of the Great Attractor of human culture, the unseen mass that pulls human cultures toward it and so limits their diversity. (1999, 25–26)

In other words, for authors such as Cronk, the many similarities in human cultures are not consistent with the idea that culture has free rein to invent any form of life whatsoever: that is, as E. O. Wilson put it, the genes have culture on a very short leash.

Cronk explains apparent cultural variation with a metaphor originally put forward by Cosmides, Tooby, and Barkow (1992) in which human minds are likened to jukeboxes, each equipped with thousands of different songs, and programmed to select a particular song depending on the current time, date, and geographical location. The variation that would be observed would be the result of accidental factors, but the total song program of each jukebox would actually be identical:

> Every jukebox in Kinshasa would be playing the same song, but it would be a different one from the song playing in every jukebox in Irkutsk. But, clearly, no culture—in the limited sense of socially transmitted information—would be involved

in generating this diversity. Involved instead would be entities that, like humans, share a universal underlying design that responds to local conditions. (1999, 33)

From this perspective, a range of possible cultural and social scenarios has been built into the human genome by our evolutionary experience, which must certainly have led us to adapt to a wide variety of ecologies and climates. From this repertoire, most of which is latent, society chooses and enacts the one that most nearly fits the current situation.

These polarized views thus give opposite but similarly unequal weight either to culture or genetics in determining the forms our society will take. In contrast, Richerson and Boyd, as is clear in the quotations cited above, argue that culture was superimposed on the genetic program, but that the genetic program did not then just go away. Rather, the two co-exist in an uneasy and conflictual alliance, each pursuing its own goals but tailoring itself to the reality that the other one places constraints upon it. As I think I have made plain by now, I am in agreement with this position, and my aim is to show how viewing things this way is consistent with, and makes sense out of, data in the ethnographic record.

Genetic Inheritance/Symbolic Inheritance is not "Nature/Culture"

In his major works *The Elementary Structures of Kinship* (1967) and *The Raw and the Cooked* (1969), Lévi-Strauss made use of an existing distinction in Western thought between nature and culture, arguing in the former work that the incest taboo, as a universal rule of human social life, represented an intrusion of culture into the otherwise natural realm of kinship, and in the latter that native South American mythology rests in large measure on the transformations of the proposition that raw is to cooked as nature is to culture (and presumably as incest is to marriage). Sherry Ortner (1974), building on the work of Simone de Beauvoir, who had also used these concepts, extended this idea to the assertion that in cultural conceptual systems around the world, male is to female as culture is to nature, with the further implication that this conceptual scheme is at the root of what she asserted to be universal male dominance, since culture is seen as superior to nature. This formulation, coming at the inception of the renewed interest in feminism and in gender issues more generally in anthropology in the early 1970s, led to a lively debate.[2] After this subsided, however, contributions using the concepts of nature and culture uncritically largely fell into desuetude, until they resurfaced again this time in the context of thinking about the relations not between male and female but rather between the human and animal, or more broadly, the "natural" world (Descola 1994, 2013; Descola and Palsson 1996).

As the debate about nature versus culture demonstrated, one of the main problems with the concept is that since both "nature" and "culture" are polysemous words with wide ranges of meaning and implication in our own language, the opposition between them lacks specificity and cannot easily be mapped onto the conceptual systems of other socio-cultural systems. Thus, "nature" may be opposed to "culture" in ways not all consistent with each other: as the innate to the learned, the uncultivated to the cultivated, the bush to the settlement, the animal to the human, the wild to the domesticated, the raw to the cooked, and so on. This is not to say that other cultural symbolic systems do not in fact make similar or identical distinctions—many of them do so very extensively. But there can be no "one size fits all" application of what is at root one of our own quite diffuse folk categories. So, for example, Gillian Gillison (1980) argues that the distinction of nature and culture cannot be applied directly to the very pronounced gender asymmetry among the Gimi of Highland Papua New Guinea, because the rainforest "is regarded as a male refuge from women and from ordinary life within the settlement. The Gimi wilderness is an exalted domain where the male spirit, incarnate in birds and marsupials, acts out its secret desires away from the inhibiting presence of women" (143). Since one of the meanings of "nature" in our own cultural conceptual system is "the wilderness" as opposed to the settled area, by this logic Gimi men, though seen as superior to women in that society, would nonetheless have to be identified with nature rather than with culture, as Ortner's original formulation would appear to require.

In the work I cited above, Sahlins (2008) makes a stronger point against the nature/culture idea, namely that the Western presumption that humans are selfish and bestial by nature, and have to be kept in check by a regulatory social arrangement, is a perspective that is peculiar to our cultural system, and a dangerously misguided one at that. Most of the rest of the peoples of the world, so he argues, do not see humans as superior to animals, or regard children as little wild beasts whose unruly and selfish natures need to be tamed by socialization; he cites several ethnographic examples in which humans are seen as inherently and essentially social rather than as asocial monads requiring chastisement and discipline to bring them around. Having shown that many hunting and gathering groups hold that animals were once human, and, from their own perspective still think of themselves in human terms, he poses this rhetorical question:

One has to ask, if man really has a pre-social, anti-social animal disposition, how has it happened that so many peoples remained unaware of it and lived to relate their ignorance? Many of them have no concept of animality whatsoever, let alone of the bestiality supposed to be lurking in our genes, our bodies and our culture. Amazing that, living in such close relations with so-called "nature," these peoples have neither recognized their inherent animality nor known the necessity of coming to cultural terms with it. (2008, 97)

In light of this antipathy to the "nature/culture" distinction as an analytic tool in cultural analysis, it is important that I emphasize that the dual inheritance model, despite any apparent similarity, is not the same as the "nature/culture" concept, in whatever form it takes. This is because the dualism it identifies avoids the wide fan of meanings in "nature/culture" and is a very specific and clearly defined and observable one: that between two forms of information transmission that are empirical facts about the real world and about which objectively valid statements can be made. Most animals, as Boyd and Richerson said, are largely guided by the genetic program alone, with its prioritization of reproductive fitness. This program does not allow for the good of the wider group to play a role in adaptation. (I will examine some examples of how some animals that live in social groups manage to do so within the constraints of Darwinian principles in the next chapter.) Humans also are self-evidently guided to a great degree by the genetic program, but in addition they are subject to the symbolic program that in many instances works to offset the demands of the genetic program. I provide supporting evidence of this later in this chapter and in ensuing chapters. There is nothing vague, polysemous, ethnocentric, or folk psychological about the dual inheritance model; it follows directly from the observed and undisputed facts of life that I enumerated in the Introduction to this work.

Nor does dual inheritance require that other cultures need to have conceptualized it in their own cultural systems for it to be valid, as some anthropologists seem to have expected of the "nature/culture" construct. It is an analytic formulation that only proposes that any vi-

CHAPTER FOUR

able human society will have found ways to resolve the contradictions between the two, not necessarily that they will have reflected on this in explicit symbolic terms. Nonetheless, I do feel it will be useful in strengthening my argument if I give an answer to the (rhetorical) question Sahlins posed, about why other peoples have not recognized the presocial "animality" of their natures and the necessity of coming to cultural grips with it. The answer is that they have, though not always by using the specific contrast of "animal/human." I will show this by giving some ethnographic examples demonstrating that we are by no means the only people to have noticed that social life juxtaposes an inherently "selfish" program, conforming to the requirements and implications of the genetic channel of inheritance, with one that imposes upon it rules for harmonious life in society, conforming to the implications of the cultural channel.

Gapun

The people of the village of Gapun, in the lower Sepik region of Papua New Guinea, interpret much of their infants' prelinguistic and nonverbal behavior as dissatisfaction and aggression (Kulick 1997). This is because of their thoughts about the nature of the self. An essential part of the self in their cultural system is what is called *hed* in Tok Pisin (that is, the Pidgin language of PNG), or as *kokir* in the vernacular. Both words simply mean "head" and refer to a fundamental sense of personal will and autonomy. Babies are talked about and treated as "stubborn, big-headed individualists. Preverbal infants are frequently shaken by their mothers and chastised playfully that their heds are too 'strong' and 'big,' and that they 'never listen to talk'" (101). The first word a child utters is held to be a vernacular word meaning, approximately, "I'm getting out of here." This word, attributed to infants two months of age, reflects adult beliefs that babies "do what they want" (101) The next words they are thought to speak are supposed to mean "I'm sick of this" and "stop it."

Hed is tolerated only in small children, of whom it is expected and taken to be the default or "natural" condition. In anyone else, hed is officially condemned, since it is used to signify "egoism, selfishness, and maverick individualism. Hed is bad. It is antisocial and stubbornly autonomistic. It is held up in stark contrast to development (*kamap*) which, as Kruni explained . . . is a group pursuit" (102). Elsewhere, Kulick describes hed thus:

This is the dimension of self which is individualistic, irascible, selfish, unbending, haughty, and proud. Villagers express and find expression of *hed* in everything from a baby's first words, to women's harangues of abuse at villagers who have offended them in some way, to the now defunct custom of murdering an innocent maternal relative to display one's anger and shame. (19)

Juxtaposed to hed is a second aspect of the self, called *save*, which means "knowledge." It is the sociable, cooperative side of a person. All people, but especially men, are expected to use their *save*, their "knowledge," to "suppress" (*daunim*) their antisocial *heds* and cooperate with their fellow villagers (19). All good qualities—generosity, placidness, cooperativeness, willingness to work hard and help others with their work—emerge from and are evidence of one's *save* (19–20).

This is, in a word, the picture of human life as one in which social norms must be imposed on recalcitrant and self-serving human raw material—much as Campbell, Durkheim, and the more recent version of E. O. Wilson have proposed; and, I might add, that is what is to be predicted from a view that one channel of our inheritance, the genetic one, is intrinsically interested only in its own priorities, while the other, cultural one does what it can to try to mold such individuals into a larger and harmonious society.

Pintupi

The Pintupi people of the Western Australian desert have as a key symbol the concept of *walytja*, or "relatedness," which recognizes the relationship of the self to various others:

The Pintupi . . . have based their culture on the concept of *walytja* as the dominant symbol of shared identity and mutual support. Official representations of Pintupi social life stress that they are "one family," or "all related." . . . The concept asserts a relationship between oneself and persons, objects, or places; it recognizes as fundamental in Pintupi life the identity extended to persons and things beyond the physical individual. (Myers 1986, 109)

Sharing, exchange, and compassion, and generalized unqualified support are the values surrounding the idea of relatedness, while those who are not walytja are regarded with fear and suspicion. However, relatedness is a quality of self that is only acquired through socialization and the overcoming of what is regarded as an autonomous willfulness:

CHAPTER FOUR

> Essentially . . . maturation depends upon the ability to recognize one's relatedness to others, and to subdue one's will in order to sustain relatedness. In Pintupi theory, this development is perceived as an increased ability to "understand." . . . However deficient in understanding, children are not lacking in will. Adults recognize and respect the willfulness of children, who are seen, even at an early age, to direct their activities by their "own idea." . . . Unlike comprehension, personal autonomy seems to be a given in human life, as reflected in the origin of the spirit outside of society, in The Dreaming. (107)

Thus, although the Pintupi self is one conceived not as a bounded entity but as formed in relatedness to other people, places, and objects, this is understood to be a desired sociality that does not simply come naturally but is opposed to, and must overcome, an asocial original "willfulness" that, if expressed by adults, is subject to disapprobation and even, if extreme, to collective punishment.

Bororo

For the Bororo, who once inhabited a large area of the north central Mato Grosso in Brazil, and who now live in smaller numbers in communities around the town of Rondonopolis on the São Lourenço River and its tributaries, "reality is constituted by the operation of the antithetical dyad, the *bope* and the *aroe*. . . . together they exhaust all of non-empirical reality" (Crocker 1985, 121). The aroe are the immaterial souls possessed by everything including humans, and they are immutable, giving structure to society and to the entire cosmos. Each clan has its own aroe, and the representation of these in ceremonial dances and in exchanges constitutes the essence of Bororo social life. The aroe provide the Bororo with "high-grid high-group" in the terms proposed by Mary Douglas (1973), that is, an elaborate classificatory system and an intensively differentiated social structure in which the ego is controlled by other people's pressure (Crocker 1985, 35).

The bope, by contrast, are supernatural beings that represent the disruptive forces in social life. The bope

> cause all things to reproduce and to die. They are therefore the principle of all organic transformation, of fructification, growth, death, and decay, the spirits of metamorphosis. As the principle of the aroe manifests itself in each human being's immortal names, his or her transcendent soul, so too the *bope* show forth in the hu-

man body, and in the organic being of all living creatures. Their manifestation is in the *raka*, the "blood," that *"élan vital"* which animates every being. (36)

It is the bope who are the source of this life force, the raka that manifests itself in the body as either semen or mother's milk. The bope are, furthermore, "considered completely licentious and lustful creatures, ungoverned by any moral principles" (50), while the aroe represent the timeless classificatory system that regulates life. Crocker resists the temptation to equate these two principles with "nature" and "culture":

At first glance one is tempted to consider them as collective representations in which *aroe* can be equated with "culture" and *bope* with "nature." This would be a considerable distortion of the subtlety and nuances of Bororo thought. Informants do sometimes say that "everything" is either "a" *bope* or "an" *aroe*. But they mean, usually, that one element or the other predominates in any given entity or class of entity. In a sense, there is one great continuum running through the universe from total "bopeness" at one extreme to pure "aroeness" on the other. (50)

But this qualification does not detract from the fact that everything is constituted by some combination of two opposed principles, one of which is immaterial, immutable, and regulatory of the cosmos and of the moiety and clan-based social system, and the other of which represents organic life, its ruthlessness, and its ability to reproduce, grow, and die.

Further, in other parts of his text Crocker does himself make the distinction between "nature" and "society." When a child is born, a balance must be struck between its raka and its "name," that is, its place in the fixed structure of the clans and their associated aroe. The former is provided by the biological father and mother, but the latter must be achieved through a ceremony from which the biological parents are excluded and instead represented by the father's sister and the mother's brother:

By utilizing cross-sex siblings of the biological parents, and by restricting the ceremonial roles to those who have not had sexual contact with each other, the Bororo at once invert and connect the two sets of relationships. This disjunction between those collaborations which produce and maintain the natural self, the *raka* [produced by the bope], and those who originate and nourish the social self, the *aroe,* is basic to all Bororo ritual involving the preservation and renewal of the individual. (67)

Thus, as the bottom line, Crocker describes the Bororo concept of the self as arising from a balanced collaboration between the natural and the social, the former associated with sexuality and reproduction and thus with "raka," and the later, associated with the aroe, representing the moral order of society that fixes, names, and places, and thus constrains, the otherwise unbridled raka. I need not stress how closely this follows the formulation implicit in the dual inheritance model.

Baining

It would not be an exaggeration to say that Jane Fajans's understanding of the social system and culture of the Baining of northern New Britain is almost entirely based on the opposition between nature and society (1997). This opposition is not conceptualized or named within Baining culture, but Fajans finds herself only able to explain essential features of Baining social life by assuming the existence and operation of this opposition. The unsocialized infant is "a natural being, untouched by social expectations." A baby is in fact a "bundle of uncontrolled natural processes, constantly carried by the parents." In addition, a child does not "understand" at first the words spoken to restrain, educate, and socialize it (89). As the child grows and becomes socialized, anything associated with "natural" processes is experienced as shameful.

The Baining person is, according to Fajans, "characterized primarily by his or her engagement in social relationships, predicated on productive labor, food giving and food taking" (114.) However, there are disruptive aspects of their relatively placid, work-oriented communal life. Ghosts, spirits, or bush creatures called *aios* can take hold people and cause them to "run wild":

A wild person might run off along a path out of the village or break straight into the bush. When this occurs, others have to go after him and restrain him. Once when Kangmani was dancing with an *akavuganan*, a piece of dance regalia with a shield atop a pole . . . he went wild and started to run out of the village toward the cemetery. Somewhere en route, he grabbed a raw taro [painful to eat] and started to eat it. The men who pursued him did not know where he had obtained the taro; he said a "ghost" had given it to him. He held it by the stalk with the leaves pointing to the ground [i.e., backwards] and started to eat it from the bottom. (128–29)

Such behavior, caused by the *aios* in Baining thought, is antisocial in the literal sense. Not only do the actors not see their action as their own, and not only are they disruptive of the fabric of ordinary social life, but they literally invert the norms and symbols regulating the social:

> The emotions and behavior that erupt in these situations resemble Mauss's "Le moi envahissant." . . . These "invading emotions" (perhaps better translated as "encroaching emotional chaos") are seen as dangerous but potent forces, which pull the person away from his or her social milieu. . . . [The state in which such eruptions occur] is a composite of traits that are symbolically meaningful as the antithesis of proper social behavior. . . . From a normal Baining perspective, reality is inverted during these experiences. People leave the social space of the village and head for the forest, but they think the forest is clear like the village; they see night as day; they see the dead as alive; they see raw, unripe, or inedible foodstuffs as good food; they fail to respond to social relationships or encounters. (133–34)

Samoa

Although the people of Samoa have no abstract term corresponding to the English "nature," they nonetheless "do classify human behavior according to its motivation in a way that does suggest our own nature-culture dichotomy" (Shore 1982, 195):

> The key concepts employed in these classifications are *amio* and *aga*. *Aga* refers to social norms, proper behavior, linked to social roles and appropriate contexts. *Amio* describes the actual behavior of individuals as it emerges from personal drives and urges. . . . *Aga* thus links behavior to society, while *amio* links it to self, to what we call "motivations." That one's actual *amio* may conform to *aga* on any particular occasion is understood as a reflection of proper teaching . . . or socialization, and the ability to "think" (*mafaufau*) properly. (195)

Although the term amio does not refer to specific acts but rather to their source in personal motivation, the word often serves as a euphemism for "impulsive, self-gratifying behavior, most commonly sexual in nature" (195).

Throughout his ethnographic work, Shore examines in detail

> the manifestations of this dualism in Samoan thought in Samoan language, social structure, and aesthetic forms. In each case, one term of a complementary opposition corresponds generally to *aga* and suggests social constraint, dignity, and subordination of personal impulse to cultural style and social control; the other term

corresponds to *amio*, implying lack of social restrain or form, and the expression of personal impulse and spontaneity. (196)

Shore is careful to point out that the aga–amio dichotomy is not identical to that between culture and nature in the Western cultural context:

Aga does indeed suggest culture, explicitly, for Samoans. *Amio*, like reproductive sexuality, only approaches nature. Insofar as *amio* and its associated symbols represent precultural forces, they do suggest nature. However, insofar as these are symbols, manifest in social institutions and coopted by culture, framed as an elaborate dual structure that is itself part of culture, *amio* is a cultural representation of nature and not quite nature itself. (196)

Shore here makes a crucial point which I need to underscore. I do not mean my formulation of dual inheritance to suggest that the genetic program is enacted in humans without the mediation of the symbolic code. On the contrary, as Shore's example shows, both the differing imperatives are encoded within the symbolic system; but some portion of what is encoded in the symbolic system expresses the priorities and imperatives of the genetic program, since the symbolic system names and encompasses all aspect of experience, even those antithetical to its own interests. This is part of what gives it its great power.

Manambu/Avatip

While I could extend this catalogue of ethnographic examples quite a bit further, I think I have established the point that many cultural systems, even ones that do not have an equivalent concept to some version of the Western nature–culture division, do explicitly recognize that human life is based on an opposition between the social and the "willful" and egoistic, the latter often further explicitly thought of as involving impulses or drives, particularly sexual ones, that can disrupt the operation of ongoing harmonious social life. I will conclude this exercise with a consideration of the ethnically Manambu people of the village of Avatip, on the Sepik River of Papua New Guinea just upriver from the coastal Iatmul. This example is of particular value for my purposes, because the ethnographer, Simon Harrison (1992), places himself quite explicitly on the side of those who, partly under the influence of Marilyn Strathern (1980, 1988, 1992), have understood many New Guinea societies as differing from the Western view of the relationship

TWO KINDS OF SOCIALITY

between the individual and society, which is said to posit that individuals exist prior to social relationships and are separable from them conceptually:

To Melanesians, individuals are the creators of social relationships, but also the creations or outcomes of social relationships that pre-exist them and into which they grow. . . . To put it differently, Melanesian societies conceive of the person as fundamentally sociable. This inherent sociability does not refer to psychological dispositions of the individual. That would be to attribute to Melanesians just another variety of our own essentialism. It is not a statement about the nature of individuals but about the nature of their relationships, namely, their inescapability. (Harrison 1992, 22–23)

The implication is that Melanesians understand themselves as interimplicated with others from before birth until after death, so that any individual person partakes in the lives of others just as others actively participate in his/her existence, through exchanges at the economic, ritual, and reproductive levels. It is of course far from being my intention to deny that such complex social and cultural understandings exist, or that they are all-encompassing and inescapable—and not just for Melanesians. However, I do want to insist that even in societies and cultures that are said to be "socio-centric" there is a conceptualization of the distinction between that part of human life that is prosocial, and another part that is assertive, self-affirming, willful, and potentially at odds with, and disruptive of, social harmony. As Shore points out in one of the passages I cited from him above, this personal, willful aspect is itself a symbolic construct, and is not (nor can it be) unmediated nature itself. But the ubiquity of the cultural formulation points to a recognition that there is a dichotomy between two realms of human existence and that many cultural systems explicitly recognize that they have to come to terms with it.

It is therefore of interest that in his own cultural analysis of the ethnopsychology of war magic in Avatip, Harrison writes as follows:

All normal human beings, male and female, child and adult, are viewed as having two basic dimensions of selfhood which I shall call Spirit and Understanding. The Understanding is essentially the ability to apprehend social obligations. It is the source of sociability, compassion, and 'amity' in Fortes' (1969) sense. Spirit, on the other hand, is the individual's life-force, an *élan vital* conceived of as the source of physical growth and health, or self-assertion and self-will and, in time, of certain mystical powers over others. Spirit is the aspect of the personality to which

CHAPTER FOUR

the men's cult specifically pertains. The ritual grades represent increasing degrees of potency which the spirits of men are assumed to gain as men pass through the initiatory system. Spirit is also the central construct in Avatip theories of violence. (96–97)

At this point in my exposition there is no further need for me to underscore the agreement of this formulation with the main points of the other such dichotomies I have sketched above. Even in this society rich with social interpenetration of relationships, and even when described by an ethnographer partial to a theoretical viewpoint stressing the difference between their system and our own presuppositions, the ethnographic facts speak for themselves: human social life is in tension between cooperative social imperatives and self-willed organic impulses. These in turn reflect the difference Boyd and Richerson proposed between a social instinct evolved to further the genetic interests of individuals and their genes and one evolved to generate altruism and cooperative behavior to others in the group beyond the immediate circle of genetic kin, over and above the principle of balanced reciprocity.

Sex/Non-Sex

It is of paramount importance for the success of genetic inheritance in a group that births occur in a population at a rate that exceeds that of deaths or at least keeps the balance even. This implies that individuals should try to optimize opportunities to procreate sexually, which in turn requires that they seek out or create and take advantage of opportunities to copulate with a fertile member of the opposite sex. (I wish there were a better word for "copulate" than the standard and by now very commonly used but still very vulgar Anglo-Saxonism, but there isn't. Let me also stipulate that for purposes of clarity in this book, I use "copulation" to refer specifically and only to potentially fertile heterosexual genital-to-genital sexual intercourse, so as to distinguish this from other forms of sexual activity.) Whatever else an individual may accomplish in life, failure to copulate ensures the failure of genetic inheritance for the individual and his/her genes, and failure to do so on a larger scale in a group spells trouble or ultimately extinction for any population, group, or species in which copulation does not thrive unless reparative action is taken. Therefore, it must be of the very highest value for the genetic program that those phenotypes that bear it be strongly motivated to copulate when they can and when the outcome is likely to be genetic fitness. Success in doing so is tantamount to winning the game of natural selection, which is, after all, a numbers game about leaving the most progeny who will then go on to do the same.

One might imagine, then, that if human social life were guided solely by the evolved interests of the genetic program, copulation would be regarded as the highest good and would be encouraged and celebrated both publicly and in individual life as the greatest of life's pleasures and the most praiseworthy of life's goals. However, this is manifestly not the case. On the contrary, while it certainly is a highly sought goal and has on its side internal impulsions that can be as powerful as any motivator in human experience, and is often the subject of myth, song, story, and many other expressions of imagination private or public, it is also, at least in many of its manifestations, widely regarded with shame, hidden away from public view, not considered a fit topic for public discourse, and expressly excluded wholesale from vast dimensions of social life. In Western societies such as our own, any allusion or reference to copulation in public discourse was and for the most part still is considered a major breach of what is acceptable; and even in societies in which discussions of sexual functions are more open, or where they are more highly valued in the public domain, or where there are specified occasions for sanctioned public sex, copulation is generally done in private and excluded from the normal conduct of everyday life in the public arena. In situations in which the dwelling or settlement does not afford adequate privacy, copulation even between spouses may be done surreptitiously in the bush.

Why should this be? If sexual experience is exciting, pleasurable, and gratifying, and if, furthermore, it is the necessary motivator that leads to the realization of the main Darwinian goal and supreme value of the lives of humans as biological creatures, then why does it so often lurk in the shadows, cordoned off as it so often is by prohibitions, guilt, conflict, shame, disgust, fear, and concerns about morality, decency, health, propriety, and/or aesthetics? Certainly no such compunctions trouble our hairy uncles the apes and monkeys, most species of whom readily copulate in full public view.

The answer of course is that human life is not solely determined by the genetic program, as is the life of other organisms, but by the interaction of that program with a second program, the cultural program, that does not operate by the same principles as the genetic one and therefore does not share its goals and priorities. As the Marind or Mbaya cases illustrate, it is enough for the cultural program that someone or other engages in genetic reproduction, in these cases, the neighboring peoples who provide them with kidnapped children to make up for their own ineffectiveness (Marind) or disinterest (Mbaya) in reproducing sexually themselves. But this freedom of culture from genetics is not restricted to these cases. On the contrary, it is in one form or another a necessary feature of any viable human society.

Margaret Mead (1949), with her usual perspicuity, wrote that "every human society is faced not with one population problem but with two: how to beget and rear enough children and how not to beget and rear too many" (224). From these conflicting necessities, she argues, it is essential for a society, and for the humans who compose it, to develop regulating mechanisms so that the society reproduces itself but does not exceed an optimum size; these regulations include such measures as postpartum taboos, contraception, abortion, and infanticide. But for Mead there is something else at work that drives these apparently anti-reproductive measures: "Behind all of these age-old devices there lies a subtler factor, a willingness or an unwillingness to breed that is deeply embedded in the character structure of both men and women. How these relative willingnesses and unwillingnesses function, at what point in the reproductive process blocks are introduced, we still do not know, but the evidence leaves little doubt that they are there" (225).

There is in itself nothing that violates the Darwinian principle of maximizing genetic fitness in the "unwillingness to breed" that Mead notes. Life history theory in the field of biology recognizes that, the goal of genetic inheritance being to optimize fitness, local environmental circumstances or other factors may not always favor simply having as many offspring as possible. A clear example would be the practice of killing all or all but one of a set of multiple births found among some hunting and gathering peoples: there is simply no way the mother can nurse and care for more than one infant at a time and also fulfill her subsistence obligations, move rapidly with the band in its wanderings, and so on. The one surviving child actually has more probability of enhancing the mother's long-term fitness than would be the case if she tried ineffectively to nurture twins.

Hill and Hurtado (1996), in an exposition of biological life-history theory in anthropology, refer to "reaction norms," that is, evolved adaptive responses to immediately arising environmental conditions that produce genetically adaptive results:

Reaction norms that produce phenotypic variation with ecological variation are expected to be adaptive because they are based in the organic machinery of the body and are therefore heritable. . . . A clear understanding of phenotypic plasticity, as produced by reaction norms, has important ramifications for all studies of human social behavior, since much of what we call culture may simply be the result of evolved reaction norms in varying contexts. Cultural variation under these conditions will *not* be genetically determined but will be adaptive, a fact that early anthropological critiques of human evolutionary ecology failed to appreciate. (13–14)

In other words, phenotypic organisms have the ability to react to their circumstances in ways not simply predetermined by genetically evolved behaviors. The choices they make regarding whether or not to reproduce, though not themselves genetically preordained by evolution, are made by a cultural capacity that has itself evolved in the course of the evolution of the organism. Since it is the product of evolution, this cultural choosing mechanism will, in the main, produce choices that will prove to be adaptive in the genetic sense.

Thus, in this view, culture is assigned the task in human society of limiting reproductive activity in the ultimate service of the adaptive purposes of the genetic program. A cultural postpartum prohibition on sexual intercourse between spouses, for example, can be seen as a way of ensuring that there is optimal birth spacing to enhance the infant's probability of surviving to reproductive age, while a woman's decision to abort might reflect her accurate appraisal of the survival chances of her fetus if carried to birth. And this is certainly a convincing argument: the ability of culture to participate in the regulation of reproduction to maximize the genetic fitness of a phenotype, or to help the group maintain optimum demographic size, as Mead proposed, would not have evolved if it were not genetically adaptive. I want to argue, however, that once established, this feature of culture became capable of disentangling itself from necessary service to the genetic channel of inheritance and so of operating on its own behalf. Why this happened from a theoretical standpoint is something I will address at a later point in my argument, but for now I will simply submit that an acquaintance with the ethnographic record leaves no doubt that this process has indeed occurred.

The evidence is the fact that in every society cultural norms supporting the life of the society demarcate times and places within which reproduction, or the copulation that leads to it, may and may not be permitted, and further restricts the range of who may engage in copulation and reproduction, and with whom, enforcing these restrictions with frequently (though not always) harsh sanctions. One may observe

CHAPTER FOUR

in any society, often demarcated quite clearly, the realms—the times, places, ritual conditions, etc.—in which sex is permitted and with whom, and others in which it is not. Regular and long-term permitted copulation (as opposed to ritually sanctioned occasions for license) almost universally involves the recognition of an enduring bond of some sort between one or more members of one sex with one or more members of the other. Usually these are heterosexual couples whose sexual relationship is recognized and regarded as legitimate by the social group at large. (I will call this "marriage," despite some problematic implications of doing so, because there is no better word for it.)[3] The marriage of one man to two or more women is fairly common; the marriage of one woman to more than one man, or the marriage of more than one man to more than one woman, also occurs though less frequently. Within the marital unit and its offspring, an incest prohibition forbids copulation of children with their parents or—usually—with each other. The extension of incest prohibitions beyond these first degree relatives is very variable from society to society.[4]

Beyond this social unit, and its juvenile offspring, which may terminate in separation, divorce, or death, but which is significantly more than a temporary consortship, there is the wider society. Whether a hunting-gathering band, a settled camp, hamlet, or village, or a more inclusive unit, this society is composed of several married pairs and involves social relations between these pairs and other pairs, as well as between and among each individual with other individuals, either as kin, or co-residents, or by means of some other sort of socially recognized tie. Any viable social system must find a way to harmonize the interests of these two kinds of basic social units: the domain of genetic reproduction and the domain in which the social ties are cultural and nonsexual. While of course there is a wide range of permitted, prescribed, tolerated, or surreptitious non- or extramarital sex in different societies, in many if not most societies reproduction is not a socially legitimate or desirable outcome of such copulations. (I am of course fully cognizant of the fact that sexual activity by individuals and in society encompasses far more than just copulation intended, expected, or allowed to produce offspring, but that is a vast topic that cannot be explored here.)

There are, then, two domains of society, one characterized by the presence of legitimate biological reproduction and the copulation it requires to accomplish it, and one characterized by the absence of these, and by the presence of mutual identifications among society members on the basis of shared symbolic features drawn from the publicly available systems of symbols constituting the culture of the society. Of these two social types one is guided by the values and priorities of the genetic program, and the other by the values and priorities of the cultural program. These correspond to the two kinds of society Richerson and Boyd proposed, and which are also exemplified in the ethnographic survey I included earlier in this chapter.

From that survey it emerged that many cultures conform to, and many themselves recognize, two dimensions of human life. One is a willful, self-interested and self-promoting, asocial individual autonomy that includes sexual and aggressive behavior, is reflected in a failure to "understand" the demands of others in society, and may be seen as the presocialized state of children. The other is self-effacing, prosocial, altruistic, nonsexual and mild, and the result of the regulation, transformation, or incorporation in developing children of the autonomous mode by the prosocial mode, in which people "understand" not only each other but the dominant prosocial morality enshrined in the public culture, enacted in ritual and myth and sanctioned by public approbation or disapproval, or by stronger forms of reward or punishment.

These different modes reflect the expected results of the genetic program and the cultural program respectively. The former places genetic self-interest, including that of very close genetic relatives, at the top of the hierarchy of goals and values, while the latter views the former as a potential disruption of social life and seeks to limit, confine, and otherwise minimize its presence in the public arena, reproducing itself not through copulation but through the transmission of cultural symbols. These two modes can both lead to social groupings, the cultural one obviously so, the genetic one as illustrated in the case of noncultural social animals who form societies that reflect the attributes of the genetic program alone, as I discuss in chapter 5. The more or less harmonious or conflictual reconciliation of these two kinds of social life is what we see in successfully realized human societies, that is, ones that reproduce themselves with some degree of continuity and stability over generations.

While this picture requires further refining, a project I pursue in the upcoming chapters, it is a first approximation of the social consequences that follow from the simple fact of dual inheritance. It has in its favor that it is theoretically logical, consistent, and sound, and that it also conforms to what is actually out there in the ethnographic roll-call of social systems.

CHAPTER FOUR

Culina as a Model System

Before going on to further develop the ideas presented so far, let me offer a single ethnographic case study as an illustration of what I have been talking about that may make it easier to grasp. In taking the Culina as a "model" I do not mean to imply that they represent the "natural," "typical," or "default" position of human society, only that their social organization and the cultural beliefs that sustain and govern it offer a particularly clear example of a solution to the problem of how to harmonize the two modes of social relations I have been discussing. Indeed, my point is not that any one society shows us the solution in ideal form but rather that all cultures can be approached with an eye toward analyzing how each one has gone about achieving the necessary reconciliation of the genetic and the cultural reproduction and inheritance programs and their social manifestations.

The Culina Indians (sometimes spelled Kulina) number about 4,000 people, speaking an Arawakan language and living in villages of 5 to 250 inhabitants along the rivers of the Brazil-Peru border region (Pollock 1985). They are organized into groups called *madiha*, which correspond roughly to local or village groups, and are identified by different animal names: the fish people, the monkey people, and so on. All members of any madiha regard themselves as real relatives—*wemekute*. Beyond the madiha boundary, other groups are perceived to be scarcely human and are characterized as dirty, untrustworthy, sexually loose, and hostile. One can infer that, by contrast, the members of any Culina madiha view themselves as clean, loyal, sexually restrained, and amicable.

We have of course encountered this distinction before, among those peoples I listed who see a similar division within their own society and indeed within their own selves. The Culina case illustrates that while this distinction is going to be found everywhere, it may not in every case be taken to describe two aspects of humans within a single society. In the Culina case, the devalued mode, that which is unclean or transgressive, immoral or antisocial, sexual, and aggressive is identified with an out-group and contrasted with an in-group seen as completely devoted to prosocial and asexual norms. In other societies, the negative values might be projected onto beings in the realm of the supernatural. Thus, societies that overtly describe themselves as social all the way down will, on inspection, be found to have conceptualized the other side of the coin, and if not ascribing it to some aspect of themselves, then imagining it in a realm beyond the borders of their own society or in another unseen realm of existence. Yet other categories useful for the projection of antisocial and aggressive thoughts and impulses are, as I will explore in greater detail later in this book, women and with somewhat more justice—*pace* Marshall Sahlins—children.

The kinship system that unites each madiha is based on the metaphor of siblingship. Their ethnographer, Donald Pollock, writes:

> The term for kinship in [the] broadest sense is *wemekute*, derived from the root *-kute* for "sibling." Culina do not in this sense conceive of the *madiha* as linked sets of siblings . . . but rather as a field of siblings, all of whom are related equally and in the same degree. This conception is underscored by statements such as . . . "we are all real siblings" to describe the madiha as an undifferentiated field of relatives. (1985, 9)

It follows from this construction of co-residence as siblingship that all the members of the madiha should observe an incest prohibition regarding each other, and that therefore the assumed "amity" of sibling sets, and of larger kinship units modeled on the sibling set, should characterize the entire community. The village is supposed to be a realm from which sex and aggression are absent or have been banished and in which mildness, cooperation, and fellow-feeling and the altruistic actions that follow from it prevail in an atmosphere of complete equality among the symbolic siblings.

But the Culina madiha is endogamous: one must find a spouse from within the village. Yet if all members of the madiha are siblings, and therefore governed by an incest prohibition, how are marriage and the resultant sexual procreation possible? The Culina cultural system solves this problem by differentiating between two types of siblings: (1) those related people who do not belong to the same set of siblings in the sense of being born of the same two parents (but are nonetheless siblings by virtue of being members of the madiha in terms of wemekute) and who are related through same-sex (genetic) siblings of their parents; and (2) those people who are not siblings in the genetic sense but are related through the parents' cross-sex genetic siblings. The latter are allowed to marry each other, while the former are not.

In terms of Culina cultural symbolism, all wemekute, as members of the same madiha, share the same blood, and they also share a relation to the souls of dead madiha ancestors who live in the present world in the form of white-lipped peccaries. At the same time, the Culina specify that biological reproduction necessitates repeated acts of intercourse, in which it is the father's semen alone that forms the fetus. After the

CHAPTER FOUR

child is born, the mother's milk, which is closely related conceptually to semen, further informs the child and helps it to grow and develop. This construction solves the marriage problem in the following way: although one cannot marry anyone who was formed from the same semen or milk, one can and should marry someone of the same blood.

By this logic, then, one's own brothers or sisters are forbidden as marriage partners because they are not only of the same blood but also partake of the same paternal semen and maternal milk. The same also applies to the children of father's brother and mother's sister, because father and his brother were formed by the same semen and milk, and pass these along to their children, and the same is true of mother and her sister. But father's sister's child is formed by the semen of father's sister's husband, and by father's sister's milk, neither of which played any role in forming or nourishing oneself. Therefore, one can marry father's sister's child or, by the same token, mother's brother's child. Thus the whole of any madiha is a zone in which reproductive sexual relations are prohibited by a general rule, which however contains a special provision for enabling marriage between cross-cousins. This is a good example of how the wider society encompasses, defines, and constrains the part that is marked off and set aside to serve the purpose of genetic reproduction.

All wemekute are expected to live by the harmonious code of conduct described above, symbolized by their shared essence in the form of blood and the common relationship to the souls of dead wemekute. But when a marriage takes place, the nature of the relationships among those involved changes dramatically. A telling aspect of this change is that at the marriage ceremony, those former siblings (via blood) who have now become affines (via the new marriage tie) are expected to whip each other. Pollock observes that "whipping may well be a formalized and socially acceptable manner in which to express aggression" (12)—a suggestion it would seem hard to dispute. No ordinary wemekute would ever do such a thing as whip another one under normal circumstances, since it violates the nonaggression ethos supposed to rule in the madiha. But the special circumstance of a wedding, that is, the creation of a new social bond based on copulation expected to produce offspring—implies that the ordinary rules that govern the madiha society are altered, and new ones in which sexuality and aggression are expressed are allowed, if also ritually restricted as to time and place and held in check by the (ritualized and contained rather than unregulated) aggression that is enacted at the ceremony that sanctions legitimate copulation.

After the marriage, the new husband must exhibit shame around his mother-in-law and must perform labor for his father-in-law. Both are shifts away from the openness and egalitarianism that normally characterize the wemekute relationship. Because there is a relationship involving reproductive sex connecting them, relations between a man and his female affines morph from the amity of siblings to the shame accompanying sexuality. The father-in-law, meanwhile, ceases to be an equal wemekute and becomes an authority figure in a hierarchical relationship characterized by the power of one over the other—again, contrary to the usual wemekute norms. The wife's brother ceases to be a "sibling" to the new husband, and becomes a "friend," that is, one not related but of a different substance altogether. The wife's brother may whip the husband if the latter does not treat the former's sister well; again, this would be forbidden in the normal wemekute form of sociality. Finally, a joking relationship is established between the new husband and his wife's sister, indicating a change from her status as just another generalized "sibling" covered by the overall amity and incest norms to that of a person with whom sex now is "in the air," and a distinct possibility, but with whom it is also out of bounds. The dilemma is solved by the compromise of permitted sexual banter without actual copulation.

The charged and marked status of the married couple now customarily persists until they have had three children. After this limit has been reached, the married couple's social status changes once again: they move out of the wife's home, where they have been living under the thumb of the wife's parents, and into their own home. At this point, they return to the status of wemekute with everyone in the madiha including their affines: the husband stops working for his father-in-law, stops sexual joking with his sister-in-law, stops feeling shame around his mother-in-law, and once again regards his brother-in-law as his wemekute rather than as his ambivalent "friend." In short, relations are now normalized back into the prevailing wemekute model because the couple's reproductive activities, now successfully accomplished, are no longer the salient feature of their relationship (though they may in fact go on to have more children). In this way, the zone of reproduction is further restricted in time: it lasts only as long as the new married couple has less than three children, and then it is socially over, though it may be reproductively not over at all. Once again, the reproductive zone is limited, demarcated, and ultimately submerged under the encompassing totality of the nonsexual social domain that links but also overrides the marital relationship in its reproductive aspect.

CHAPTER FOUR

In summarizing the overall situation, Pollock comes to the following conclusion:

> Culina social organization presents a dynamic in which a set of relations exemplifying (and evaluated in terms of) "siblingship" as a code for conduct are transformed into affinal relations, and which are finally re-transformed into metaphorical "sibling" relations, one might say, when the social consequences of affinity are fulfilled. . . . Affinity emerges for the Culina as an uncomfortable phase in a social process or developmental cycle, one which conceptually contradicts the norms of interaction which Culina presume to characterize "kinship," or in a broader sense, humanness or personhood. Throughout, it is "siblingship" which remains the constant, both as a general model of social interaction and as a field against which affinity becomes marked. (1985, 14–15)

What this brief outline of Culina social organization shows is that there are indeed two clearly defined sorts of social systems, one, the madiha, encompassing the other, the married nuclear family, of which there are several constituting the whole madiha. There is no need for an outside observer to parse them in imposed analytical terms, since the Culina demarcate them themselves conceptually, institutionally, and behaviorally. One of these, the madiha with its ideology of universal egalitarian kinship framed as "siblingship" prescribes pacific, friendly, cooperative relations among all co-residents, but does so on the condition that there be no intrusion of sex, reproduction, aggression, or power imbalance to disrupt the peace. This form of sociality thus represents the principles inherent in the cultural channel of information communication: it unites all who are equal and identified with each other by their common access to the public symbolic system that encodes the wemekute ideas. They are, indeed, all kin by virtue of shared symbolic substance. At the same time, it privileges amicable relations among members and devalues sex and aggression, which are seen to be a necessary evil that must be tolerated but needs to be cordoned off from public social life if the system is to maintain an even keel. Behind this is what Richerson and Boyd call the tribal social instinct.

The "archaic," genetic system, on the other hand, was evolved for success in the evolutionary competition for (inclusive) reproductive fitness, in which self-interest and competition, including violent competition, are available weapons, and reproductive sexual activity is the ultimate prize that alone assures the good outcome. The family, including its affinal extensions, embodies the characteristics that one would expect from such a system, though somewhat hemmed in by social constraint and cultural ritualization: copulation between marriage partners, manifestations of shame, power differentials, the possibility of disruptive nonmarital sexuality, and (controlled) aggression in the form of whipping.

What has been demonstrated, then, is that the consequences of dual inheritance that I have analyzed in the foregoing pages have clear real-world manifestations. One could, of course, argue that I have cherry-picked the Culina because they seem to reflect the implications of my model particularly well. (This would of course be a serious misunderstanding of my method.) It is true that I picked them as an illustration for just this reason. What I must now show in the remaining pages is that this is no fluke, and that it is a general rule that human societies must find some way—not the Culina way, but some way—of reconciling the sexual program of the genetic channel and the non- or antisexual program of the cultural channel, so that they both survive and go on to inform future generations of the society in question. There are very strong reasons why this integration is so hard to achieve and consequently so rare in the world of living organisms that reproduce sexually, and I show that it is the symbolic system itself that, by its particular characteristics as an external means of information storage and transfer, both creates and resolves the difficulty.

FIVE

Society beyond the Genetic Program

Several animal species have managed to create something we would call a social life without the benefit of symbolic culture, which I have argued is essential to the capacity of *Homo sapiens* for the distinctively human form of social organization.[1] And I have characterized the genetic channel of inheritance as one that pursues competitive aims intended to favor one's own DNA, in contrast to the prosocial character of the cultural, symbolic program. How then do the social animals manage a social life, if this is the case? This was the problem the field of sociobiology was called upon to solve. It did so with the concepts of inclusive fitness and balanced reciprocity, along with a few other ancillary principles. It is a premise of this book that explanations employing evolutionary and sociobiological concepts alone are inadequate to account for the observed facts of human society in all its diversity, and that the dual inheritance model is a more successful way of understanding human society as it is represented in the ethnographic record. To make this case, in this chapter I place the human form of social life in the context of that of some other social animals in order to see what is unique about human societies and to show how the symbolic channel of information transmission is instrumental in enabling this unique system to thrive and to produce widely differing manifestations that nonetheless reveal an underlying regularity.

Conditions for Society among Some Non-Human Animals

Most animals that reproduce sexually do not have anything we would want to dignify with the term "society." They may live in groups, often enough, but many of these are better seen as "populations" in the evolutionary sense: collections of individuals all pursuing their own reproductive self-interest, without significant or organized interactions with other conspecifics that benefit anyone but themselves (or their very close genetic kin) in terms of fitness. Since the evolutionary process involves the winnowing of phenotypes via environmental selection operating on inherited variation in the population, the individual organisms bearing one type of gamete must be in competition with all others bearing the same kind of gametes in the quest to unite their gametes with those of the opposite sex. What may appear to be "altruistic" behavior will, from this perspective, turn out to serve individual inclusive reproductive self-interest: if it did not, it would never have been selected, since the competition for mating opportunities is the decisive factor for the future chances of any and all. Apparently altruistic behavior may, on this account, also result from the expectation that the altruistic assistance will be reciprocated by the individual who benefited from it.

Some species among the primates and the canids for example, however, have managed the feat of creating social units that actively cooperate in ways that solve survival and reproduction problems, that benefit group members generally, and that nonetheless do not violate the expectation that each individual operates on the basis of a strategy designed by genetic evolution to maximize its own reproductive fitness and in doing so to out-compete fellow group members of the same sex. The advantages that individuals gain from cooperative behavior range from mutual grooming, as is found in many primates, through the building of friendships and coalitions that can enhance an individual's relative rank in a dominance hierarchy, to group defense of territory, mobbing of predators, coordinating hunting (as found especially among the canids), and raiding the territory of other groups so as to expand one's own foraging range (observed among chimpanzees).

Given that the individuals doing the cooperating are, with the exception of very close genetic relatives, all in competition with each other for reproductive advantage, and hence mating opportunities, mechanisms must have evolved in the relevant species to assure that at least in some circumstances mutual benefit outweighs individual competitive self-interest. The potential for any individual to freeload, defect, or otherwise cheat in any truly "social" endeavor must somehow not be allowed to undermine the evolution of the possibility of at least a minimal level of coordination, cooperation, and perhaps even friendly fellow-feeling with other group members.

As we have already discussed with regard to humans, the successful accomplishment of effective social life entails the exclusion, limitation, containment, and/or control, to at least some degree, of sexuality and of aggression within the cooperating group. Otherwise the competition for mating opportunities would lead to conflict among group members that could disrupt or undo cooperation. Since it is cooperation that is itself a leading adaptive benefit for the participating individual, its loss would also put individuals in the group at an adaptive disadvantage. How then is it accomplished?

One strategy is for the cooperative group to contain only a single mating pair. This is the case, for example, among wolves (Mech 1970). The hunting packs that these animals form, which give them a great advantage in bringing down prey cooperatively, are actually large extended families of genetically related kin, mainly the juvenile offspring of the single adult mating couple. Sometimes there are adult pack members who are not direct offspring of the mating pair but are genetically related in some more distant way; they usually play a minor ancillary role in reproduction, though they do help in group hunting. What makes this system work is the (virtual) exclusion of all but one male and one female from breeding, by virtue of the reproductive immaturity of the offspring who constitute the pack. Whatever adaptive mechanism is in place to prevent too close inbreeding, which would have deleterious fitness outcomes, prevents sexual competition within the group. Finally, when maturation impels a juvenile to seek its own opportunities for mating, it leaves the pack and finds its own mate, and the cycle begins again.

Canids also are well known to have dominance hierarchies that further regulate breeding within the group. It is thanks to this mechanism that some adults other than the breeding pair can be included in the hunting pack. The auxiliary adult occupies a subordinate position to the alpha adult of the same sex, and has far fewer mating opportunities available to it. One has to assume that for the subordinate male, the choice is either to have no mating opportunities at all or to have at least the possibility of a few by helping out the pack. Although overt aggression within the group is kept to a minimum, dominance hierarchies, established through both displays and limited physical contests, serve to express and also to channel and contain aggression associated with competition for mating opportunities, since they are established by real but non-lethal fights.

Dominance hierarchies are very well developed among many monkey species that live in multi-male, multi-female groups. In most of these groups whatever cooperation there is takes the form of mutual grooming, friendships, coalitions, and sometimes group defense. Thus, to take one exemplary case, that of the red colobus monkey of central Africa,

> within the group there is a pronounced dominance hierarchy, which is expressed through priority of access to space, food, and grooming position. The adult males are dominant to the other members of the group. . . . The most dominant male does by far the most copulation in the group. Adult males are groomed most and groom the least, whereas adult females groomed others more than they were groomed. There is a strong cohesion among adult males of the group as evidenced by their united effort during intergroup conflicts and by their long-term membership in the group. (Struhsaker 1975, 27)

Thus it appears that in this and similar species, social cohesion among the males is achieved with the aid of the dominance hierarchy that limits sexual competition once rank order has been established, leaving one male in the position of monopolizing the great majority of mating opportunities and therefore of achieving the greatest possible reproductive advantage, and of consigning the other adult males to subordinate status. This is apparently no bar to their contributing to group efforts such as fighting off rival groups, as is also the case among wolves.

When we turn to the great apes, we find an interesting array of social types (Smuts et al. 1987). At one end of the spectrum are the orangutans, who have no social life to speak of whatsoever. They live almost completely as solitaries, the sole exception being a female with her juvenile offspring. Adult males are hostile to each other and to females, and mainly steer clear of them entirely, except during a brief breeding period in which the males seek females to inseminate. The periodicity of female sexual receptivity and ovulation, and its relative rarity, create long periods when sexuality is off the table.

Gibbons come together in mating pairs for long-lasting unions in which both partners contribute to the care of the offspring. However, there is no extended or meaningful interaction between mating couples, which remain the sole social form in this species. There is no higher level group bringing the mating pairs, or their individual members, together in a larger cooperative grouping. Gorillas, similarly, live in small groups usually with a single dominant adult male and his one or more long-term female consorts, together with their juvenile offspring. Occasionally, ancillary adult subordinate males may join the group. As with the gibbons, there is no higher or more inclusive social grouping that coordinates these mating units.

In all these cases, it appears clear that the competition among same-sex adults precludes the formation of anything other than a mating pair, which may be temporary as in the case of the orangutans, or long lasting, as among the gibbons and gorillas. Nothing resembling Richerson and Boyd's tribal social organization is present at all, hence there is no need to explain how social life is possible over and above that necessitated and enabled by the requirements of genetic reproduction. These animals, then, conform to the expectation that cooperation in most non-human societies stops at the boundaries of the mating pair and first degree relations.

This leaves our close relatives the chimpanzees and bonobos, both of whom live in multi-male, multi-female groups. These species present the opposite picture from that of the gibbons and gorillas: whereas the latter have durable mating pairs but no multi-member group ties, the chimpanzees and bonobos have multi-member group relations but no long-lasting mating pairs. The chimpanzees are capable of a good deal of social cooperation and seem to possess at least rudiments of empathic understanding of and feeling for other group members. They can hunt cooperatively, and conduct both aggressive raids and defensive operations in competition with other groups in adjoining territories, though the degree of chimpanzee consideration of others is a matter of debate.[2] Adult males also share meat among themselves after a successful hunt. Several mechanisms mitigate overt disruptive mating competition among them. These include the estrus cycle that delimits periods of female sexual receptivity, creating periods in which reproduction is not at issue and therefore male competition is rendered irrelevant. During these non-reproductively active periods, there is general copulation among males and females of the group in a so-called "scramble competition," since no reproductive outcome is at stake. But during the period of receptivity, the well-established and enforced dominance hierarchy lessens overt competition by assigning each male a rank order that determines his access to mating opportunities. Chimpanzees also have extensive mechanisms for reconciliation and peace-making after conflicts, and senior group members, often respected older females, serve peace-keeping functions as well (de Waal, 1989, 1996, 1998). On the other hand, they need these because they also display relatively frequent intragroup aggression that is occasionally life-threatening or even lethal.

The bonobos, by contrast, are quite peaceful and enjoy a more harmonious group life than the chimpanzees, though they do not form long-term mating relationships either (de Waal and Lanting 1997). There are dominance hierarchies in both sexes, but the relationship between the sexes is less unequal than it is among chimpanzees, among whom adult male coalitions can wield control. Among the bonobos, females dominate males, and indeed the most important social relationships linking the group into a cooperative unit are female-female dyads, who bond to each other through sexual play. In general, the level of overt sexuality practiced largely for pleasure and social relating rather than just for reproduction, without much internal conflict surrounding it, is a distinctive characteristic of bonobo life.

The creative evolutionary leap made by *Homo sapiens* was the integration of the social unit formed by lasting mating pairs, such as that found among gibbons and gorillas, and the social unit formed by multi-male, multi-female groups like those found among the chimpanzees and bonobos. Furthermore, humans accomplished this synthesis of the two systems without recourse to either the periodicity of the estrus cycle or dominance hierarchies, though they probably must have built upon, refined, and enhanced prosocial emotions and aptitudes inherited from evolutionary predecessors and observed in lesser degree among other great apes with a common ancestry.[3] How they managed this will concern us later in this chapter, but before turning to that theme I want to single out for a more focused discussion one particular species of primate which, though not as closely related to humans genetically or evolutionarily as the chimpanzees or bonobos, nonetheless offers particularly instructive lessons for the study of human social organization, namely, the hamadryas baboons.

Hamadryas Baboon Social Organization

The hamadryas baboon is one of five species of the genus *Papio* found in various parts of Africa. The hamadryas species, *P. hamadryas,* lives in an arid semi-desert region of Ethiopia, surviving largely on the flowers and beans of the acacia tree (Abegglen 1984; Kummer 1971; Swedell 2006). Unlike most baboons in somewhat more favorable environments, who live in bands numbering from 10 to 100 individuals, the hamadryas baboons have a three-tiered social organization. The smallest unit, and most closely bonded, is the one male unit (OMU), consisting of a single adult male, his long-term female consorts, their juvenile offspring, and occasionally a "follower," that is, a subadult male hanger-on, who is tolerated by the adult male up to a point but is prevented by him from copulating with any of the females of the OMU. A typical such group, consisting of about five or six members, is just the right size to exploit a single acacia tree during a day's foraging.

The next highest level is what Kummer calls the "band," a rather loose federation of two or more OMU's that habitually forage together during the day, having a high tolerance for each other's presence. Abegglen (1984) has further distinguished a unit called a "clan," with the presumption that they may be close genetic relatives. Finally, the "troop" includes many bands and may number up to several hundred individuals. The troop spends the night together clinging to cliffs high enough to protect them from predators, among which the leopard is the most dangerous. The hamadryas baboons are diurnal, and when morning comes they set out as a troop toward a foraging area, then break up in to bands (or clans), and ultimately dissolve into OMU's, each climbing an acacia tree and eating its products on the spot. There seems to be little or no food sharing, even within the OMU.

In addition to the three layers of the OMU, the band, and the troop, there are singleton male adults, though never single adult females, and also groups of unattached males. In these male groups amity prevails among the group members, who affiliate with and groom each other. No such amity characterizes the relations between adult males who possess "herds" or "harems" of females; it seems they are wary of each other, but even more importantly they are very much preoccupied with keeping an eye on their bonded females and making sure these do not stray far from their control. The male hamadryas baboon is twice the size of the female, and can be quite ferocious, and thus can easily dominate his herd of female consorts. This dimorphism has presumably evolved to enable a male "mate-guarding" strategy, although as we shall see a unique and very interesting evolutionary development among them has also addressed the problem of male-male conflict over females in another way. The ferocity of the adult males serves not only to control the females but also to protect the band from predators such as wild hunting dogs during daylight foraging hours. A foraging troop posts sentinels to scout the approach of danger, and adult males will mount a joint defense or counterattack against would-be predators. Thus, one can say that there are real mutual benefits to joint action among troop males.

In his important paper on rational preselection in the process of evolution, Christopher Boehm (1978) relied extensively on Kummer's field data on the hamadryas baboon. Boehm argues that humans can influence their own evolution by rationally preselecting behaviors that are likely to be adaptive based on purposive behavior guided by memory, knowledge of the terrain and of resource availability based on experience, and goal-oriented planning. He further posits that this adaptation, now so highly developed in humans, must have had pre-adaptations in the evolutionary past, and presents the case of the hamadryas foraging expedition in support of this argument. Citing an eloquent description by Kummer of how a troop descends from the cliffs in the morning and decides in which direction to set out for their daily travels in search of food and water, Boehm shows that experienced older adult males direct the movement of the troop by positioning themselves facing a certain direction, after which other males orient themselves in the same direction. Boehm concludes that these behaviors indicate systematic coordinated group organization that draws on and benefits from the old baboons' experience of the countryside and knowledge of where the best resources are likely to be found under present seasonal and weather conditions.

The same set of behaviors indicates that despite their general stand-offish attitude toward one another when it comes to their females, the adult males who head OMU's are capable of cooperation and each one benefits from the coordinated group behavior via the protection it offers of mutual defense. The various tiers of hamadryas society seem geared to the different needs these animals face: the troop provides nocturnal protection on the cliffs and daylight protection when the troop as a whole is on the move, and also acts as a repository of shared knowledge about the advisability of different foraging options that benefits each troop member. The band or clan corresponds to the normative group size of the other species of baboons and is probably the "default" from which the unique hamadryas social structure evolved, large enough to provide protection while on the ground and small enough for individuals to become habituated enough to each other so as to be able to affiliate on a regular basis. The OMU fits nicely in an acacia tree, seated on whose branches the baboons are relatively safe from predations by land-bound predators such as the hunting dog pack.

If the males are on the one hand jealous guardians of their females, whose mating potential they monopolize, and on the other are capable of at least some level of mutually beneficial cooperation and coordination, how are potential conflicts managed so that the group is not unduly disrupted by mating competition? In other baboon species—from which the hamadryas species probably represents a late evolutionary divergence selected for when they moved into their present harsh environment—male-female consort bonds form, but only while the females are in estrus. Among the hamadryas, by contrast, these consortships, involving one male with one or several females, have become permanent:

> In the hamadryas male, a high motivation to associate with females merely had to be extended to the anoestous female to make him the intolerant permanent mate he now is. To own more than one female, however, required an innovation, the herding technique. The repertoire of baboons already included its behavioral components, that is, brow-raising and neck-biting. The hamadryas, however, are the only baboon species that built these behaviors into a tool for keeping females close by timing their attack to the female's behavior. Anubis [baboon] males never herd their females even when they consort with them. (Kummer 1971, 99)

In other words, the hamadryas baboons evolved a permanent pair-bond (or set of pair-bonds with one male and several different females) within the context of a large multi-male, multi-female troop. We could say that this arrangement is evolutionarily "on the way" toward the basic human social arrangement, except that neither the troop nor the band are particularly strong social units compared with the OMU. However, a key step has been taken in extending close and exclusive affiliation between a male and a female that does not simply correspond to the period when the female is sexually receptive and fertile. That it seems to be based mainly on threats and fear as opposed to mutual affection does not negate the fact that it establishes a system in which the reproductive unit and the larger unit are synthesized by a method other than that in which the highest ranking members of each sex are the ones who get to do the reproducing to the exclusion of the less fortunate or well-endowed.

But this new system needed further adjustments before it could work:

> On the side of the males, the new social system also introduced a new problem. It created males who, already aggressive, were now highly possessive of females. Whereas anubis females may provoke male competition for a few days each month,

hamadryas females become permanent incentives. When 30 new females were introduced to a colony of about 100 hamadryas baboons in the London Zoo, all the old males tried to secure females and, within one month, killed 15 of them in competitive fights over their possession. This event, though provoked by an unnatural manipulation, points out the risk in evolving possessive males. (100)

This passage is particularly important since it highlights the very real problem posed by male competition for sexual access to females, which when not socially controlled can be massively and lethally destructive.

This brings us to the very important question of how the hamadryas baboons solved the problem they had created for themselves by bringing into their midst a full-time temptation to competition for reproductive sexuality among the males. This is where the story gets even more interesting. On the basis of experiments performed in enclosures built in the wild, Kummer concluded that male hamadryas baboons, quite uniquely among non-human primates, had evolved an inhibition that prevents them from aggressively attempting to possess a female who was already in the possession of another male. Kummer observed that when a pair-bond formed, it was respected by all the other males. This was not due to lack of overall sexual interest on the part of males in general. On the contrary, if a male is captured and removed from the troop, his females are taken over at once by rivals. If the captured male returns, the new possessor will not give up his females without a fight to determine who will possess the herd.

This last observation raises a key question:

Fighting ability can obviously decide who will own a female, but if it is the new possessor who wins, why did he not attack the original one and conquer the female long ago? Fighting power is clearly not the whole answer. If it were, we would expect that a few superior fighters would appropriate all the females of a troop, and this, in reality, is not true. In the Erer-Gota population, no less than 80 per cent of all adult males owned females. (103–4)

To explain this apparent conundrum, Kummer put forward the hypothesis that "a male does not *claim* a female if she already belongs to another male" (104). He and his colleagues then tested this hypothesis: A male and female were placed in an enclosure, while a second male outside the enclosure was allowed to watch them "consorting," that is, mounting, grooming, and exhibiting the "following of the male" behavior by the female. (In other baboon species, when males attack females the latter withdraw, whereas among hamadryas baboons

CHAPTER FIVE

they approach and follow the attacking male.) After fifteen minutes of this, the second male was introduced into the enclosure with the now bonded pair:

> He behaved rather strikingly. He refrained from fighting over the female and even looking at the pair. Most of the time he sat in the corner opposite the others, looking away from them. At intervals, he looked at the sky, inspected the well-known countryside as if something were moving there, or aimlessly fiddled on the ground with one hand. His social behavior was strongly inhibited. (104)

Further experiments supported the hypothesis. The conclusion Kummer reached was that "the inhibiting effect of the pair Gestalt is the stabilizing factor that protects pair bonds and lends the one-male group a certain immunity" (105).

The implications of this finding for the understanding of human social behavior were not lost on Kummer: "Independent of each other, man and hamadryas baboons both evolved stable family units linked together in larger bands, and both presumably arrived at this type of society while living in open country" (106). His hypothesis about this is that some unknown enhancement of sexual attraction led to the observed shift from temporary consortships during estrus to long-lasting pair-bonds (which in humans would be connected with the loss of estrus and the innovation of concealed ovulation). The difference between the two species would lie, for Kummer, in the fact that whereas in humans it was females who evolved to become more attractive to males (by having permanent sexual receptivity), in hamadryas baboons the males evolved to become more attracted to females (during anestrous periods). In order to cope with the disruptive threat caused by this increased force of sexual attraction, meanwhile, an inhibition had to arise to protect pair-bonds from competition by rival males. Kummer is quick to assert that he does not mean to suggest "that humans have or should have such inhibitions!" (107). But clearly they do, in the rules governing the institutions we call by the admittedly inexact term "marriage," and in other characteristics as well. The big difference is that the hamadryas baboons dealt with the problem by evolving the necessary inhibition by genetic means, whereas humans did so largely by means of cultural symbol systems (though these may of course be derived from genetically evolved dispositions), locating the inhibition in the norms and values obtaining in the shared communal public space of social life. A bridge is provided between the two methods by the fact that our big brains, which allowed humans to develop the symbolic capacity necessary for the creation of cultural marriage rules, are also instruments for inhibiting our impulses, as Michael Chance (1961, 1962) first suggested in positing the capacity for impulse control as a key evolutionary advantage leading to the rise of hominids.[4]

The Cultural Revolution

To create basic human social organization as it empirically exists, then, perhaps the biggest hurdle that has to be overcome is the disruptive potential that is present when adult males try to form cooperative groups. This disruptive potential is inherent in the genetic program, which motivates each individual to do whatever it takes to out-compete all rivals for sexual access to fertile females and the resulting reproductive advantage. It is also certainly true, as Hrdy (1981) has shown, that there is competition among females in primate societies as well, but it is not so fierce as that among males. As we saw in the case of hamadryas baboons, 80 percent of adult males have consorts, leaving 20 percent of males without reproductive opportunities, but all females are in a position to fulfill their reproductive potential—there are no unclaimed mature females, and they are all kept in some stage of childbearing about as often as nature allows. There is thus less for the females to be competitive about. In fact, the hamadryas are quite atypical in allowing so many males to have mating opportunities: among many monkey species, for example the red colobus monkey, with their multi-male, multi-female groups, the alpha male alone accounts for over 80 percent of the copulations with fertile females, leaving his many rivals out in the evolutionary cold. We have also seen, in the example from the hamadryas baboons in the London Zoo, that this competition, when aggravated by artificial tampering, can quickly turn very destructive and is waged in deadly earnest. (That the competing males in that situation killed the females rather than each other is a telling part of the story with implications for human behavior that might be illuminating, but I leave this matter to one side here.)

We do not of course usually observe such deadly competition in the wild since a society, to have survived, must have found some solution or other to the problem. In many animal species males simply avoid each other altogether, or have evolved ritualized combats to minimize the lethal potential of their conflicts. But in those that develop social living with some real measure of mutual amity or at least tolerance,

CHAPTER FIVE

and some degree of cooperation or mutual benefit, mechanisms are in place to allow the males to coexist in a viable degree of harmony. These include dominance hierarchies, advertised frequent or lengthy periods of female infertility, mechanisms of reconciliation and peacekeeping, and, in the case of the hamadryas baboons, some evolved inhibition that prevents males from trying to fight for already claimed females and encourages them to respect the ownership of females by males who would otherwise be rivals.

As is the case among the baboons, and among most other mammals, there is more, and more potentially harmful and disruptive, competition for mating opportunities among the males than among females. This is because of their asymmetrical reproductive strategies determined by biology: while there is an almost inexhaustible supply of the male contribution to sexual reproduction, spermatozoa, even when there are only a few males present, and these can be obtained practically on demand, the female contribution to sexual reproduction, ova and the nine months of gestation, are a scarce resource in much more limited supply. The assumption that it is male competition that requires cultural managing is another of the fundamental premises of the argument of this book. The evidence to support it is supplied by the ethnographic descriptions of the efforts socio-cultural systems make to minimize and contain male competition that I present throughout much of the remainder of this book.

Going back now to the ethnographic example of the Culina, we can see a major difference with any other primate or other social mammal in that in this particular human society the processes leading to individual genetic reproductive fitness, the zone of sanctioned sexuality that sustains it, and the impulses that motivate it are opposed, contained, and encompassed by the wider tribal social system in which they are embedded. Whereas among the hamadryas baboons the troop and the band are relatively weak and ineffective forms of social groups in contrast with the OMU's, among humans the equivalent higher level has most frequently gained the upper hand and become the dominant position, into which the sexual reproductive program, represented by the "OMU" of the married couple, is a potential disruption carefully kept within symbolically prescribed limits and treated as something set apart and often even dangerous. It may even be viewed as an aberration, an eruption of dangerous and powerful but necessary forces into the normal smooth functioning of social life, as the Culina case illustrates.

What has happened, then, is a reversal of what at first must have been the domination of the genetic program over the competing system it presumably created in the first place to give itself genetic reproductive advantage. In observed human social life, however, the cultural, symbolic channel of information often takes precedence and treats the genetic program as something to serve its own interests, rather than the other way around. Presumably, to be sure, the cultural capacity, like the capacity for social learning, for memory and the ability to learn from experience, and other intelligent traits that many animals share with humans, evolved genetically to be in the service of furthering the reproductive advantage of the genetic program. Yet in human society, it seems that the sexual program has frequently been made subordinate to the cultural program. How could such a reversal of values have come about?

I have already made the point that the invention of a second system for the creation, storage, transmission, and inheritance across generations of information vital to the survival of individuals and of the societies they compose represents an adaptive advantage of untold magnitude—a fact proved by the rampant domination and exploitation of the globe by *H. sapiens*. Furthermore, the advantages of collective social life, from mutual defense to reciprocity that provides resources in times of need to the enhancement of information by dispersing it among many individuals and encoding it in media external to the brain, are likewise of practically unlimited adaptive potential. Once again, that it was possible to achieve this is proved by the simple fact that it has happened. The hard part, from a Darwinian point of view, is to imagine how these obviously advantageous features could have been achieved in the face of the obstacles placed in their path by the self-serving character of the imperatives that prevail in the normal course of natural selection working blindly on the genes. This latter process would, according to the theory, have snuffed out any incipient form of prosocial inclination in the constituent members of a group by favoring those not so inclined, who would out-compete their uncompetitive group mates in the competition for fitness.

An important point to bear in mind in trying to overcome this conceptual difficulty is that originally, and in many if not most ways to this day, prosocial impulses and the capacity for culture that enable them do in fact serve the interests of the genetic program. Those able to use cultural symbolism can attain a level of foresight and intelligence, through the manipulation of external symbolic models of real life situations, that can outdistance all the competition, and will in turn presumably spread their genes in the same way other beneficial genetic advances are disseminated. Again, the cultural capacity would not have evolved if it did not first and foremost serve the interests of the genetic program. Once in place, however, its ability to unite non-kin in cooperative groups would have begun to hold its own against the tendency of the genetic program to lead to defection and cheating, and as it gained ground it would itself become a selective factor that would lead the genetic program to evolve in the prosocial direction of tolerating genetic competitors. After all, if the hamadryas baboons could do something to neutralize aggressive mating competition among potentially fierce jealous males, so, one would suppose, could humans.

I have already mentioned how measures enabled by cultural instructions that might seem to oppose sexual reproduction may actually favor it. Thus, infanticide, abortion, sexual abstinence, and contraception, which are learned and culturally transmitted procedures (though not present in all cultures), can be seen to allow the reproducing parent(s) to optimize rather than crudely maximize offspring. People using these measures are presumably measuring the costs and benefits of investing nurturing energy against the chances of any individual infant's survival and reproduction, based on such considerations as birth spacing, the apparent robustness of the child, current climatic conditions, resource availability, and so on. In these cases, the imperative to reproduce is opposed by an inhibition that urges against reproduction. But this latter inhibition is actually acting in the long-term genetic interest of the individual who applies it.

Similarly, in the case of the institution of marriage, cultural rules and norms both prescribe and prohibit some potential spouses, and provide positive sanctions for respecting and negative ones for transgressing the inviolability of marital bonds. (That these sanctions are only partly successful does not obviate the point that by and large, and normatively, marital pairs are by definition not supposed to be challenged.) These culturally created and enforced inhibitions serve the interests of the genetic success of individuals in the society by allowing them all access to the great adaptive advantages provided by human social and cultural life. This is illustrated in the case of the hamadryas baboons, who observe just such inhibitions. These, in turn, allow for the creation of long-term, permanent male-female pair-bonds set within a (relatively weak, to be sure) larger kin group, the clan, which is presumably exogamic; and a still larger society, the troop (weaker still, but not negligible) that encompasses many OMUs, bands, and clans.[5]

In cases such as these, it seems clear that it is in the interests of the genetic program to evolve a mechanism that opposes its own evolved motivations to reproduce in the higher interest of overall long-term fitness considerations. But one then has to imagine, on the basis of the empirical ethnographic evidence, how in many instances this provisionally oppositional factor, whatever it might be, once called into existence, might come to liberate itself to a greater or lesser degree from its condition of servitude to the genetic program and become oppositional to the genetic program not only for the latter's ultimate benefit but on its own account.

To illustrate how the genes might have designed something that seems to oppose them only to serve a longer-term genetic interest, I draw an analogy with a well-known episode from literature. Jon Elster (1978), in discussing "rationality," has used the example of Ulysses and the Sirens, from *The Odyssey* of Homer. In that episode, Ulysses is able to hear the sublimely seductive song of the Sirens without meeting the inevitable fate of any sailor who does so, which is to be lured to one's death. Ulysses has his crew tie him tightly to the mast of his ship and orders them not to release him no matter how much he pleads for them to do so. He then orders the crew to plug up their ears with beeswax so they won't hear the song and themselves be tempted to steer toward the Sirens, and then row past the Sirens' island. The result is that he and his crew survive, and he gets to hear the Sirens' song without succumbing to the fate of all those who preceded him.

True, this story is not exactly on all fours with the idea I am trying to convey, since precisely what Ulysses possesses—foresight and the ability to plan for adverse eventualities that he can anticipate—are what genetic evolution is lacking.[6] Nonetheless, since students of biology are willing to credit blind chance and natural selection with mimicking reason and achieving results that appear to have been preplanned in all sorts of other cases, then perhaps we can do so here for heuristic purposes: the genetic program, like Ulysses, creates a situation in which its own wishes and impulses are inhibited, but with the higher end in view of getting what it wants after all, despite an apparently destructive obstacle lying in its path. Since what it "wants"—the adaptive advantages of cooperation and external information storage and communication—cannot be had according to the limits set by its own *modus operandi*, it has to counter and oppose those limitations in the interest of transcending them.

The tale of Ulysses and the Sirens illustrates how the genes might benefit from creating something that restricts itself—the bonds that tie Ulysses to the mast and hold him fast even when he desperately wants to be freed—in order achieve their ultimate goal of surviving a threat to their existence. But culture has gone further than that. To illustrate the next step in the process, I turn to the current fantasy, realized in works of imagination such as the film *The Matrix* or the television series *Battlestar Galactica*, to the effect that the machines living humans have created, whether depicted as computers or robots, have (in the future world) taken over control of their former human masters and now hold them in servitude as they once served humans. My argument is that this is in fact what has to a certain degree happened with symbolic culture: that this non-living external cultural prosthesis that the human genetic program devised to give itself a tremendous technological and social adaptive advantage has seized control and turned the tables, making its former master into its creature, treating it more or less as humans treat their domesticated animals.

Imagine the DNA, at an early moment of the evolution of life, as an absolute master, giving out orders via the RNA to the humble phenotypic protein, which does its bidding and follows its instructions. Since the DNA is blind and receives no feedback from the outside world, this is a fairly ineffective system, successful only because in the primordial era, chance had billions of single-celled beings to work with and eons of time within which some lucky hits could happen to come up and manage to allow the phenotypes to survive and reproduce themselves. No wonder that what has happened did happen: some DNA developed a way to give instructions for the phenotypes it governed to be able to receive information from the environment in the form of perception (in the broadest sense), enabling them to make better-informed moves in the competition for survival and reproduction. We might imagine that, as in the juke-box metaphor to which I alluded in an earlier chapter, organisms at this level are constructed with a number of different programs, like records in a jukebox, and that environmental conditions trigger which one of these is put on "play." Such an organism would obviously stand a better chance of survival than one that was blindly preprogrammed in advance.

The DNA might then go even further, once multicellular organisms had evolved, and instruct the organism to observe other organisms, and copy them, or copy those that appeared successful, as in "social learning" models, which various creatures have employed. Once again, this would surely represent an evolutionary advance. Now imagine that the DNA goes one step further still and creates a phenotype with a brain and nervous system that can wire itself up and rewire itself creatively on the spot in response to external conditions, and, manipulating external symbolic representations, can evaluate situations on its own and decide what would be the best course of action, sometimes drawing on a stored repertory but also sometimes inventing novel strategies on the basis of metaphoric and metonymic leaps of imagination. The DNA would in effect say to this nervous system with its perceptual apparatus: "You know more about what's going on out there in the world than I do, and you know what I'm trying to do. Use your superior intelligence, and the powers I have given you of foresight, memory, reason, and so on, to devise the most effective strategy to further my interests; then act on it." One can easily imagine that the newly created superior intelligence system might one day see no reason not to serve itself, rather than its creator, the genes.

Let me translate this perhaps too abstract line of reasoning into a concrete historical example by way of yet another analogy. The early Islamic polity consisted of various Arabian tribes united by fervor for the new religion. That unity began to fray only a short time after the death of the Prophet. As it made the transition from being a confederation of tribes with conflicting loyalties to a unified political entity, the Islamic state needed some principle other than fractious clan affiliation upon which to found the necessary military dimensions of its political power. This it found in the institution of military slavery, in which non-Arabs, mainly Turks, were taken in, not as tribes or clans or hordes but as individual slaves composing a non-tribal army loyal only to the Caliph. The ninth-century Caliph al-Ma'mun created the original slave regiment called "Mamluks" (the Arabic word meaning "slave"), which set the precedent for future Islamic rulers. One such ruler, the last sultan of the Ayyubid dynasty in Egypt, created a Mamluk regiment in the mid-thirteenth century, recruited from Qipchak Turks living across the Caucasus mountains. Some of these were captured, but most were purchased as young boys from their families. No one could become a Mamluk except by starting out as a non-Arab slave: the Mamluks could not pass on their status to their children, but instead took in young Qipchak (and later Circassian) boys, converted them to Islam, trained them in the arts of war, and inculcated in them strong group loyalty.

These Mamluk armies were responsible for the successful defense of the Egyptian sultanate, but in the course of safeguarding the sultan's regime, and realizing that they were in fact the strongest element within the state, they overthrew their Arab overlords and seized power themselves, appointing one of their number, the military hero Baybars, as the new sultan. The Mamluk Sultanate, recreating itself over time not by genetic reproduction but rather by the recruitment of young

CHAPTER FIVE

Turkish boys through purchased slavery, went on to rule Egypt for most of three hundred years, until defeated by the Ottomans.

I will discuss the Mamluk system of succession in further detail in chapter 8, but for now I introduce them as an example of an external "means" created by a ruler to win battles for him and provide him with the advantages of power, that is, to serve his own interests; but which, once created, by virtue of the very fact that it is the ruler's best means of controlling his destiny, decides to use that ability to do away with its master and use the military advantage for which it was designed to take over power and run the state itself. The piquancy of this example for my present argument is enhanced by the fact that the Mamluk seizure of power represents the replacement of the principle of the reproduction of the sultan and the ruling elite by genetic procreation with the principle of recruitment and generational succession by nonprocreative means, that is, by capture and increasingly by the purchase and ensuing resocialization of foreign-born boys into a society of individuals without ties of genetic kinship to anyone and with ties of cultural kinship only to each other.[7] Something like this is what I envision must have happened in our evolutionary history when the cultural program assumed (partial) mastery over the genetic program. The genetic program did not disappear, indeed it could not, since it remained the only source of fresh human phenotypes. In just the same way, the Mamluks could not reproduce themselves solely by cultural means but used the Qipchak and Circassian people, themselves reproducing sexually, as a source of new members to be recruited into a society united by cultural uniformity and not biological kinship.

Finally, to give this historical example a more generalized philosophical grounding, let me allude very briefly to the Hegelian dialectic of the Master/Slave relationship, as interpreted by Alexandre Kojève (1969) in his influential reading of Hegel's *Phenomenology of Spirit*. In Kojève's interpretation of Hegel's famous parable, the Master and the Slave have emerged as, respectively, the winner and loser of a fight for recognition by the other. This fight was originally supposed to be to the death, but the loser, out of fear of death, chooses slavery instead. The Master now puts the Slave to work and enjoys the fruits of the Slave's labor:

The Master forces the Slave to work. And by working, the Slave becomes master of Nature. Now, he became the Master's Slave only because—in the beginning— he was a slave of Nature, joining with it and subordinating himself to its laws by accepting the instinct of preservation. In becoming master of Nature by work, then, the Slave frees himself from his own nature, from his own instincts that tied him to Nature and made him the Master's Slave. Therefore, by freeing the Slave from Nature, work frees him from himself as well, from his slave's nature: it frees him from the Master. In the raw, natural, given world, the Slave is slave of the Master. In the technical world transformed by his work, he rules—or at least one day will rule—as absolute Master. And this Mastery that arises from work, from the progressive transformation of the given World, and of man given in this World, will be an entirely different thing from the "immediate" Mastery of the Master. The future and History hence belong not to the warlike Master, who either dies or preserves himself indefinitely in identity to himself, but to the working Slave. The Slave, in transforming the given World by his work, transcends the given . . . ; hence, he goes beyond himself and also goes beyond the Master who is tied to the given which, not working, he leaves intact. (1969, 23)

If in this passage we substitute "the genetic program" for the Master whose domain is (in this Hegelian discourse) Nature, the raw, the "given," and who is ruled by the instinct of self-preservation (within which we would want to also include preservation of the genes over generations), and understand the "Slave" as the phenotypic human armed with the ability to go beyond the genetic program by means of actual engagement with the world via perception and action, and able to construct symbolic systems for the creation, preservation, and transmission of experiences that are represented in shared social space and that allow for successful adaptation ("work"), then this is the picture I would draw of how the cultural program, originally set up by the genetic program to help it preserve itself, comes by its nature to have the capacity to preserve and reproduce itself, using its original "master," the DNA, as its means.[8]

So much for parables and abstract models. Let me now proceed by looking at marriage in societies with elementary social structures to see how a few simple rules governing the realm of sexual reproduction in human culture extends the hamadryas pattern and solves the problem of male competition in the human context.

Marriage in Elementary Social Systems

The Culina, who I used as a model case in making an argument for the separation and opposition of the tribal and the sexual/reproductive aspects of society, and the apparent subordination of the latter to the former, are not hunters and gatherers, of whom there are only a tiny handful of groups left in South America (as is true almost everywhere). The rest of the indigenous groups on that continent are to some degree cultivators (and now perhaps workers in the wider local, national, and global economies), and as such they represent an evolutionary development in human society that is very recent, since cultivation became a way of life only a few thousand years ago. That development often leads to social characteristics markedly different from those that characterize hunters and gatherers. It has sometimes been suggested that some of the latter live in some version of what human societies may have looked like throughout most of evolutionary time, before either cultivation or domestication were practiced. This stretch of time that hominins (including humans) spent as foragers lasted millions of years and therefore represents the transitional zone between precultural primate social organization and neolithic social systems.

But hunting and gathering societies themselves must have differed widely in prehistory, as they do today, depending on environmental, demographic, and cultural factors. If they were anything like today's foragers, some were nomadic, some settled in villages, some hunted collectively following great herds, others hunted alone in tropical jungles, some had elaborate kinship systems, others had next to none, and so on. Putting the Kwakiutl of the Northwest Coast of America, the Walbiri of the Central Australian Desert, and the Inuit of the North American Arctic, to take some random examples, all in the same basket greatly tests ones willingness to generalize about "hunters and gatherers."[9] Furthermore, one cannot uncritically assume that today's hunter-gatherers give us an accurate picture of any real prehistoric society. Add to that the facts that many of such groups are in fact refugees or a defeated group who once were cultivators but have been displaced, and that almost none survive without some meaningful interaction with neighboring cultivators, pastoralists, or urban or town dwellers, or live on state reserves, and one hesitates to use the category at all to say anything about our evolutionary past, which must have seen the rise and fall of a great variety of different adaptations as our ancestors moved into new and different environments over the course of those many millennia.

Nonetheless, contemporary hunter-gatherers do have the advantage for our purposes of illustrating for us at least some of the strategies groups using this adaptation can employ to organize themselves, survive, and reproduce as populations and as socio-cultural entities. Some among these groups have very small populations and only a minimal amount of internal organization, formal or otherwise, so that they show us what is possible for a society at the most elementary level. I will focus on one such group here, again in the spirit of taking it not as representative of all but rather as a single particular case that illustrates some of the points I want to make. The question I will be addressing is this: what minimally had to happen to get from the hamadryas model of social organization to that of relatively uncomplicated human hunter-gatherer groups? The hamadryas had already solved some of the problems of integrating mating pairs into a larger social group without the benefit of symbolic culture. Furthermore, hominins are only very distantly related to baboons. Baboons do not literally represent an antecedent, but a case of parallel evolution, as Kummer noted. It seems likely that the hominin line developed from an ancestor shared with the chimpanzees and bonobos that lived in multi-male, multi-female groups with affiliated groups of related males practicing male philopatry. How did hominins get from "there" to "here"?

Fortunately there is a solid body of thought already in the literature to help understand this. Bernard Chapais, in his recent important contribution *Primeval Kinship* (2008), asks us to perform the following thought experiment: in a hierarchical social organization with a male dominance system, the hierarchy emerges because some males are stronger or smarter or luckier than others, and the desired resources—specifically mating opportunities with females—are distributed accordingly. But what if all the players are equal in their ability to compete? If one starts from a baseline of hamadryas-type "multiharem" organization, and competitive abilities are defined as equal among all the males, then would the best strategy be a scramble competition such as occurs among the highly promiscuous chimpanzees? This is not likely because, as we have seen, hamadryas baboons are intensely possessive of their harems; so the result would be destructive rivalry and general mayhem. Therefore, they should continue to build pair-bonds; and if they did so, given equal competitive prowess among them, this would lead to a situation of monogamy. Chapais's logic is that any male attempting to defend more than one female would be challenged by other, equally powerful males with no more to lose than he had. The outcome "would be an egalitarian distribution of females among males—generalized monogamy—because it is the arrangement that minimizes conflicts and hence the costs of aggression" (177).

The reason Chapais brings up the consequences of a situation of equality of competitive ability among the males is because he believes that is exactly what happened when humans acquired weapons. And we know they did this early on since weapons are indispensable for human hunters, whose organic endowments are not designed for killing animals of any size:

> Armed with a deadly weapon, especially one that could be thrown some distance, any individual, even a physically weaker one, was in a position to seriously hurt stronger individuals. In such a context it should have become extremely costly for a male to monopolize several females. . . . Because any man *can* make tools and form coalitions, generalized polygyny was bound to give way eventually to generalized monogamy. (177)

In short, the ability of men to form coalitions, and presumably, possessing symbolic communication techniques, to talk in order to plot and plan effectively, and of any (male) individual or coalition to inflict grievous harm or death on any other, leads to the result that, outside of a dominance hierarchy, the only stable organization is an equal distribution of women among the men on the restrictive but fair principle of "one to a customer." Needless to say, monogamy among humans coexists with various forms of polygamy, at least for some in the society; and in many societies opportunities, illicit or condoned, for nonmarital copulation are normative. But the underlying principle of marriage as a sexual and reproductive lasting bond that everyone in a society can have seems to have the effect, empirically as well as theoretically, of reduced competition and increased social harmony (Henrich, Boyd, and Richerson 2012; Ingham and Spain 2005).

Christopher Boehm (1999) also proposed that the escalation of potential destructiveness made possible by weapons is what allowed hominid hunter-gatherers at some point in their evolutionary history to suppress the dominance hierarchy principle altogether, and enforce the only viable option, individual male-female pair-bonds integrated into a larger social community:

> After weapons arrived, the camp bully became more vulnerable. Such political equalization could have had meaning as a preadaptation for an egalitarian cultural revolution, particularly if one considers the *combined* weapons of a group of rebellious subordinates being directed at a *single* too-aggressive alpha. The latter could be readily eliminated at a relatively safe distance or driven from the group with little immediate risk to the rebels. (177)

A comparative ethnography of the Hadza of Tanzania and other nomadic "immediate return" hunter-gatherers—those with the least complex social structure even among foragers—by James Woodburn (1982) lends further support to this dynamic:

> Hunting weapons are lethal not just for game animals but also for people. There are serious dangers in antagonizing someone: he might choose to simply move away but if he feels a strong sense of grievance that his rights have been encroached upon he could respond with violence. . . . The knowledge that . . . weapons have indeed been used for murder in the past, the dangers of conflict between men over claims not only to women but more generally to wealth, to power or to prestige are well understood. . . . In normal circumstances the possession by all men, however physically weak, cowardly, unskilled or socially inept, of the means to kill secretly anyone perceived as a threat to their own well-being not only limits predation and exploitation; it also acts directly as a powerful leveling mechanism. Inequalities of wealth, power and prestige are a potential source of envy and resentment and can be dangerous for holders where means of effective protection are lacking. (436)

We may thus conclude that one of the important roles of "marriage"—a socially recognized and respected male-female long-lasting union that entails sexual reproduction—is that it solves the problem of organizing potentially competitive and dangerous males, armed as they are with lethal weapons, into a cooperative community, thus keeping the peace. The people themselves recognize this, as Woodburn notes, and the fear of what would happen if equal opportunities for marriage for all men were not respected is a good deal of what enforces the system and ensures (relative) compliance.

It is true that viewing the role of lethal weapons in producing social harmony creates an image of society as not unlike a Mexican stand-off, in which peace (insofar as it exists) is only maintained at the point of a gun. But one can certainly imagine that once pair-bonding began evolving into marriage, genetic changes might well have led to a more irenic approach to group life. Nonetheless, the staunch individualism and insistence on autonomy and noninterference that prevails in most hunter-gatherer and many horticultural groups, along with the high rates of homicide, lend credence to the idea that social harmony does depend to some degree on the maintenance of measures to contain potential lethal competitive violence among the men.

Woodburn, like many other writers about hunter-gatherers, or at least the nomadic, immediate-return foragers represented by groups such as the Hadza, the Mbuti, the Ju/'hoansi (!Kung), and others, stresses that they share a spirit of independence, self-sufficiency, non-

CHAPTER FIVE

interference, and a stance that communicates, as Fried put it, "Don't fool with me" (1967). The aversion of such groups to intragroup violence, sometimes quite extreme, as among some of the forest groups of Malaysia, the injunction to share any sizeable resource, usually in the form of a large killed game animal, and the negative sanctioning of anyone who tries to deviate either from the ground rules of the group or from the ethos of equality suggest that the egalitarianism of these elementary societies is not simply nice behavior on their part, but, as Boehm asserts, a powerful shared communal strategy designed to keep any tendency to hierarchy and dominance in check. The universal tendency to envy and resentment of success, to which Woodburn refers, results in a communal commitment to sanctions ranging from malicious gossip to ostracism and even death.

One major piece of the puzzle is still missing, however. Classic thinkers from Tylor to Leslie White and Lévi-Strauss emphasized that the key to human social organization, differentiating it from other non-human systems, is exogamy, leading to the creation of pair-bonds in the framework of a pattern of exchange. Building on these insights, Robin Fox, in his ambitious integration of evolutionary, psychoanalytic, and structuralist thought *The Red Lamp of Incest* (1980), argued that what was new in human kinship was not exogamy per se but rather alliance, that is, the social relation created between the relatives of the principals in a male-female couple; or in other words, the affinal or in-law relationship that holds not only between a man and his wife's relatives or a woman and her husband's relatives, but between larger exogamous kinship ("descent") groups to which the spouses belong.

In their influential article "The Human Community as a Primate Society" (1991), Lars Rodseth and his collaborators elaborated on the points these earlier writers had raised, armed, however, with new and much better data on primate social organization than had been available to their predecessors. Like Robin Fox before them (and like me), they were struck by the fact that hamadryas baboons seemed to have managed to have merged long-term pair-bonds with a wider community structure. They also accepted the argument of Abegglen (1984) that hamadryas OMUs were clustered in clans of two or three or more representing a lineage of related males. Arguing that these clans foraged together and provided each other protection, and that their modal pattern was the transfer out of the clan of females upon maturity, Rodseth et al. suggested that the hamadryas baboons have what amounts to kin-group exogamy together with kin-group alliance, but without any direct connection between these two features, that is, without a recognized link: the females who transfer out retain no cognitive or affective links to their natal group or relatives. These authors recognize, further, that when females enter a herd, the other clan members are inhibited from any competition, so that the "in-law" relation is of necessity also a nonsexual one. This all seems almost human—hamadryas baboon society is a "halfway house" toward human social organization, as Rodseth et al. put it.

What is missing are affinal alliances that continue after pair-bonding is established, and that go both ways. That is, in hamadryas society, a female leaves her brothers and initiates asexual relationships with, or at least recognition and toleration of, the male clan mates of her mate. But this does not forge a link between her new associates and her natal family. What human society does, so the alliance model argues, is to establish a permanent affinal tie with implications of amity and mutual recognition and respect between the exogamous groups from which the two members of the new male-female marital pair come.

The hamadryas pattern depends on continuous spatial proximity, because the harem must be protected and watched over by its male "owner" at all times. There is no way a female in his herd could, for example, go home for the weekend and visit her parents and siblings. Humans in elementary society, however, follow a fission-fusion pattern in which families regularly separate to perform their complementary daily labor, most usually hunting or fishing for the men, gathering for the women. One evolutionary development can tie all this together, as Rodseth et al. demonstrate, and that is symbolic communication, especially language. With language, families from which a member has transferred out to form a pair-bond can remember and remain in touch, at least sporadically, with relatives. Similarly, as Deacon points out (1997), symbolic communication allows for the creation of categories of persons, so that relations among them can be regulated by knowledge of the symbolic system rather than relying on perceptual recognition and thus on immediate presence and spatial proximity. It is a cliché, indeed, of the anthropological literature that such societies hasten to assign the anthropologists who come to study them a place in the symbolically constructed marriage and kinship system so as to know how to behave toward them. Thanks to this new communicative skill, humans became "unique in the extent to which both males and females form affiliative relationships, with nonkin as well as kin, both within and between groups" (Rodseth et al. 1991, 232). Symbolism and the expansion of prosocial bonds thus coevolve and implicate each other—much as I have been arguing, but with the striking consequences I have also proposed.

With this lengthy preamble, I am now ready to look at one particular case of a group with an elementary social system. In his response to Rodseth et al., Robin Fox speculated that the hominid transition occurred during the era of *Homo erectus*, "with hunting bands established, rudimentary language in place, and competitive polygynous 'allocation' of some kind going on in a kinship context" (1991, 243): "In this case why not a context close to that of the hamadryas with their savannah-dwelling adaptations? An exchange of mates between 'clans' could produce a 'Dravidian' version of elementary kinship, and would thus fit my scheme quite nicely" (243). Following Fox's lead, then (though disregarding the question of whether his proposed evolution took place in the era of *H. erectus*), let us take a look at an elementary society that is not only of the Dravidian type but is actually Dravidian, namely, the Hill Pandaram of Kerala Province in South India (Morris 1982a, 1982b). An examination of this hunting and gathering society will help shed light on the question of whether it is the case that it is alliance of recognized kin groups that makes the leap from hamadryas to human social organization.

Hill Pandaram Marriage

Brian Morris remarks that his reports on Hill Pandaram society are often met with the same sense of disbelief from anthropologists that had greeted Woodburn's initial descriptions of the Hadza:

> Here was a society in which people placed a high value on individual autonomy and movement, where social life was not elaborately structured and ritualized, and where kinship was not the all-pervasive matrix that anthropologists had come to associate with tribal communities.... An earlier article of mine on the Hill Pandaram, which indicated their lack of interest in systematized knowledge, and their lack of ritual classifications, led some to suggest that I had somehow discovered a group of 'non-human humans.' (1982a, 452)

In examining the extremely basic character of Hill Pandaram social life, I am far from arguing that the group is a survival from early human history, or even that they represent the earliest form of human society; still less do I mean to imply that they are themselves a missing link between hamadryas society and more complex human social forms. No such historical chain of course exists. I focus on them because they illustrate one of the forms human social life can take when its formal and institutional features are reduced to a minimum.

The Hill Pandaram are a largely endogamous, self-defined, and externally recognized ethnic group occupying the mountainous forests that lie in the border region separating Kerala from Tamil Nadu, in the Western Ghats of South India. They survive mainly by collecting forest produce, most of which—edible roots, nuts, and palm flour—they consume, but a significant portion of which—honey, resin, herbs, and spices—they trade with the lowland peoples, from whom they receive, in return, rice and other staples such as tea and coffee, metal utensils and tools, and muzzle-loading guns. The latter are used by the men for hunting some of the game animals that inhabit the forests in relative profusion, guns having replaced the bow and arrow in historical times. But women and children also hunt for smaller game, usually with digging sticks, while men also freely participate in the gathering activities. In addition to hunting with guns, the men specialize in trade with the settled people on the plains, and in the gathering of various kinds of honey, which is avidly consumed and also traded with settled neighbors. Beyond that, there is a high level of equality between the sexes, reflecting the overall ethos of autonomy, noninterference, and resistance to any sort of authority. Gathering rather than hunting is the predominant mode of adaptation.

The kinship terminology system is, as one would expect given their location and ethnic affiliation, Dravidian, a scheme that emphasizes the distinction between cross- and parallel-sex kinship links to ego, dividing up ego's relatives into kin on the one hand and (potential) affines and/or marital or sexual partners on the other.[10] The preferred marriage is to a cross-cousin, though it may be a classificatory member of that category rather than an actual mother's brother's daughter or father's sister's daughter. Marriage is at once very casual—there is almost no ceremony, and begins simply when a couple begin to cohabit, and often breaks up to be followed by other marriages. But it is also very serious, in that once married, a couple is expected to have sex only with each other, and men are rivalrous with each other and jealous of their wives. The Hill Pandaram have a reputation among the lowland people of being promiscuous and as allowing some unions others would regard as incest; but while there are scattered instances on record that might support this view, people in this society are generally respectful of marriage and many unions in fact last for a lifetime.

But marriage does not take place as an exchange between exogamous kin groups—there are no recognized exogamous kin groups—and the most that is expected is that the wife's father grants permission to the prospective husband to visit her. No goods are given or received at marriage. On the basis of an examination of the census of actual marriages, Morris is able to propose that there may be an incipient moiety system between certain localities, which appear exogamous in practice, with spousal exchange regularly going on between them; but this is not recognized or institutionalized in any way, and in practice marriages are usually matters of individual choice. Married couples and their small children live in shelters made of leaves; each one has a fireplace and each person is responsible for providing his or her own gathered food. Meat from sizable game, however, is shared among neighboring families.

The married pair is the strongest social unit, and there is no overarching collectivity—they have no lineages, clans, or totemic groups—except the ethnic affiliation of the group itself. However, families typically cluster in camps of two or three marital pairs. Since the Hill Pandaram are highly nomadic and shift location mainly because that's just what they do, localities have only a temporary and ad hoc existence, and the camp composition may shift over time.

The main conceptual bifurcation of society is determined by the fact that, in virtue of the Dravidian kinship terminology system, for each ego half of the population are designated consanguineal kin, or can be reckoned as such, while the other half are considered affines and, if of the same generation and the opposite sex, potential mates or sex partners. The Dravidian kinship terminology basically divides all people relative to ego by sex, by relative generation, and by whether they are classificatory kin or affines. When a person marries, however, the spouse's siblings begin to be addressed as "sister" or "brother" rather than using affinal terms. This presumably is intended to indicate that they are no longer available as sex partners by the spouse, but must be viewed and treated as if they were consanguineal kin and thus off-limits for sex.

What minimal ceremonial life there is revolves mainly around women's sexuality, especially the puberty of young girls. During the time of the female puberty ritual men may not gather honey. The same prohibition applies to a husband when his wife is having her menstrual period. There are no puberty rituals for boys; the girls' puberty rites and funerals, which again are quite minimal, are the only occasions on which kin beyond the immediate family come together. There are no communal rituals of any other kind. What little ritual symbolism there is in Hill Pandaram culture, Morris states, is related to the sexual division and the marriage classes created and implied by the kinship terminology. The ceremonies that do exist have three main components: the separation of the sexes, the removal of women at puberty or during menstruation from food preparation, and a taboo on honey collecting by men. Some men also refrain from sex on the day before a honey collecting expedition. By way of analysis of this symbolism Morris suggests that "the productive aspect of the sexual division of labour seems to be deliberately played down in order to highlight the procreative function of sex. . . . Many hunting and gathering societies regard hunting and sexual intercourse as mutually antagonistic. For the Hill Pandaram honey-gathering seems to have the same functional significance" (1982b, 157).

Honey-gathering in fact plays a key role in Hill Pandaram life. Honey is not only a major source of income (in the form of goods) in trade; it is a favorite food and ingredient, an eagerly sought special treat; excitement reigns when it comes into season. Gathering it is also the main task exclusively conducted not only by men but by cooperative groups of men, usually groups of two or three: one person has to climb the tree, while at least one other has to remain on the ground and man the ropes holding the tin containers in which the raw honeycomb will be lowered. In this respect, as Morris suggests, honey gathering does seem to play a role similar in certain respects, at least conceptually, to that played by hunting in other societies that place more importance on hunting than do the Hill Pandaram. We know that in many such societies hunting is actually thought of in sexual terms, so that, for example, to dream of a sexual adventure may be interpreted as presaging success in the hunt. It is not hard to see how honey gathering could have similar symbolic sexual significance, since it involves men taking something sweet from a source that is somewhat inaccessible and surrounded with potential danger in the shape of angry bees. So it is possible to imagine that when women's biological procreative function is made salient, as in puberty rites or during menstruation, symbolic "sex" is tabooed for the men. This would suggest and reflect the antagonism between the genetic channel of inheritance and the symbolic one.

Children are cared for very affectionately when small, and the whole community of the camp looks after a child, to the point of allowing other women besides its mother to nurse it. By five years of age, however, children are free to run off by themselves, and are actually capable of performing most of the collecting activities necessary for survival. They may even, in keeping with the high regard for individual autonomy, decide to leave their parents and live with another family for a while. By the time they are eight or so they are fully self-sufficient. (This would not be the case if hunting were an essential source of sustenance; but in this case, where gathering is the default source of nutrition, anyone, including children, can effectively provide for themselves in this resource-rich and sparsely populated environment.)

Morris argues that despite the fact that social life is fragmented, with marriages and parent-child relationships subject to rearranging themselves without much ado, dyadic relations between husbands and wives, parents and children, real siblings-in-law and close friends are affectionate, warm, and emotionally rich.[11] Indeed, the whole of Hill Pandaram society, in Morris's view, is composed of networks of close dyadic relations, with little by way of formally organized social structure above that level.

Some writers (Paige and Paige 1981, Thoden van Velsen and van Wetering 1960, Otterbein 1970) have argued that a building block of elementary human society is the fraternal band or interest group, also characteristic of chimpanzee society. Chapais (2010), likewise, reasons that male familiarity was a distinctive feature of proto-hominid life, shared with chimpanzees, and perhaps the common ancestor, and would have laid the groundwork for peaceful cooperative groups among hominids including humans. This may be so, but it is not the case among the Hill Pandaram, where brothers tend to be rivalrous and suspicious of each other's intentions. This is because their wives are also "affines" to their brothers and thus temptations to a sexual liaison, despite the official respect for the boundary of marriage and the rebranding of wife's sister as "sister." The latter custom, indeed, must very likely have been instituted as an attempt to curtail fraternal rivalry over wives. Furthermore, men, once they are married, refrain from too intimate contact with other women, for the same reasons. While the state of marriage is indeed usually respected, sex is understood in this group to be a powerful motivator that can and often does circumvent or undo these boundaries.

With whom, then, do men in this society cooperate? The answer also explains the nature of Hill Pandaram camps: the two or three married couples living together in a camp correspond to the families of the two or three men it takes to form a honey-collecting party. And by far the preferred partners for such ventures are real or classificatory brothers-in-law. This clearly follows because the wife of the brother-in-law is male ego's own sister in classificatory terms—a consanguineal kinswoman in terms of the kinship reckoning—and so poses no danger of disrupting his marriage. We thus see how important a role in structuring the social order the major male cooperative enterprise, honey-gathering, actually plays. We also learn that while no corporate groups of fraternal bands engage in formally recognized spousal exchange, individual men do preferentially ally with brothers-in-law in an exchange system that is ad hoc and one-on-one, based on the kinship nomenclature itself, rather than one involving anything resembling corporate kin groups, fraternal interest groups, or philopatry.

Before examining how this system matches any predictions or assumptions we might have been led to make on the theoretical bases discussed in the previous sections, let me hasten to add that it may well be argued, as Richard Fox (1969) and Peter Gardner (1966, 1972) have done for the South Indian hill tribes generally, that their very rudimentary social system may indicate that they might be better thought of not as isolated and coherent societies but rather as either "castes" specializing in forest collection within a much wider economy and society, or as "refugees" under intense pressure from the surrounding settled cultivators, or perhaps as both. In either case, their system would not represent any sort of real inherent hunter-gatherer type but would be a "devolved" result of fitting, in an inferior or marginalized position, into a much larger post-agriculture system. As I have said, however, I do not in any case consider them to represent the most elementary form of human society in any historical sense, but rather take them as a case of how, even when they are seen in the light of intergroup relations, humans living in a forest rich in edible fauna and flora that their small numbers cannot exhaust are able to construct a way of life with few of the structural constraints that characterize more complex, usually permanently settled, societies.

That said, we can see that the view that human society entails formal exchange between kin groups or any form of kin exogamy at a group level is not quite right. This feature may have been historically important in the original evolution of proto-human and human society for all I know, but it is clearly not necessary to the formation of a stable human group capable of sustaining and reproducing itself over time under minimal social conditions in a plentiful resource base. Like the hamadryas baboons, Hill Pandaram society is built around pair-bonds, and, as Chapais theorized would happen when weapons are introduced as equalizers, these have by and large been realized as monogamous dyads (though there were some polygynous families and one polyandrous family in Morris's census). There is always the threat

CHAPTER FIVE

of rivalry, jealousy, and the break-up of one dyad and the formation of another, but the marital bond itself, while it lasts, is respected as more or less inviolable from within and from without. Alliance with brothers-in-law on an individual-by-individual basis reflects the effects of the division of society into affinal and consanguineal halves. As in a moiety system, the whole society is bifurcated, but in this case the bifurcation is created by the kinship terminology in relation to each individual, rather than being recognized as forming two corporate groupings.

The main structural innovation in this minimal system over the hamadryas model, therefore, as Rodseth et al. suggested and as Robin Fox predicted, is a culturally defined kinship system of the "Dravidian" type that divides the population for any ego into affines—potential sex- and/or marriage partners—versus consanguines with whom sex and thus marriage is taboo. This is an important innovation, however, not because it sets up systems of dyadic exchange between groups, for it does not do that, at least in this case in any consistent or recognized way, but rather because through symbolic nomenclature it automatically rules out half the population as possible sex partners and makes alliance with one or more brothers-in-law possible from within that half. Combined with the expectation (not always met) that one should marry someone from the same generation, which rules out two major sectors of the remaining eligible mates, those in the senior and junior generations to ego, the result is that about 5/6ths of the members of the community of the opposite sex are excluded from the start as potential partners for either marriage or more casual sex.

This, I think, is the really critical difference between a very elementary human group and the hamadryas baboons, who obviously do not designate available and unavailable mates symbolically or otherwise; it is only respect for an existing pair-bond by each male that results in a stable social formation. A viable human group, by contrast, needs an asexual grouping principle, and this is achieved by the symbolic means of designating every member of society in relation to their sexual or asexual status in relation to any other. In this society, the number of potential sexual partners is far outstripped by the number of people with whom one's relations must be asexual, thus constituting the larger society as an asexual one within which marital bonds are then tolerated. As with the Culina, it is thus the broad extension or cooptation of incest avoidance, itself probably originally based on an evolved aversion to sex among first degree genetic relatives, that makes the elementary social system of the Hill Pandaram viable.[12]

What the human system does that the hamadryas system does not do, then, is to suppress any urge by a male to sexually approach any female not by means of an innate inhibition, nor merely by means of fear, caution in the interests of self-preservation, or learning from previous experience, nor yet again by any other physical constraint, but rather by the application of a rule, embedded symbolically in the linguistic code, that is public and known to all, that each individual knows how to apply to his/her universe of social others. Knowing that a particular person one encounters is either a consanguine or an affine defines the relationship in a way that has no correlates in the physical or genetic characteristics of that person, but is solely the effect of a cultural definition.

Terrence Deacon (2012), in the course of an argument far too complex to reproduce here, about how order, life, and mind emerge in biological evolution without violating the laws of the physical universe, rests much of his case on the concept of constraint. Constraint, by limiting the degrees of freedom of options within any process, thereby increases the degree of orderliness in the system and creates the conditions for the emergence of organization: "In colloquial terms, constraint tends to be understood as an external limitation, reflecting some extrinsically imposed factor that reduces possibilities or options. So, for example, railcars are constrained in their movement by the location of tracks, farmers are constrained in what can be grown by the local climate, and citizens are constrained in their behaviors by laws" (192). Deacon goes on to argue that constraint can be intrinsically as well as extrinsically imposed. We might then say that the constraint on sexual freedom for hamadryas males is extrinsically imposed by the threat of conflict with other males similarly motivated, but has become intrinsically imposed insofar as the genetic system itself has adapted to this danger by developing an inhibition mechanism that constrains the baboon's behavior from within. In humans, the constraint is likewise an adaptation undertaken to avoid continuous conflict among males over mating access to women, but it is encoded in the symbolic system, not in the genetic system. We might guess, however, that the motivating force that compels people to obey the cultural rules is the displacement or extension of sex-avoidant emotional reactions to certain members of the opposite sex, originally evolved to prevent close inbreeding, into situations far removed from any such danger but latterly useful for allowing men to live and work together in relative harmony.

Constraints on sexual behavior reflecting culturally encoded and transmitted rules mark the beginning of human as opposed to animal

CHAPTER FIVE

social organization and pave the way for more and more constraints resulting in more and more complex forms of social organization. Culture, in this aspect of its operation, performs the function of suppressing, countering, inhibiting, or channeling and containing sexual impulses—and thus organizing them into a system capable of operating over time without undue conflict. The interests it is serving are not (necessarily) those of the genetic fitness of any individual but of the social order itself, within which those individuals will live and reproduce. This is because, as I have said, culture needs genetic reproduction to take place somewhere, so as to generate recruits for its own survival over time; but it is the latter—its own continuing integrity as a system—that it is "pursuing." Individuals within the system benefit from a compromise: they agree to the limitation or constraint of their sexuality in return for the tremendous adaptive advantages of living in a society organized by cultural rules and employing symbolic encoding of knowledge. And as we have seen, even if individual genetic fitness falls short, as it does in the case of the Marind, the Mbaya, or the Mundurucu, whether by design or accident, there can be other ways to replace the population than by generating them sexually oneself.

I argued earlier that when culture established itself as a necessary source of instructions for the development of a fully operational human in society, it is unlikely that the genetic instructions that had preceded it evolutionarily just disappeared. In the case of sexuality, on the contrary, it is obvious that if anything it often enough has redoubled its efforts to motivate people in the face of the constraints imposed by the cultural rules. The exponential growth of the human population of the world is testimony to this. But since the great advantage of the cultural channel over the genetic one is its flexibility and capacity to change under changed environmental, demographic, or other conditions, one can picture the actual situation as, once again, a compromise: the genetic program retained fertile copulation as its highest priority, but instead of designing an innate system of reproduction, including a social organization consistent with it, it instead delegated to culture the task of organizing sexuality according to symbolic instructions that could vary with circumstances. In order to do this, culture, in turn, needed the capacity to inhibit, suppress, or otherwise contain and constrain reproductive sex—or, frequently, sex altogether. The genetic system must have granted it that power, just as Ulysses granted his men the power to tie him to a mast as they passed the island of the Sirens. But, like the Egyptian Mamluks, the servants have, on many occasions, decided to act on their own behalf instead of that of the master; so that one very frequently encounters, in the ethnographic record, the phenomenon, too often simply taken for granted, that the sexual, reproductive dimension of life is not only constrained and sequestered off from the social life of the wider community but is actively devalued. It is to the containment and devaluation of the genetic system by the cultural system that I now turn in the next chapters.

SIX

The Asymmetry of Cultural versus Genetic Reproduction

When Jane Collier and Michelle Rosaldo (1981) began a comparative investigation into the cultural construction of gender in "simple" societies, that is, hunter-gatherers and hunter-horticulturalists, they expected to find that Man the Life Taker would be balanced by an image of Woman the Life Giver; they assumed that women would be recognized and indeed prized for taking the lead role in the procreation of new lives for the community. They were surprised to find that this was not the case:

> Instead, men and women are defined by a system opposing uni-(male-male) and heterosexual models of reproduction and social relationship. Men are celebrated for their skills as providers, and providing—especially through hunting—is characteristically associated with men's ability to order, energize, and nurture both social and natural worlds. Women's rituals, by contrast, have much less to do with the creation of life than with health and sexual pleasure. It is not as mothers and nurturers that women win ritual status, but rather as sexual beings. (276)

I leave to one side the question of whether or not their characterization of women's rituals in elementary societies is accurate. The reason I mention it is to emphasize that biological reproduction is not only not celebrated in elementary (or many other) societies, as some may have expected or perhaps hoped to find, but rather matters relating to motherhood and birth are frequently conceptualized as dangerous, polluting, repulsive, shameful, or in other ways of problematic, and even as threats to the integrity of society and of human life and health. At this point in my own argument, I hope this situation is not surprising. It is a consequence of the different agendas pursued by the genetic and the cultural programs. The birth of every individual as a helpless organism from a woman's womb is perhaps the most fundamental scandal challenging the integrity of the claims of the cultural system to independence from biology, and as such it is very widely (though not universally) culturally devalued in one way or another. In this chapter I focus on some ways in which human social systems denigrate, devalue, restrict, or otherwise oppose the biological process by which new human phenotypes are created to populate and thus reproduce those same social systems.

No Sex Please, We're [New] British—Baining

We have already encountered the Baining of New Britain as described by Jane Fajans (1997). They are not hunters and gatherers but swidden cultivators, but they do exhibit many of the features often associated with the most elementary societies. The essential unit of social coherence is the married couple and their young offspring; no higher level of integration has much valence or salience over and above this one. There being no segmentation, hierarchy, or political authority, an egalitarian ethos prevails and great care is taken not to create distinctions in property, wealth, or prestige among individuals or marital units. The marriage ceremony is "quick and perfunctory" and "virtually free of symbolic content" (59). Indeed there is little by way of ceremonial life, myth, or ritual symbolism altogether, and explicit expression of concepts or values underlying their own practices are not articulated in ordinary discourse, or even very much in response to questioning from the ethnographer. Fajans herself admits that the apparently featureless and boring character of this culture has been a bar and a disappointment to previous ethnographers, and a challenge for her as well. What the Baining seem most interested in is humdrum horticultural work day in and day out, and what they value above all else is the hard labor, and indeed the actual sweat of the brow, that household couples devote to making their gardens grow.

The underlying logic of the Baining ethos opened up for Fajans when she began to grasp the central importance of adoption, as opposed to biological procreation, in this society. This represents the type case of the Baining preference for socially produced phenomena over those that are produced by natural means, that is, in this instance, without the transformative effect of human labor:

> Work takes the concrete forms of social activities (e.g. gardening, adoption, and socialization). What these activities have in common is that each transforms elements that are initially classed as natural (e.g. virgin forest or an unsocialized child) into social products (e.g. food, fire, families, and socialized persons). Work is a process that alters the materials it touches: it transforms them into social products. . . . Gardening, cooking, adoption, socializing children, and preparing dance regalia, for example, are all processes that utilize natural materials for the creation of social products. They are, at the same time, the activities that form the core of Baining life. (8)

The importance of adoption in Baining society reflects a view that children produced by copulation are less desirable than ones acquired by a social interaction with the donor family. Underlying this idea is the equation of sex with what is natural and thus of little value, unless transformed into a social product by human effort. Though the married couple is the key unit of Baining social life, the marital pair is understood primarily as a cooperative effort in labor, not as a sexual relation:

> Intercourse is an activity about which the Baining feel comfortable only within the confines of the marriage bond, and not all that comfortable even then. Work need not take place only in marriage, but given the sexual division of labor, marriage is perhaps the most efficient way of organizing subsistence activities. These latter are activities that the Baining value as social . . . and in which they take pride, whereas sex is shameful, natural and isolating. . . . Acknowledgment of sex is extremely shameful. Since sex is permitted only within marriage, marriage is considerably tainted by this association. The considerable cultural emphasis on the other activities (gardening, cooking, etc.) serves to balance or override the natural, shameful aspects of marriage. (59)

This disparagement of sex, and thus of marriage as the one institution within which it is allowed, carries over to attitudes toward children. Children are loved and valued, but "relations within the family are filled with shame, tension, and denial." Therefore, the preferred method of obtaining offspring for the family is via adoption. While the rate of adoption is about 30 percent—lower, Fajans notes, than is the case elsewhere in Oceania, citing Carroll (1970), Levy (1973), and Brady (1976)—and almost every family is involved in adoption as either a giver or a taker, the real importance of adoption is its centrality in cultural logic: "The act of adoption is itself a social act. An adopted child is said to be one's 'true' child. Parents call their adopted children by kin terms . . . [and] refer to their [adopted] children in conversations with others by these terms. . . . Informants also say that adopted children should be favored with more or choicer food" (63). By the same token, mother's milk is disparaged as "nothing," in opposition to garden food, because of the same distinction between what is natural to biological reproduction and what is transformed by human work.

Consistent with this logic, the Baining prefer plants that are not grown from seed:

In their gardens the Baining cut down all the wild, natural vegetation of a plot, and plant domesticated crops in its stead. It is their labor that produces and cultivates these plants. This is particularly true since the majority of Baining crops are grown by vegetative propagation rather than from seed. These crops are grown by cloning. The plants are already sprouted or developed and are transplanted from one garden to another. They are analogous to adopted children who are transplanted from one household to another. Forest crops, which grow from seed, are more like biological children. Here again the Baining replace a substance they see as natural with one they consider socially produced. (69)

Although the distinction between "nature" and the "social" is not explicitly articulated by the Baining themselves, it is hardly possible to deny the implicit presence of this distinction in the face of empirical observations such as these. In this society, while of course copulation occurs between marriage partners, it is surrounded by shame and denial, and its result, biologically produced offspring, is devalued in favor of "transplanted" children acquired by social exchange, that is, by adoption. Even naturally born children are only valued insofar as they undergo socialization and feeding with garden food produced by work rather than by the mother's "natural" substance. In this system, relative equality is maintained between the sexes by the expedient of devaluing not only the mother's role in procreation but the very existence and necessity of procreation altogether. At the same time, the marital couples that form the main elementary units of society are linked to a larger system by the exchange not of spouses but rather of children.

Adoption emerges in this light as a form of exchange that serves the tribal dimension of society in its efforts to overcome the isolating characteristics inherent in the imperatives of the genetic program, and to spread responsibility for reproduction throughout the society beyond any particular marital pair. In this way it is analogous to other forms of exchange that have a similar effect, such as marital exchange and gift exchange. In the same way that one must not reproduce with one's own close relatives, or eat one's own pigs, as the Melanesians say, so one should not raise one's own children but allow them to circulate.[1] The phenomenon of "partible paternity" (Beckerman and Valentine 2002), whereby it is thought that successful conception requires the semen of multiple men, serves a similar function in negating self-interested "paternity certainty," associated with the genetic program, and requiring group participation in the production of children at both the genetic and cultural levels. All these forms of reproductive exchange and redistribution can be seen as cultural efforts to offset and undercut the imperative toward paternal investment in one's own genetic offspring only, by literally confusing the issue so as to make paternity *un*certain and to thus require the sharing of parental investment among the men, robbing them of a basis for potentially disruptive competition.

I do not mean to suggest, any more that I have with other examples, that the Baining system has any privileged place in understanding human social life; I have chosen it rather because it seems to me a particularly clear case in which it is easy to see how the cultural program systematically restricts and deemphasizes the genetic program upon which it nonetheless depends, and works with it to overlay genetic reproduction with a system of cultural reproduction that is more highly valued. That enables the formation of a wider community over and above the mating pairs that constitute it (weak though that community is).

The one spectacular instance of communal ritual in Baining society, the masked dancing, while almost completely devoid of any cognized mythic foundation or local exegesis, is an examplar of the core values of the society: forest products, transformed by men in the bush into cultural artifacts in the form of masks, are brought into the social arena of the public space of the community and displayed, and then left to revert to nature by rotting. It is not the symbolism constituting the ritual, or about the ritual, but the ritual itself as a symbol, that conveys the key meaningful schema that organizes Baining social life. The ritual says, in effect, that society arises out of nature, flourishes for a while as persons (masks), and then reverts to nature.

Meanwhile, On an Other Part of the Island—Kaulong

The Baining occupy a region at the northern end of the island of New Britain. The southwest region of the same island is the home of the Kaulong people studied by Jane Goodale (1980, 1983). As will be seen, their social and symbolic system is very different from that of the Baining, but it accomplishes, by entirely different methods, the same feat of devaluing, marginalizing, and containing sexual reproduction within a larger asexual social matrix. Among the Baining, most adults are in fact married; but the sexual and reproductive dimensions of marriage are hidden away, shameful and inappropriate to mention, and counteracted by the practice of adoption that involves married couples in a system of child exchange that offsets the atomism of a society otherwise composed almost solely of isolated marital pairs. Among the Kaulong, there is "little separation of men and women in public or private spheres of life, which contrasted with a noticeable separation between individuals who [are] married and those who [are] not: a separation both physical and symbolic" (1980, 119).

In this system, celibacy as a way of life is highly valued, and marriage is seen quite literally as a fatal condition, best postponed or even avoided if one could help it:

Men are quite literally scared to death of marriage (and sex). Young men often repeated the phrase 'I am too young to get married and die.' Or they said 'When I am old and ready to die, I will get married and find a replacement for myself who will bury me.' . . . Sexual intercourse is not only considered intensely polluting for the male; it is also considered 'animal-like' and should take place away from the *bi-* [central clearing] and garden in the forest. Neither gender should show overt interest in copulation. Married couples who spend long periods of time together and away from the public area of the *bi-* may be suspected of 'inhuman' preoccupation with marital sex. (133–34)

The Kaulong (in pre-contact times) subsisted on taro, domesticated pigs, forest products that could be gathered or hunted, the produce of tropical trees of various kinds, and fish. Residence was flexible but centered around a hamlet in a cleared space (*bi-*) in which usually about a dozen people lived. The central feature was a hamlet house in which deceased affiliates of that hamlet were buried. This house was maintained and occupied by unmarried men who affiliated with that hamlet. Married couples and their children affiliated with the hamlet lived in garden huts some distance from the hamlet house. The cleared space, the hamlet, and the hamlet house were said to have been founded by a brother-sister pair who in turn sprang from a tree. Unlike the Baining, the Kaulong looked to forest plants that propagate themselves without the need for copulation as an ideal model for human life. A person might affiliate with any hamlet in which s/he could trace descent through either a paternal or maternal line—a person's sex was not particularly important in their system, as would of course logically follow for people who valorize asexual reproduction. The chief duty of a person was to replace the old fruiting trees around the hamlet's cleared space with newly propagated ones, and likewise to replicate him- or herself with at least one biologically produced child, preferably but not necessarily of the same sex. Doing so was clearly fraught with danger, since the Kaulong consider females to be inherently beset by sickness in the form of menstruation and childbirth, and believe that their weakness is contagious. As women gradually have their own substance reduced by a series of childbirths, men are subject to contamination from sexual contact with their wives' genitals and will ultimately die of it.

Marriage is for life—there is no other option except celibacy. No sex other than between a married couple is tolerated. Men are expected to be passive in the face of the women, who are encouraged to be the sexual aggressors and initiators of courtship. Once a couple has had sex they are married for life, and their main task in life is to produce their replacements by having biological children (adoption is sometimes practiced, but an adoptee does not count as a replacement). Sexual relations are so shameful, however, that married couples live in isolated huts near their gardens, away from the hamlet clearing and hamlet house. The latter is for unmarried people only.

In addition to producing a replacement for oneself by sexual procreation, the other main goal of Kaulong life is to enhance ones own vitality, made evident through a glowing physique. This, in turn, is produced by success in networking and trade; an aspiring "big person" (I phrase it this way since the role is open to either sex) utilizes ties to as many hamlets as s/he can claim affiliation with, and, if successful, attracts a number of clients eager to be associated with a successful person. Since all the aspects of subsistence can be carried out by a single person, there is no need to marry to achieve high status and the power, prestige, and wealth it brings with it. The only restriction is that women cannot perform the clearing of forest trees; but they can call upon consanguineally related men to help them and can therefore do without husbands if they choose.

THE ASYMMETRY OF CULTURAL VERSUS GENETIC REPRODUCTION

Married couples are not only separated from the main settlement and public arena spatially; they also have to begin using a new "dialect," since not only are they forbidden to speak the name of their spouse, but they cannot use any word that sounds like it or is related to it in any way. Furthermore, they are required to observe a great many prohibitions in regard to their affines. It seems clear, in other words, that marriage, and the sexual activity that defines it, are shameful, indecent, marginalized, and set apart—and also necessary for the production of replacement children.

In contrast to the husband-wife pair, the brother-sister pair is the one that is publicly approved. While men are rivalrous and jealous of their brothers, as women are, to a lesser extent, of their sisters, cross-sex consanguineal ties, especially between mother and son but more importantly between brother and sister, are marked by warmth, intimacy, and lifelong devotion. Since this relationship is regarded as self-evidently asexual, brothers and sisters may delouse each other, eat together, and sleep together (not in the sexual sense) with no opprobrium. This relationship endures after the woman's marriage, in which, as I have said, the sister's husband has to observe prohibitions—chiefly against any hint of aggression—toward her brother, and to show subservience to him.

Since the goal of an unmarried, asexual existence, while not a problem for achieving individual success in life, is incompatible with the ideal of producing a biological child to replace oneself, a tension is set up between brothers-in-law over the matter of whether the father or the maternal uncle has the greater rights and influence over the latter's children. A man can at least partially resolve this tension in the following remarkable way: when a man died, it was customary (in pre-contact times) for his wife's brothers to strangle her so that she could continue the marriage in the afterlife. This of course left the couple's children orphans; they could then be raised by their mother's brother as his own. He might then marry his sister's daughter to his own son, and in this way forge a compromise in which he could reproduce indirectly, and vicariously transform his asexual relationship with his sister in his own generation into an affinal, sexually procreative relationship in the next one.

Another compromise was to be found in the practice of allowing older brothers to marry but expecting the youngest son to remain celibate. The reason for this is that since the married brothers would die early from the sickness caused by contamination from copulation with their wives, it was important for one of the brothers to stay alive by not subjecting himself to the lethal effects of marriage so that he could take care of the orphaned children of his married older brothers after the latter died and their wives had been strangled at their graveside and buried with them. As for whether there is conflict between a man's youngest brother and his wife's brother for control over the children, it seems that the stronger hand is held by the wife's brother, since in all respects the asexual relation trumps one that is polluted by sexual activity and biological procreation.

Goodale sums up the situation in Kaulong society thus:

The Kaulong have two main concerns or desires for personal achievement; 1) immortality through reproduction of identity, and 2) self-development through production and social activity. Because immortality requires human beings to engage in animal-like sexual intercourse in contrast to spontaneous plant-like reproduction through cloning, human reproduction is physically separated from plant and tree replacement and production, and takes place in the forest away from garden and central clearing. Self-development is displayed in the central clearing. It is in the pork distribution and singing that all may see the tangible and intangible results of extension of self. (1980, 139)

The dream of a human society that reproduces itself without marriage or sex, on the model of the forest itself, has to give way to an allowance for sexual reproduction, but only at great cost. While marriage can and must be undertaken for the reproduction of the hamlet clearing, through the replacement of trees around the clearing and the replacement of deceased people by sexual means, almost every bad thing that can be said about reproductive sexuality is present in this cultural system: marriage, and its defining characteristic, copulation, is a sickness, it is lethal, it is polluting, it is disgusting, it is shameful, it cannot be mentioned, and it separates one from others by prohibitions, spatial isolation, and other restrictions such as the ones on language. It stands in stark contrast to the brother-sister relationship, which is asexual and hence free of all the ills that beset marriage, protected by the assumption that sexuality between cross-sex siblings is simply an unthinkable impossibility. It is important to note that the negative evaluation applies to sexual procreation, not to women per se: while women as sexual beings are contaminating and dangerous because of their role in childbirth and menstruation, as individual entrepreneurs or as asexual sisters or mothers they are very highly regarded by the men.

The means culture has at its disposal to oppose sexuality and reproduction, so well illustrated in this case, may be seen from this example to involve mobilizing the gamut of the negative emotions. These include: dread of sickness and death (it is dangerous and will kill one); shame (it marks one as animal-like and excludes one from the main public arena of social life, the *bi-*, and makes one the butt of disapprobation from the community); disgust (it makes one behave like a pig rather than a human, and thus arouses revulsion); fear of contamination from the pollution of menstrual blood; fear of lowered status and the loss of power that are entailed in the restrictions placed on one vis-à-vis one's affines; and whatever the emotional barriers are that act, in concert with socially established incest prohibitions, to inculcate an effective aversion to sex with a sibling.

Some but not all of these emotions have equivalents or precursors in non-human primates; they are all self-regulatory methods that are either created by life in society or are innate dispositions that require reinforcement and elaboration from symbolic sources to become effective, or probably some combination of both. The dread of death is accentuated in a creature with the foresight culture enables in humans. Shame is an emotion foreign to non-humans, involving as it does being seen in a negative light by others in the community in the public arena—something that does not exist as such without human society and culture. Separation from the community and the loss of power, status, face, or honor likewise are terrors that only humans, with their immense dependency on the regard and support of community and fear of its loss, can experience (I return to this topic in chapter 10). Disgust may have evolutionary roots in aversion to things likely to actually transmit disease, or to eating dangerous or non-nutritious things, but it can be extended by cultural symbolism to other arenas as well, and the functions humans visibly share with other animals, mainly alimentation, elimination of waste, and sex, are prime targets. Reduced power and status of course occur in non-human societies with status hierarchies, but the idea that sexual behavior results in such a narcissistic and social injury seems distinctively human. In humans, there is an official cultural norm that devalues procreation and sex, but these latter nonetheless motivate people thanks to the continued influence of the genetic program, whether sanctioned by the social value system or not. Shame and loss of honor, face, or reputation thus arise from the gap between the imperatives of the two systems. Status exists in any hierarchical society, including non-human ones, but "prestige" is a human phenomenon requiring socially shared and understood symbolic markers. I will explore the phenomenon of prestige further in the Conclusion to this book.

Finally, aversion to too-close inbreeding would seem to have its roots in pre-human genetic dispositions that would have been necessary to prevent the deleterious consequences that follow from it. Exactly how this is acquired in humans, and how it comes to be so profoundly motivational, is a subject that has troubled anthropologists and others for well over a century; I will just take it for granted here that it self-evidently happens.[2] This is what makes it possible for brother and sister in Kaulong society to share intimacies that would be punishable by banishment or death in any other situation. The extension of the incest taboo to people beyond first-degree relatives, however, is a human faculty requiring the capacity for symbolization that places large numbers of people in the category of those with whom marriage and/or sex is either forbidden, or unattractive, or both, and thus paves the way for relations of amity and asexual affection to prevail among them.

We have seen that the Hill Pandaram kinship system places large numbers of cross-sex people in the forbidden category by the classificatory system that divides the whole society between ego's affines and ego's consanguines. Likewise, among the Culina, the whole of the community is regarded as having asexual amity according to the idea that they are all real siblings; in that case, the tension between the asexual community based on an incest prohibition, and the necessity of marrying endogamously, resulted in the restrictions and changes in behavior placed on marriage and its consequence of biological reproduction. Among the Kaulong, a similar idea holds sway: "A *bi-*, place of orientation and affiliation, is the only geographical space which is owned on a permanent basis. The cognatic group sharing in the ownership also share an identity as *poididuan* (glossed as 'all related' or in Melanesian Pidgin as either *barata* 'brother' or *bisnis* 'business')" (1980, 122).

The means employed by the Kaulong for simultaneously suppressing reproductive sexuality and at the same time allowing it to happen at least enough so that there is demographic replacement include the following: separating the categories of married and unmarried people; allowing the latter to inhabit the community center, while separating off and placing at a social disadvantage married couples; postponing marriage (for men) until late in life; making a distinction between older brothers, who do the sexual reproducing, and the youngest brother, who remains celibate; and valorizing the asexual brother-sister relation in myths of origin as well as in enacted social reality, while allowing asexual consanguines to produce affinal pairs in the younger generation by giving mother's brother authority over his sister's husband and thence over her children.

Different as it is from the other cases I have examined, one can also see that the Kaulong example is similar to them in that it illustrates one of the ways in which a social system resolves the conflict between the valorization of asexual reproduction and the necessity of undergirding it with sexual reproduction, while at the same time protecting the integrity of the community formed by the former through the restriction and devaluation of the latter.

High Caste Nepalis

The problem of how to reproduce as asexually as possible, while bowing to the necessity of actually allowing sexual reproduction to take place so as to avoid social death, occurs in many societies around the world. We have just seen some of the ways the Kaulong deal with it. I will now turn to another example, a very different one, that of the high Brahmin and Chetri (*bahun-chhetri*) castes in the Kathmandu Valley of Nepal. From a sociological point of view, there are of course vast differences between a small-scale foraging and swidden society in Melanesia with little by way of formal structure, on the one hand, and a large, complex, agricultural Hindu society in a partially urbanized nation state with distinctions of caste, clan, lineage, ethnicity, and wealth that are absent from societies like the Baining or Kaulong, on the other. However, the solutions to the problem of how to reconcile the ideal of asexual reproduction that reflects the values of the cultural channel of information transmission with the need to find a way to manage sexual reproduction of the genetic channel are, if not the same, then, in some ways at least, strikingly similar.

The Brahman-Chetri conceptual system places the highest value in life on achieving liberation from organic existence and the mortality that it entails (Bennett 1983). However, there are two antithetical ways of doing this, reflecting the contrast between the system of ancestor worship built into the caste and lineage system, on the one hand, and the religious ideal of complete liberation from the cyclical round of existence and endless rebirth in the material world altogether, on the other. In the system of the married "householder" way of life, a premium is placed on the continuity of the agnatic patriline and the worship of the ancestors. After a well-lived lifetime in this regime, a man can be born in the realm of ancestors himself where his sons will perform ceremonies for him back in the realm of the living. For the *sanyasin*, the ascetic, on the other hand, complete rejection of the organic life leads to true liberation from mortality rather than the conditional immortality of the one who lives as a householder. For the latter, immortality is assured only by the reproduction of sons through marriage, whereas for the ascetic, reproduction through sex, like any form of sexual activity, must be foresworn, with the end of the cycle of being born over and over in a material body, and union with the absolute, as the spiritual reward.

While some people do become true ascetics, the vast majority of village and town dwellers choose the householder path, in which one must forge a compromise between the ascetic ideal that holds sway in Hindu ideology and the imperative of marrying and having sons. A solution is found by acting first on one, then on the other in a temporal sequence in the course of a lifetime. As O'Flaherty (1973) puts it in her magisterial treatment of asceticism and eroticism in Indian thought: "The tension which is manifested in metaphysical terms as the conflict between the two paths of immortality, between Release and *dharma* of conventional society (in particular the *dharma* of marriage and procreation) appears in social terms as the tension between different stages of Hindu life" (78).

These four stages, for a male, are that of a celibate student, a married householder, a forest-dwelling renouncer who may however maintain his marriage with his wife, and a sanyasin, who has renounced everything. In two out of these four, the first and last, the ideal of complete celibacy is upheld; in the second, the householder values prevail, while in the third an uneasy compromise is in effect. However, in practice, the last two stages are hardly ever realized, and so the practical task is to integrate devotion to the ideal of the continuity of the patriline with the ideal of renunciation, including renunciation of sexuality, in the ongoing course of everyday village life:

Ritual purification is extremely important in village religion as a means by which the opposing values of asceticism are symbolically incorporated into the householder's life. Ritual purity, attained by obeying the rules and performing the ceremonies of conventional religion, is at least on one level a metaphor for ascetic purity attained through renouncing samsara [material existence] and performing harsh austerities. Both kinds of purity require discipline and set limits to the bodily and egotistical desires of the individual. Significantly, the strongest rules for maintaining ritual purity have to do with the regulation of sex and eating, both areas where the desires of the flesh are strong. (Bennett 1983, 44)

The problem of reconciling asceticism and eroticism is built into the conceptual system surrounding the idea of patrilineal descent and agnatic solidarity, since each current *gotra* (an exogamous clan-like agnatic descent group) is said to be descended from one of the seven sages or *rishis* who lived in mythic times. But if these mythic first ancestors were all extremely pure ascetics, how could they also be the patriarchs of a patriline? The answer to this conundrum is supplied by the mythic tradition concerning the wives of the rishis, who, though sworn like their husbands to a life of austerity, succumbed to sexual desire, deluded their husbands, and seduced them, thereby weakening their spiritual purity but also establishing the patrilines in the process.

In an ideal world, from the Brahman-Chetri perspective, patrilines of perfectly pure men would reproduce without having to resort to sexual procreation. Bennett suggests:

The patriline's ideal self-image could best be represented by an unbroken line of celibate holy men, such as is found in the mythological lineages of the North Indian Nath cults. There, as elsewhere in Hindu mythology, male offspring are produced by a series of miracles instead of by normal sexual intercourse between men and women. Fertility, and thus continuity, is achieved without the loss of purity entailed by the householder's involvement with women. (1983, 126)

However, since this ideal is not to be achieved in the real world, householder virtue must be something short of complete asceticism. Bennett lists four strategies for achieving rebirth in the abode of the ancestors, if not true and final release from the round of rebirth. These include the continual maintenance of purity through ritual cleansings; the accumulation of merit through good works, such as by giving alms; devotion to the gods through the diligent performance of worship rituals or *puja*; and the production of male offspring. A continuous agenda of life cycle rituals, especially death commemoration rituals, daily pujas, and regular communal religious festivals, works symbolically to weave together these (actually incompatible) values—values that can easily be understood as representing the competing aims of the cultural and of the genetic channel of inherited continuity.

This brings us to the matter of the role that women play in mediating this system, since they are the necessary accessories by whose agency the patrilines of (conditionally) pure agnates actually perpetuate themselves. As the story of the bad behavior of the wives of the rishis makes clear, Hindu ideology is quick to lay the blame for the intrusion of sexuality into the realm of purity at the feet of women. Not only are they ritually impure in comparison with men, and socially subservient in most (but not all) ways, but it is also thought that it is in-marrying affinal women who sow dissension among the brothers of the patriline by seducing them away from their group loyalty to each other and egging them on to compete with each other. Therefore, many ceremonial events stress the cooperation and solidarity of the agnatic kin, while women are relegated to the sidelines. Thus for example, during the festival of Devali, which centers around the performance of the blood sacrifice of goats and chickens, the men of the *kul*, or local lineage group, reunite, temporarily undoing the tendency to fission that breaks the lineages up into smaller units. "The oft-repeated sentiment that brothers would get along if only their wives didn't set them against each other is here enacted. For on the afternoon of Devali, the kul temporarily reestablishes the single joint family from which all the participating household units originated. There are no women—only classificatory brothers" (1983, 136). Thus far, the women who have married into a family or lineage to ensure its perpetuation through the birth of sons, are, beyond that desired function, understood as dangerous sources of friction and agents of the undoing of agnatic solidarity and the purity it would otherwise maintain.

There is, however, an entirely different side to the story. These same women, who as soon as they are formally betrothed leave behind their natal patrilineage and, as wives and then mothers, become full members of their husband's patriline, are also daughters and sisters in their natal lineage. Their virginity in youth is essential to the family's honor, and so it is carefully guarded; this will bring them to the pivotal moment in their lives, in which they are given as virgin brides to a family whose status is ideally higher than their own. When this is achieved, the young woman has fulfilled her obligation to her natal family, for whom she remains the pure and asexual person she was before her wedding. For it is only as a member of her husband's patriline that she engages in marital sex and biological reproduction. Thus she avoids infusing any pollution associated with sex and procreation into her own natal patriline.

In their natal families, these young as-yet unmarried women are the objects of reverence that reverses the normal hierarchy whereby men outrank women and senior people outrank more junior ones. In their own natal households, these women are sacred, and the most senior men show them ritual respect, such as by washing their feet on the occasion of their marriage. Since in this system the family of wife-takers outranks the family of wife-givers (the opposite of what happens in many other systems), even if there is no actual status difference between the two parties to a marriage, the marriage itself creates an imbalance whereby the groom's family now outranks the bride's family. The ideal of hypergamy for the bride and her family is upheld. In order for the bride herself to effect this, she must be a highly prized gift (*dan*) from the one family to the other, and she is raised to be exactly this.

The marriage ceremony itself comprises a series of steps whereby the bride, who begins the event by being in a state of purity and ritually outranking everybody including the groom, reverses her role and, by eating food that the groom has already eaten, thus defiling herself, and by allowing the groom to touch her breasts, becomes sexualized and hence subservient to her husband. She will remain so for the rest of her life—in the husband's lineage (although as she produces sons her value as a mother of the lineage becomes more appreciated there). But in her own family, she remains the pure, sacred, and beloved virgin daughter who was given as a gift; she does not defile her brothers through her married sexuality since in that capacity she is now acting as a member of her new lineage and bringing potential contamination and discord only to them—as well as giving them the sons they want and need:

> The sexual and procreative roles which are felt to endanger the purity of the patrilineal group are exclusively associated with affinal women. Hence the bride, so strongly identified with her procreative role, must be protected by the strictest social and ritual means in her *ghar* [husband's household]. The daughter, on the other hand, is categorically shielded from any association with sexual roles. To her *maiti* [natal household] a woman is always perceived as the pure virgin *kanya* [prepubescent girl]. There is no need for the harsh restrictions of the *ghar* because, from the point of view of her *maiti*, a daughter has no sexuality to be controlled. (Bennett 1983, 241)

The equivalent moment to marriage in the woman's life cycle is the initiation ritual for young men, the *bartaman* that gives them status in their patriline and caste. The boy is ritually transformed from the initial male life stage of celibate student (*brahmacarya*) to that of real or potential householder. Although in former times young men might actually be sent for a period of study with a guru, in present times the celibate student phase is enacted in the initiation ritual itself. The young boy is first treated as if he were a chaste student, and introduced to a modified form of asceticism congruent with married life. Since in Hindu thought ascetic practice actually heightens ones virility, there is no contradiction between imposing "some restraint and control on the natural man" and his future life as a householder. This ritual enactment of celibacy will in principle actually lead to his likely fertility in his married life. During the first parts of the rite the boy's father has aided him ritually in becoming an ascetic brahman, during his hair cutting and investiture with the sacred thread, for example:

> But now the father's role is reversed. The father becomes anti-ascetic and begins preparing his son for life as a married householder. . . . This need to balance the opposition between asceticism and fertility helps explain the father's double role in the bartaman. As "great teacher" . . . and representative of the ancient rishis, the father leads the boy into a life of celibacy and Vedic study. But when the boy's asceticism threatens to become too extreme during the alms-begging ceremony, the father, as representative of the patriline, tries to lead his son back to householder values. (68)

As in the Kaulong case, then, the Nepali situation presents us with the picture of a contradiction between the wish to perpetuate a descent line and the fact that the ideal form of this descent line—the succession of brother-sister pairs in one case, the all-male solidarity of the agnatic patriline in the other—cannot perpetuate itself via asexuality and has to rely on sexual reproduction somewhere in the system. This in turn leads to the need to integrate sexuality and asexuality in an overall system that accommodates both but also clearly separates them conceptually and in concrete ways as well. In the Kaulong case celibacy was associated with early life stages with a man who would eventually marry and with the youngest brother in a sibling set, with the public and socially valorized sphere of social life, and of course with the consanguineal cross-sex relations dominated by the incest prohibition, chiefly those between mother and son and between brother and sister. Marriage was regarded as dangerous, shameful, animal-like, and separated from the public arena in space, through symbolic markers such as a new dialect and by a subservient position to the wife-giving affines.

Among the Brahmans and Chetris, asexuality is attributed to a consanguineal woman from her natal family's point of view, not just during her years as an actual prepubescent girl and virgin but in perpetuity, due to her complete formal transfer into her husband's patriline. Men begin physiological adulthood as celibate students (now only represented briefly during initiation, but previously an actual life stage) and, in initiation, are reborn with the sacred thread both as (modified) adult ascetics but also has potential fertile householders who will perpetuate the patriline. In the wedding ceremony, women are transformed from pure asexual virgins to somewhat defiled potential mothers from the groom's family's perspective, while keeping their virginal status from their natal family's point of view. In both cases, it is clear that the ideal would be reproduction without sexuality altogether; but, since this is an impossibility, reproductive sex is introduced into the system but surrounded by restrictions and boundaries and thrown into a devalued light in comparison with the higher value given to celibacy and asexuality.

Merina

That "raw" human infants must be transformed into socially "cooked" socialized members of society is an anthropological truism. Sometime around or after birth, a new human organism is given a name, placed in a kinship matrix, often given a ritual of incorporation into the community, and otherwise assigned a place in the universe of the symbolic system. Subsequent efforts are often made, through maturational initiation rituals, to fully complete the transformation of beings created by sexual reproduction into fully cultural members of the community. The final outcome is usually that they may now—as in the case of the Brahman-Chetri boy initiate—play their part in sexual reproduction, under a regime of social, religious, and other constraints. (I expand on the theme of transforming a baby human organism into a social person in chapter 8.)

There are so many good examples of this whole process in the literature that it is hard to represent it by a single one, but I will choose to look at the case of the Merina of the Malagasy Republic, for whom male circumcision at an early age transforms the little boy from a biological creature of his mother's womb to a member of the endogamous descent group into which he has been destined by his birth to his particular pair of parents (Bloch 1986, 1989).

Merina religion and social organization center in large part around the maintenance of and worship at the ancestral tombs of both members of the ancestral couple of the descent group. Rights to membership in this descent group pass without distinction through both males and females. This can happen because the descent group is, as the ethnographer Maurice Bloch puts it, "suprabiological in that it overcomes all discontinuities created by biology, including sexual differences." It is thus not a group in which membership depends conceptually on sexual reproduction but instead rests on the nonmaterial, nonbiological

CHAPTER SIX

transmission of shared substance separate from that which produces the biological organism: "The Merina descent group is represented as reproducing not through biological generation but through superior mystical means. The higher form of reproduction is by blessing (*tsodrano*), the mystical transmission of life-giving virtue through the generations" (1989, 155–56). Clearly, then, the Merina system distinguishes between a biological and a cultural form of inheritance across generations, consistent with my overall theoretical schema in this book, viewing them as operating along separate parallel tracks. In the descent group, women and men are of equal conceptual and ritual value; no distinction is made between the sexes in this context, as is exemplified in the fact that the tombs in the ancestral cult are those of both the patriarch and the matriarch. This reflects in concrete terms the difference I have proposed between genetic reproduction, which relies on the complementary of the sexes and their necessary sexual union, and cultural reproduction, in which sex and the distinctions it entails between males and females are irrelevant: mystical blessing has no need of any reference to the genitals and gametes of those who inherit and transmit it. It can thus travel through a descent line without fear of violating either incest or purity restrictions, and thus fulfilling, albeit only at a "mystical" (i.e., symbolic) level, the dream of asexual reproduction that also was seen to be motivating both the Kaulong and the Nepalis.

Since the ancestral cult is oriented to past, not future, generations, there is no need for sexuality to enter into it as it would if it were going forward toward the future and as-yet uncreated generations. Exiting from life may often enough be conceptualized as a kind of new birth, but it doesn't require anyone to be actually impregnated. (Actually the situation is more complex, since the funeral rituals of the Merina do involve some unrestrained and ordinarily prohibited sexual matings; but that is a separate matter [Bloch and Parry 1982]. I return to this issue in chapter 8.)

However, when it comes to the production of new members of the descent group, the asexual, unisex character of the descent group alters dramatically. While it is true that, as members of the descent group carrying the mystical blessing, women are of no less value than men, and are not distinguished from them, in an ideological strain running counter to that one women are seen as distinctly inferior beings, precisely because of their involvement with sexual reproduction and especially birth, which is seen as a purely female process: "Natural procreation, interpersonal links, houses, and women alone are . . . pre-

and anti-descent. Women in this representation are categorically inferior. . . . It is possible to refer to women, children, and slaves by the same term, *ankizy*" (Bloch 1989, 157).

Birth and the rituals that surround it take place inside a model of a house built inside the actual household; only women are present. Circumcision, by contrast, which occurs when the boy is about two years old, takes the boy out of the house, the private sphere associated with femaleness, and into the public arena outdoors; from the "divisive world of women, of the home, of matrilineal kinship and of biology, to be received into the unity of the eternally undivided descent group, where the division between men and women has ceased to exist" (158–59). One might suppose that the circumcision ritual would, therefore, be one in which the descent group participated without regard to the difference between the sexes. However, what happens is quite different. While the circumcision rite is going on outside in the house, inside the house "women crawl about on the floor, humiliating themselves by throwing dirt on their heads, an action the Merina consider most polluting, and that represents the lowliness of individual kinship and birth; outside the undivided descent group is represented by men *only*, and not men and women, as the ethos of Merina descent blessing would require. On exiting the house and being welcomed into the arms of the descent group, represented only by men, the boy is greeted with the cry 'he is a man!'" (159). It seems, then, that when the ritual focus is turned away from the ancestors, the world of existence after death and the past, and turned toward the creation of new life, the distinction between biological and mystical (symbolic) reproduction is retranscribed as a distinction between men and women, whereas in the other context that distinction is ignored entirely. By ascribing birth solely to females, the males are able, in contrast, to appear as the bearers of the mystical blessing, which they enact by performing symbolic ritual actions rather than by actually extruding the infant from within what the Merina consider to be the most polluting part of their bodies as women do.

However, the story does not end there, because, as Bloch goes on to show, the male ritual symbolism requires the incorporation of symbols and materials that are very explicitly gendered as female. Plants and animals associated with the "living mother" are necessary for the efficacy of the ritual, because in the end it is necessary to inject impure but powerful vitality into the symbolic procedure. The object of the ritual is, after all, a living biological organism, who has to have been produced by sexual means, and this means including the female element. However, these potentially polluting and dangerous symbols and the femaleness they represent cannot be included in the ritual proceedings without themselves being subjugated to male mystical power: "The contradiction is, however, modified and to a certain extent lessened by the fact that these matrilineal, natural entities must first be broken, crushed, or chopped by the [male] elders at various stages in the ritual before their vitality can be added and so passed on" (1989, 159–60). This pulverizing of the female elements by the male officiants would seem to represent the ambivalence felt toward them: they are necessary, but only after having been crushed and reduced to mere unformed stuff under the control of the men.

Whereas in the Nepali case the ambiguity of women as both pure, asexual members of the natal patriline and as sexual, dangerous wives of the affinal patriline is resolved by the fiction that maintains their purity in the former by transferring their sexuality wholesale into the latter, in the Merina case women display the same two-sided aspect, but in this case they are pure, asexual members of the descent group when oriented toward mystical things, death, and the realm of the ancestors, but defiled sexual beings when involved in sexual reproduction. The descent group can be thought of either as a sexless, mystical unity where the distinction between the sexes doesn't matter, or as the pure male members separated from the defiling females. In the Hindu case the males operate in the context of an agnatic ideology of patriliny, in which daughters and sisters marry out, but in the Merina case, the descent groups, which are much larger than lineages, marry in, so of course the solution must be a different one.

Igbo of Afikpo

I began this chapter with a discussion of the Baining, who exhibit an across-the-board aversion to sex altogether, preferring reproduction by adoption to biological procreation. I deliberately chose a rather extreme example of the opposition between the cultural realm and that of sexual reproduction to illustrate my point about the two parallel tracks. In claiming that cultural symbolic systems are inherently opposed to those dimensions of human life that pertain to sexual reproduction, I obviously cannot mean that they abolish the latter completely; if that were the case, society would not be possible. What I mean to argue is rather that to understand the forms taken by social life and the various cultural symbol systems that surround and sustain it, one must observe the compromises, strategies, and symbolic sleights of hand that are necessary to make the two partially incompatible systems seem to cooperate at least minimally.

Needless to say, there are many societies that do not have an ideology as overtly antithetical to sex and reproduction as the Baining or the Kaulong. On the contrary, a great many are enthusiastically supportive of biological fertility within the public discourse. This does not mean, however, that the conflict between the genetic and the cultural modes of reproduction is absent, only that it is resolved in a manner different from the one that prevails in societies that have more explicitly negative attitudes toward sexual reproduction. Nor does it mean that they are actually better at reproducing sexually: the case of the Marind illustrates that point, since in that society people were doing everything they could, within the context of their symbolic system, to enhance procreation, only to undercut their own efforts in ways that were not transparent to them.

Many African societies, which are segmented into descent groups, regard reproduction as the highest value for individual men and women as well as for the lineage or clan. Among the Igbo of Nigeria, reproductive fertility is one of the very highest values, since it is that which strengthens both the matrilineal and patrilineal descent groups that exist side by side in their system of double descent. Nor are they troubled by a general horror of sex, as among the Baining or the Kaulong, or by a competing ideology of ascetic transcendence, as among the Hindu Nepalis, or of mystical blessing as among the Merina. Indeed, Ottenberg opens his monographic study of initiation among the Igbo of the Afikpo Village-Group with these observations:

> It is essential to have children at Afikpo. For a man it is particularly important to father sons in order to have status in society. Through sons men can acquire important titles. Without them a man is shy to speak up at public meetings and he finds it hard to be a leader and to hold a religious position. . . . And without daughters to marry off a man has difficulties cementing relationships with friends and other families. A childless man will marry and remarry, and take on additional wives to try for offspring. A woman marries at a young age, in the middle of adolescence. To have offspring that survive is the most important success in her life—nothing else matches it. . . . Both sexes are very conscious of the need to produce children for their descent groups, males particularly for the agnatic line, females for the uterine one. (1988, 1)

CHAPTER SIX

It could not be clearer, in other words, that as individuals and as a society, Igbo people value sexual reproduction highly and apparently without visible opposing ideas from within the symbol system. Yet it is just as certainly the case that there are many constraints placed upon reproductive sexuality in this society that serve to keep it contained by cultural values that demand that things be done within certain normative boundaries and in a certain prescribed order. The violation of these constraints brings on dire consequences which, by their very severity, indicate the strength of the conflicting forces of the desire and demand for reproduction and the social mechanisms shaping and regulating it.

Thus, for example, all boys are circumcised at an early age, and girls undergo genital excision, likewise in early or middle childhood. While this is not ritually elaborated, it is essential since the reshaping of the genital organs to conform with culturally valued qualities is essential for a later entry into both ritual and sexual life. The succession of stages in the boys and men's secret societies that mark the lengthy process of making boys into adult men eligible to be married and to take a place in the ritual and social life of the community requires that the initiand already be circumcised. Indeed, copulation itself is thought to require the surgical modification of the genitals in both partners. Beyond that, there are very severe repercussions if boys have sexual relations before being circumcised, or if girls get pregnant before undergoing genital excision. One of the most important supernatural spirits, Ale, the Spirit of the Ground, is deeply offended by such occurrences, and in former times at least, the punishment was harsh: a girl who got pregnant before her clitoridectomy was forced to abort, and the household goods of her mother were destroyed; her father had to perform a ritual to appease the ground spirit, and the girl was made to live in a special place until the birth of her child if she refused or was unable to abort. Derisive songs were sung about her by other women, sometimes for years afterwards. Early performance of genital excision was thus a measure often undertaken to make sure these things would not happen.

The consequences of improper births are even more dire. A woman who gives birth to twins, or who delivers a child feet first, disgraces the entire community; she is isolated from the village for a long period, her husband is required to pay for and perform a complex ceremony, her home is torn down, and she is permanently marked as impure, ineligible to perform sacrifices at a group shrine. Clearly, these anomalous births are understood as evidence of some sort of fault in or by the mother.

Igbo society was marked by a strict separation between the sexes, and men were dominant in the public spheres of society. Beginning in early middle childhood, young boys moved into boys' houses and then into boys' secret societies, preparatory to being initiated and joining the man's secret society. These were "secret" only from women, since all males in the community belonged to them. While girls were married in early adolescence so that they would begin their reproductive lives right away and before any untoward pregnancies could occur, males had to go through a very lengthy transition from childhood to adulthood before they could marry and father children. Premarital sex was not harshly forbidden for young men, though it was for young women, who were in any event already married by the time they would take an interest in it; and extramarital sex was not approved of, though it certainly occurred. The big constraint on reproductive sex, as opposed to these more recreational forms of sexuality, was the division of the society into senior men who could marry, father children, and thus join the communal social and ritual life, and the young men before completing initiation, who, while they were able to engage in hidden affairs with married women, were not allowed to be legitimate fathers of children who would perpetuate the descent group.

Sexual abstinence was expected during rituals of initiation and on other occasions, and there was a two-year postpartum taboo on marital sex. In short, while fertility was the highest goal, it was expected to be achieved only within cultural constraints well guarded by public opinion, actual sanctions, and supernatural sanctions as well. It was only by carefully observing the proper symbolically constructed order of things that the "natural" function of genetic reproduction could be managed. The emphasis on the control of women by men was salient, and corresponded to the view that women's sexuality was a potentially disruptive force in society.

This case raises the question, as do the Kaulong, Nepali, and Merina data, of what relationship there is between the opposition between cultural symbolism and genetic reproduction, on the one hand, and the opposition of men and women, or male and female gender ideologies, on the other. It is to this contentious and important question that I now turn in the next chapter.

SEVEN

The Society of Men

When the Manambu men of Avatip, a community in the Sepik River lowlands of northeastern Papua New Guinea, went on a raid against an enemy village (we are speaking of the era before pacification), they were accompanied by a *simbuk*, or war magician, who would chew a bespelled mixture of ginger and lemon and spray it from his mouth (Harrison 1993).This magic would render the enemy paralyzed and stupefied, and thus easy to kill. Then just before the attack itself, the simbuk bespelled each fighter with the most powerful form of war magic, a leaf of a variety of croton which each warrior had to keep on his tongue until the end of the encounter (it was fatal if swallowed): "This magic could only be administered after the men were well away from Avatip because, once bespelled, they became capable of killing anyone, even, so men stress, their own wives and children. Men speak of this magic as having induced in them a state of dissociation in which they become capable of extreme, indiscriminate violence" (95). When they came to after the raid was over, they themselves were astonished by the havoc they had wrought: "They claim not to even to have perceived the enemy corpses until the magic was removed and 'our eyes became clear again, and we saw all the fine men and women we had killed'" (95).

This bit of ethnographic data may raise a variety of questions, but the one on which I wish to focus at present is this one: once the warriors are bespelled and obsessed with an impulse to commit indiscriminate homicide, why don't they kill each other? After all, they have had to take precautions so as not to kill their own families. Presumably they would not be in danger of killing their wives and children unless they were in the grip of a blood lust so indiscriminate and so total that they would kill anyone who crossed their path; so why were their comrades in arms exempt if their own families were not?

The ethnographer does not give us a direct answer to this question, but he does point us in a likely direction by devoting considerable space to a discussion of the men's cult among the Manambu people of the community of Avatip, in which, as in so many similar institutions throughout Melanesia and indeed the world, men are initiated into progressively higher grades. The men's cult is the focus of Avatip religion, though (not atypically) it is an ideology representing only the senior initiated males to the exclusion of women and juniors. The cult has two major projects: the promotion of natural fertility, and war (even in post-pacification times preparedness for war remains a preoccupation, though competitive sports have been substituted for lethal rivalry). All the men in the community of Avatip are divided into five war-magic divisions, each headed by a simbuk. In pre-contact times, various forms of violent play, sometimes inflicting fairly serious wounds, were supposed to prepare the men for head-hunting raids against neighboring "outsiders." It is my supposition that it is through socialization in the cult with its war magic and its violent "team" activities that a degree of solidarity is forged among the men that is stronger even than the ties that bind them to their families, and protects them from each other's own homicidal fury in battle. It is not insignificant, furthermore, that while the ties associated with genetic reproduction—those to wife and family—are vulnerable in times of magical dissociation, the cultural ties, forged through joint participation in symbolically constructed rituals and practices in the men's cult, are apparently not.

I have been arguing throughout this book that a primary problem to be overcome in the formation of human groups is the rivalry among males for reproductive advantage entailed in the system of genetic inheritance. Human societies must find a way to deal with this situation, and one of them is the creation of male solidarity. Men's cults clearly seem to be institutions designed to do exactly that. Though they are not by any means universal, they are sufficiently widespread to represent a recurring strong tendency that can be observed throughout the ethnographic record. Of course, other strategies exist as well, but the phenomenon of the men's cult seems like a good place to begin an investigation of how cultural symbolism overcomes the male tendency to disruptive competition and enables cooperative, amicable, or

CHAPTER SEVEN

at least stable human societies to flourish. Before I turn to a consideration of men's cults, however, I must make a small detour to examine a couple of other potential obstacles to the formation of cooperative social groups.

The "Free Rider" and "Cheater" Problems

Male-male competition is not the only roadblock on the path to social living. In fact, I have elevated it here to somewhat more prominence than it is sometimes given in many evolutionary considerations of how social life evolves. Two closely related and commonly cited problems that arise from the theoretical modeling of evolutionary processes are the problem of the "free rider" and the threat posed by "cheaters." According to the theory, any attempt by people to cooperate is undermined by the fact that if an individual adopts the strategy of benefiting from the fruits of cooperation, but without contributing anything to the group effort, that strategy will flourish and spread, since by reaping the rewards without effort this strategy naturally outcompetes the strategy of putting in effort to group projects at the expense of one's own fitness. Specifically in the realm of reproductive success, by opposing any effort at regulation or constraint, the one who cheats and transgresses the rules will be pursuing the winning strategy, because s/he will reproduce more than the others, and so the genes encouraging this behavior will spread at the expense of those of the nice guys who seem perennially destined to finish last. Therefore, it is argued, real cooperation, and the altruistic sacrifice of some individual prerogatives in favor of the benefits offered by group effort, will always in the end succumb to cheating and freeloading, which in evolutionary terms will eventually prevail as the genes for such behavior spread with the reproductive success of the cheaters and freeloaders. It was precisely because of the failure to imagine how these problems could be overcome that it was necessary for evolutionary theory to propose such mechanisms as inclusive fitness and reciprocal altruism to explain the existence of what should, according to theory, have been impossible, like the flight of the bumble bee.

More recently, many evolutionary thinkers seem to have realized that in many animal societies, to say nothing of a society made up of individuals with a human level of intelligence, cheating and free riding would in fact be fairly easy to spot, and it would be in the collective interest of the cooperators to negatively sanction the miscreants. Of equal importance, as I would argue, is the fact that the commitment of a group to shared cultural identity (as opposed to genetic identity or close relation) can counteract the disruptive tendencies of individuals who may try to take advantage of the system. A certain amount of freeloading and cheating will certainly take place in even the most well-running society, but a cultural value system to which the majority subscribes most of the time can keep this within limits and prevent it from undermining the whole social system itself. The passion for equality and its flip side, envy, which are such prominent features of elementary societies, represent the evolved individual endowments, emotions, and motives that effect this socially regulation.

Avatip

Having begun this chapter with an example from the Sepik community of Avatip, I will continue a discussion of men's cults by examining this particular case in a little more detail. Although in reality regional totemic clans connect Avatip with other communities and form the basis of trade and other peaceful interactions, the ideology of the men's cult affirms that all the men of Avatip are united in a single collective entity that is sharply cut off from the world beyond the village, with the inhabitants of which Avatip exists in a perpetual state of war. It is this ideological construction that is enacted in the ritual life of the men's society. Here then is one answer to the question of what happens in the men's society to the violent competitiveness inherent in the masculine strategy of the pursuit of inclusive fitness: it is suppressed within the group and symbolically directed outward collectively toward "others" whose very existence is seen as posing an existential threat. That these outsiders are perceived as a danger is at least partly the result of the fact that they now bear the burden of the community's own disavowed aggression within itself of each against the other. The inclusivity that would spare close relatives in terms of genetic fitness has been replaced by symbolic identity with others who share the same—very local— ethnic distinctiveness. Among the Culina this inclusivity is actually expressed in the idiom of genetic kinship, by designating all community members as "siblings." But even without such a designation, such as at Avatip, the principle is the same: putative harmony within the group, enmity and chaos outside it.

One of the two main functions of the men's society at Avatip is to promote the fertility of the important food sources, principally yams and fish. There are sixteen subclans arranged into three overarching descent groups, one of which is very small and marginal. The result is a "near-dualism," which approximates to a system of exogamous moieties. One "moiety" is considered the autochthonous one, and is responsible for the magic and ritual that promote the fertility of fish in the local lagoons, while the other moiety is responsible for the yams. These two groups are viewed as linked by uterine kinship, in which each group "gives mother's milk" to the other, that is, they provide reciprocal nurture to each other. Fish, it should be noted, occur naturally, whereas yams are the product of human cultivation; thus the moiety system symbolically encodes the two components of dual inheritance.

Several points need to be teased out here. One is the ideology of overall reciprocity among the men, each group belonging to one or another half of the overall society, neither of which could survive without the contributions of the other. This means that the solidarity created among the men is partly a generalized unity, but is also expressed as a relation of difference resolved by interdependence and mutually beneficial exchange.

A second point is that one division of men is said to be autochthonous while the other is not, creating an asymmetry within the otherwise reciprocal symmetry. This asymmetry exemplifies the opposition Lévi-Strauss (1967b) noticed in the Theban myth cycle of ancient Greece. The Spartoi were those families descended from the dragon's teeth sown by the founder of Thebes, Cadmus, while the others originated in the more usual way, that is, by means of sexual reproduction. This widespread mythic construction states in simplest form the opposition between the two tracks of inheritance.

A third point is that while the cult is exclusively male, the idiom of reciprocity is framed in terms of uterine kinship—the two moieties are like "sisters" born from the same mother, and they provide each other with nurture and care symbolized through the metaphor of "mother's milk." Here we encounter a theme that will become much more explicit as our exploration of men's cults expands: that a central aspect of the all-male group is an aspiration to the status of being female, of being like "mothers"; or treating the cult itself as being like a mother. Herdt (1981), for example, in his well-known analysis of the men's cult of the Sambia, in highland Papua New Guinea, underscored the ways in which some of the ritual symbolism of the cult evoked femaleness. We have also seen how in the Merina circumcision ritual, the men, representing mystical as opposed to sexual reproduction, must incorporate materials representing femaleness into their ritual lives. Of course the fact that the value the men are promoting through their cult activities is "fertility" is the ultimate tip-off to the fact that what men are enacting is the appropriation of what are often seen as female powers, that is, fertility and reproduction.

A final point to made about the data I have presented so far is that the ascendency of the men lies in their control over "religion," that is, over the rituals and magical procedures that are believed to ensure fertility as well as success in war. Each subclan, in addition to being a kinship and ritual unit, is also a landholding entity; but these are connected insofar as landowning entails control over the myriad spirits who dwell in different territories. It is these spirits, who may be addressed and manipulated ritually, who actually produce the desired fertility by supernatural means.

In the first stage of male initiation, the boys were scarified on their backs, shoulders, and thighs, and then shown the secret flutes and taught to play them. "Flutes," or some form of wind instrument, are a feature of men's societies around the world, and their sound is represented to the women and uninitiated as the voice of spirits whom the men conjure within the forbidden confines of the men's house. The men of the village are also divided into five war-magic groups, each with its own simbuk, as we have seen; and the young men are also divided into an upriver and a downriver group, which engage with each other in competitive and often violent games. This aggressive dualism is a complementary parallel in another key to the nurturing " mother's milk" dualism of the fertility cult.

The second stage of initiation is a yam festival, and the third, which is no longer performed, in former times required that a random human victim from beyond the community be killed and cannibalized. The building of a new ceremonial house also required a homicide and the eating of the victim (today pigs are substituted for humans). War was a religious or ritual enterprise in which all the spirits of the Avatip pantheon fought alongside the human warriors—or perhaps it would be better said the other way around. Wars were understood to be between rival men's cults and their superhuman protectors, not between individual men or groups of men with a quarrel. Success in war was not the result of fighting prowess but of effective ritual magic, and defeat, injury, or death was taken as evidence that a taboo had been broken or a spirit offended in some way.

CHAPTER SEVEN

For the ethnographer, the twin themes of killing and fertility are "verso and recto of a single system of ideas" (Harrison 1993, 86), depending on whether the ritual life is seen in relation to insiders or outsiders:

> The fundamental prescription of the ritual system is mutual nurturance between insiders, and violence towards outsiders. It is by means of this absolute distinction . . . that the ritual system defines the boundaries of the polity symbolically. The role of the cult-spirits is therefore a double one. They are the sources of the fertility of the community's land and other resources. At the same time they are the supernatural weapons that the community uses against outsiders. (87)

If we look at this system from the perspective of dual inheritance, what appears to have occurred is that instead of each individual male being in competition with every other male for reproductive advantage, as in the genetic system of inheritance, the competition has been moved up a level to that between communities of men against all other communities of men. These communities, in turn, are formed by the mutual identification of each with each other and with all, thus achieving cooperation that in the genetic realm only occurs among closely related genetic kin. It is the work of the men's cult to transform their interests symbolically from each male to the collectivity of males, through the mechanism of socialization through harsh ritual and joint participation in magical activities and in war.

This allows us to reconceptualize the dynamics by which it is sometimes said that harmony is maintained within the group by the expedient of "projecting" or casting out each individual's aggression and then imagining it as coming from an enemy. This is of course a real phenomenon, as I myself argued only a few paragraphs back; but what has also happened is that the definition of the "self" whose self-interest is being enacted is enlarged so that it includes not just close genetic kin but also cultural, symbolically constituted kin, that is, all the males of a community united through the ritual symbolism of the men's cult. The cult is an emergent social reality formed by identification of the men with each other through shared identification with the cult and its spirits and ritual symbolism; it can therefore be taken as an undivided unity rather than as a group of disparate individuals. Thus the cult acts as a unit above and beyond its constituent members in the same way an organism acts (and reproduces) as a unity over and above the lives of the cells that constitute it.

Religion and Ritual

The question of the "cash-value" of religious truths (to use William James's helpful phrase) is not my concern here. I do however hold to the view that religion is, as Durkheim (1965 [1915]) understood a hundred years ago, a social fact, or more correctly, *the* social fact par excellence. Religion is the means by which social groups are constituted as entities, that is as emergent realities composed of individual humans as organisms are of cells or cells of molecules. This view is becoming more and more widely accepted even within evolutionary circles, where it is grasped that human society depends on religion to enable people to get along, cooperate, and resist the impulses arising from the "selfish" imperatives of the genetic program, much as Campbell had argued back in the mid 1970s, or as D. S. Wilson has argued more recently (2003).

Why should it be that it is religion that plays such a key role in enabling societies to form out of collectivities of individuals? The Avatip case—plucked largely at random from the ethnographic record—gives us some insights into this question. First, there is the fact that religious knowledge, enacted and taught in the context of the ceremonial life of the men's cult, is shared by all (adult male) members of the community. And whereas genetic information, requiring the union of two sets of chromosomes, can (except in the case of identical twins) at best be shared only about at 50 percent with first degree relatives, and in rapidly declining percentages as one travels further out the kinship chart, symbolic knowledge, such as that conveyed in religious instruction, can in principle be shared close to 100 percent by lots of people, insofar as everyone learns the same set of meanings encoded in symbols.[1] This produces a unity of identification and of purpose that genetic relatedness cannot achieve. Once the men have passed through the initiations and performed the prescribed rituals together, they are bound, as if they were clones, as members of a greater whole within which no conflict need arise, since they are all identical, like the cells of the body which can act in concert because they all share the same DNA. Indeed, conflict is thought to be impossible among those who have shared in the same ritual symbolism:

It is, for instance, forbidden at Avatip to kill anyone who has taken part in the community's initiation rituals. To do so would be a serious ritual offence, and any aggressor would automatically be killed by the cult-spirits. . . . Just before an important initiation took place toward the end of 1978, some Western Iatmul visitors asked to be allowed to enter the initiatory enclosure . . . but were refused. As the men explained to me, 'later we might need to fight and kill the Western Iatmul again, and then we would die.' (Harrison 1993, 87)

In other words, if the Western Iatmul men had taken part in the initiation, they would by that token have become symbolic kin of the men of Avatip. If such a kinsman's life were taken, the Avatip men who killed the Western Iatmul would die, because the sanctity of the men's cult is protected by supernatural beings who have the power to kill those who introduce conflict or difference into the men's cult. These spirits, in turn, are representations of the collective power of the united men, against which the individual man is defenseless. Because symbolic relations can be total in a way that genetic ones cannot, so that the latter always imply the necessity of conflict, even within the nuclear family group, the former can trump the latter. This consideration helps explain why the bespelled warriors of Avatip might kill their families, with whom they do not share ritual identity, and to whom they are related genetically either partly or not at all, but not their fellow fighters with whom they can identify completely by means of symbols rather than genes.

At this point in my argument, I think it would be helpful to address a theoretical issue that may be a stumbling block for some readers. One of the problems that confronts any explanation of behavior that relies on symbols as real factors in motivating people is the question of how mere words or bits of wood or stone can influence a material organism with powerful drives, desires, or impulses rooted in physiology and chemistry and deriving from the biological, genetic program.[2] The answer is that symbols do not exercise their influence through powers inherent in them, but by acting as transformers, attracting the potentially disruptive physiological forces and redirecting them toward more benign, symbolic ends.

This is the case inside the men's cult. We have just seen that nonviolence is maintained within the cult by a threat of lethal retaliation for violation of the rules. These threats are attributed to the spirits. But since the spirits are only ideas, represented by inanimate ritual symbols, how can they enforce anything? They do so because they are (or become through ritual action) symbolic receptacles of the men's own repudiated aggression. The symbolism of the ritual system is oriented toward violence directed at others, it is sustained by nonlethal but real aggressive competition among moieties of men, and it is inculcated by initiations in which the initiands are given painful wounds, and which (once) culminated in homicide and cannibalism; in short, the ethos of the cult is all about violence and death. Since the system requires that those within the group never kill each other, their violence must be relocated elsewhere, and we have already seen that outside groups are experienced as "enemies" because they reflect back to the cult members their own disavowed violence.

But another way in which competition among the men is deflected is by assigning it to the spirits, who are warlike, amoral, and chaotic, even though the system they regulate is peaceful and ruled by morality and taboos. This follows naturally if one assumes that precisely what is being disavowed in order to create social harmony are blood-thirsty, unregulated, and selfish impulses. These impulses are assigned to the spirits and put to good service as the guarantors of discipline and obedience within the cult group. The collective power of the group of united men draws its potency from the aggression it suppresses from each and reassigns to the whole, symbolized as the spirits, which they themselves form through the power of symbolism. This is the sort of phenomenon that from an evolutionary point of view enables groups to act as wholes in a process of competition and group selection.

Finally, while the ceremonial life is exclusive to men, its internal ideology (as opposed to the external, warlike one) rests on ideas and emotions taken from the realm of femininity, procreation, and motherhood. The ideas of the giving of nurture and of mother's milk, as well as of reciprocal service in which each moiety of men provides for the other by increasing the fertility of different food, is realized in the recreation within the group of men of the secular division of the complementarity of men and women. While along one axis the men split up into upriver and downriver halves to practice fighting and prepare for war, the fertility rituals divide the society of men along a different axis, that of descent groups and complementary moiety-like groupings, which perform ritual services for each other in a symbolic construction of peaceful and fruitful cooperation—such as that which takes place when men and women copulate and produce offspring. This symbolic remaking of the dualism that characterizes sexual reproduction in the ritual organization of the men's house corresponds to the fact that it is primarily the men whose tendency to rivalry and aggression must be neutralized internally and redirected elsewhere, where it will not disrupt the community. This potentially lethal rivalry among males armed both with hostile tendencies and with deadly weapons is what needs to be mastered in order for a human society to exist; and the disciplining and redirection of it is achieved through the shared, communal mutual participation in a ritual life that creates unity among the men.

This unity, which makes society possible, is a matter of individual survival to each member of the community, since it is only in and through society that each can live and flourish. Therefore, it is in the interests of each man to uphold the pact that protects each of them from each other and from the chaos that would ensue without the existence of the religious system. It is true that as men (and women too) at Avatip mature and distinguish themselves in valued activities, they also become less bound by rules and conventional morality, and themselves take on some of the self-assertive, individualistic, and forceful attributes of the spirits. As we saw in chapter 4, each person at Avatip is socialized with Understanding, which entails hearing and obeying the parental and social rules. But individuals also have their own internal Spirit, which, initially constrained by Understanding, later on grows beyond these constraints and develops as the person ages and performs well in the various meaningful arenas of Avatip socio-cultural life: "As people age, their Spirits gradually assimilate or fuse with the ghosts of their ancestors and, in the case of men, with the spirit-beings associated with the male cult. . . . The most potent curses and blessings are those of very ancient men and women, who tend for this reason to be regarded almost as sacred objects" (Harrison 1993, 108).

The fully realized person has a quality translated as "sharpness," implying that s/he cuts his or her own path, as indicated by enhanced autonomy, and has an ability to overcome pain and hardship and to require less to satisfy physical needs. As Harrison puts it, instead of "hearing" (as in "obeying"), they speak. But this enhanced spiritual power does not make men break free of the cult; rather, it enables such men to become its leaders, sharing some of the power of the spirits to enforce the order within the group that ensures harmony to all, and taking on some of the asocial features more fully developed in the amoral image of the cult's guardian spirits.

Men and the "Supernatural"

Almost all societies have cultural beliefs about a realm or realms of animate reality and existence beyond the tangible one of ordinary perception. And almost all also believe that the present life we enjoy in the physical body is preceded, paralleled, and/or succeeded by other forms of existence not directly observable through the senses. Why? It might seem on the face of it, at least to some, that there is no compelling need for such cultural ideas, and, as we have seen, some evolutionary thinkers see it as a problem to explain why such ideas have persisted given that insofar as they do not present a true picture of the actual world they are self-evidently maladaptive and should have been eliminated by selection in favor of more rational, sensible, materialistic, and down-to-earth worldviews.

But further reflection suggests that the apparent perception of a nonmaterial reality, and the sense that it is superior to the material one, has both a sound experiential reason as well as a good social reason for existing. This other reality is the cultural system itself, composed of symbol systems. These have a material existence in the world, as I have argued in chapter 3, but they also are inscribed in the mind (or brain) of enculturated individuals, where they may appear in consciousness (or in dreams, hallucinations, visions, trances, and so on) without the presence of an external material referent. These conscious experiences do have a material existence, in the neurons of the brain, but they are not directly perceived as such. Consciousness does not, for example, normally perceive words or ideas entertained in the privacy of one's own thought as having a material existence, though our contemporary science assumes that they do so nonetheless.

So it is understandable why people believe there is a realm that seems to be beyond the realm of the material: because there appears to be one. Not that it really is immaterial—I have emphasized the materiality and "reality" of symbols—but it may be experienced as if it were immaterial. It is the world created by external symbols, and their corresponding encoding in our neurons, that constitute culture, which is produced by and in turn produces the social reality within which individual people exist. That world is different from the realm created by genetic reproduction: although other higher non-human sentient beings have some control mechanism that can be called a "mind" in the very broad sense, they do not create and live in a perceptual and conceptual world made of symbols that enables a truly collective as well as an experienced inner life. This realm seems "supernatural" even from our own cultural point of view, because for us the "natural" world is that which corresponds to the perception of an external material reality. In many other societies, corresponding concepts of "supernature" do not exist: for them, spirits, ghosts, and so on are parts of the real world, the entire conceptual and perceptual cosmos as it is constituted by the symbol system. That symbol system is perfectly real, whether or not the things it is believed to represent are or not.

What all these considerations make possible is the cultural assertion that there is another life, or another form of life, or another dimension to our present life, besides the one we are experiencing in the everyday world, but which does correspond to a dimension of real experience. Since our organic life is brought into being by sexual reproduction, whereas the other is a product of symbolic construction; and since it is a requirement of our social life that the symbolic system exercise some effective control over the urgings of the genetic system; and since it follows further from this that there is a motive for the symbolic system to portray itself as superior to and to denigrate the sexual genetic aspect of life; then it follows that this "other life" is portrayed as in some sense higher and better than the one given to us by our birth into a physical body by means of our parents' copulation. Nor is it very difficult to make that case, since our organic life is to a great extent beset by dangers, dilemmas, anxieties, pains, conflicts, disappointments, frustrations, losses, sorrows, illness, and inevitably decay and death. Since the symbolic world is neither born nor dies, because it is not itself organic (though of course it depends for its own "life" and enactment on living people to produce and be produced by it), and so appears not to be subject to the ailments, physical and mental, of the physical life, it is not hard to see how it comes so regularly to be favored as an option for where to put one's cultural priorities.

But more than that, this orientation to another world may well be conceived of as essential for society to exist at all. One of the sources of unhappiness with the genetically produced world is that, unless something is done to prevent it, it leads to conflict among males over reproductive opportunity and advantage. Without a cultural symbolic system creating equitable or at least workable rules constructing and enforcing marital and other social rules or norms surrounding copulation, society would, it is (rightly) feared, descend into chaos. It is, in other words, felt to be the fault of sexual reproduction that there are so many disagreeable aspects of life in the world. But if there is another kind of life that is not dependent on sex for its existence and that life is available to us, now or later, then surely the best strategy for people, and for men in particular, who are the source of the most violent competition, would be to pursue that other life in that other realm to the degree possible, and to confine sexual reproduction to the degree necessary to assure that reproductive continuity is assured and that there are live bodies to continue to learn, embody, enact, and transmit to others and to future generations the symbol system that is so highly valued.

If then the big problem for men is the potential of interminable deadly fighting over reproductive success, and since such success requires that they copulate with women, then it is not only copulation but women who can easily come to be devalued. It is not only that by giving us birth they usher us all into the world of organic existence and the inevitable travail, sorrow, and death to which it leads, but it is also that they, and their reproductive organs, are essential to men in order to realize the reproductive success that the men's genetic program enjoins them to seek. They thus appear to be the source of men's unhappiness insofar as necessity resigns men either to chaotic conflict or to more or less unwelcome societal constraints on their sexuality. If men could do without sexual reproduction, and, by association, without women altogether, they could then dwell in that other world that seems to exist in a bodiless spiritual dimension, without the disadvantages of the present embodied life. This might all seem abstract and even arbitrary, except that it accounts for the striking preponderance in the ethnographic record of societies in which there is very pronounced status and power asymmetry between men and women. This is an undeniable and obvious dimension of much of human society, and calls out for explanation.

In such societies it is very frequently the case that the symbolic system as a social process is culturally associated with men and the genetic program with women. This association is neither necessary nor universal, as was originally suggested by Sherry Ortner (1974), but I am in general agreement with Ortner's position that men are to "culture" as women are to "nature" at least in a great many though not in all cultural symbolic systems. But clearly in light of my earlier criticism of the "nature/culture" dichotomy as a theoretical concept, I think her formulation is tenable only if "culture" and "nature" are understood much more precisely as the imperatives of the symbolic program and of the genetic program, respectively. I will show in chapter 9 that there are cases in which this metaphoric equation does not hold, and in which female subordination is not a feature of the social system. That discussion will demonstrate that the operative distinction is not that between the genders but that between the channels of inheritance. But I do believe the metaphoric equation of men with the symbol system and culture, and women with reproduction and genetic inheritance, is one that lies ready to hand and is easily made in a great many cases. So let me look still more closely at why that is so.

CHAPTER SEVEN

Men Together versus Men and Woman Together

It is self-evident that the furtherance of the genetic program requires as an essential link in its ongoing chain of replication that there be copulation between a man and a women.³ How to organize the wider social world, however, is more complicated. One way to think of it is to define it as a realm from which sex, or at least fertile copulation, is absent. This includes, for example, the Culina system which defines all community members as siblings covered by an incest taboo. In the Merina system, there are two modes of reproduction, the sexual and the mystical; in the former case the difference and inequality between men and women is emphasized, while in the latter men and women, regarded as asexual, mystical descendents of the ancestors, are not distinguished and are equal.

At the far other end of the scale is the reproductive system of the Na people of southwest Yunnan Province in China (Cai Hua 2001), in which mating occurs without marriage, and in which copulation is widespread, but with a rule assigning the social location of children to the mother's matrilineal extended household within which a very stringent incest taboo prevails. Here one might see a structural reversal of the Culina case: among the Culina, the nuclear family is a realm of sexual license between the spouses, embedded within a wider social group, the madiha, that is asexual. Among the Na, the opposite is the case: the family household is an asexual group, headed by a brother and sister rather than a husband and wife, while between and among households sexual license is the norm.

In a society such as the Mbaya, sex is pursued for pleasure only, while reproduction is assigned to people from outside the community whose children are then "recruited" forcibly into the group. In some systems it is quite possible for the realm of culture beyond the reproducing couple to encompass both sexes. This can occur either because sex is ruled out altogether, or because it does not matter whether reproduction takes place as a result of nonmarital copulation. In some cases, such as the Mbaya, unwanted offspring are eliminated by means of abortion and infanticide. On the Micronesian island of Yap, where depopulation was a serious problem in the postwar years, investigators found that women up until their thirties preferred to abort rather than give up sexual freedom (Schneider 1955). If Linton (1939) is to be believed, Marquesan women also preferred sexual attractiveness and freedom for sex to reproduction.

A very instructive case is that of Ireland, which went from being a country with very high rates of celibacy without much nonmarital sex but with high fertility for married people, in the nineteenth century, to one in the contemporary world in which premarital sex is the norm, with the proviso that birth control is practiced and reproduction does not take place until marriage, and then only at low rates (Salazar 2006). The introduction of safe, easy to use, and effective birth control has made it possible for the people of the modern (Western) industrialized world, including the Irish, to create a public realm that allows men and women to participate together, with the understanding that while there may be a good deal of copulation among them, unintended pregnancy can generally be avoided through contraception (or more controversially abortion). This mixed-sex arrangement was not possible in the era before reliable contraception, that is, most of human history.

Another way to deal with the problem of integrating reproduction into the tribal society, especially in societies in which prevention or control of birth is not possible, difficult, or strongly sanctioned, is the segregation of the sexes. In order for a society to manage with total separation of the sexes, it must be in some relationship with another part of the population, or another population entirely, either within or outside the group, that supplies new members. So in most cases, sexual segregation is partial, with the result that individuals (or at least men) in the society have a dual allegiance: one to the reproductive unit, and one to the wider tribal society. In some societies the public symbolic sphere allows for equal access to both sexes, but these are a small minority and, absent effective birth control, require special conditions. I will deal with some systems in which women play an important social role beyond the reproductive unit in chapter 9. In the present chapter, I am examining the very common situation in which the symbolic sphere is dominated by men (as the Avatip case clearly illustrates), with the result that men in these societies are divided in their commitments to a sexual reproductive life, on the one hand, and to an all-male group, either asexual and homosocial or in some cases involving overt same-sex sexuality, ritual or otherwise, on the other.[4]

This pattern is so widespread and common, in one form or another, and shows so many striking similarities among groups across oceans and continents, that one might be tempted to see it as the default human social arrangement, from which other arrangements are deviations. I do not believe this is in fact that case, but the thought is also hard to dismiss out of hand given the weight of the evidence in the ethnographic data. The best argument against its being the default position is that while aspects and traces of it may appear among hunters and gatherers, it is largely absent in the most elementary foraging societies.

In any event, as Leslee Nadelson (1981) has suggested, it appears that the most salient conceptual distinction in many societies is not that between the sexes but that between heterosexual and homosocial institutions for the men. The women in these societies are usually wholly relegated to the heterosexual sphere, which they conceptually imbue with a symbolically "feminine" connotation, but the men participate in the heterosexual sphere as husbands and householders while also participating in the all-male men's cult with a distinctly misogynistic ideology:

> I suggest the usefulness of thinking about male/female distinctions as emanating from an underlying dichotomy of homosexual and heterosexual. . . . In the Mundurucu case, the importance of the men's house and the extensive separation of the sexes make the model of two parallel social systems, one homosexually composed and the other heterosexually composed, relatively easy to detect. . . . Because of the necessity of considering "maleness" and "femaleness" as symbols to be analyzed rather than natural categories to be taken for granted, the sheer simplicity of the homosexual/heterosexual opposition has the appeal of elegance. . . . Furthermore, when the "shun women"/"seek women" motifs that emerge in the Mundurucu material are tied to homosexual/heterosexual counterparts, we are better able to understand culturally created manifestations of conflicts within individuals, for the demands of one system must violate the demands of the other. (240–41)

There are in fact many different attitudes and stances the all-male group can take in relation to sexual reproduction, which is, as we have just seen, of inherently ambivalent value to them. But the balance varies. In many societies fathering many children sexually is a source of great prestige—one thinks of many African or Middle Eastern groups. In others, such as the Kaulong and in several other Melanesian groups, copulation is approached with grave trepidation regarding the dangers supposed to be posed by exposure to females and their polluting or even lethal influence. Many societies presuppose the superiority of whatever it is the men do in relation to reproduction and society to whatever it is women do. Since women appear to have the major role in the production of biological offspring in terms of time, energy, effort, and the dedication of body parts and functions to it, and since it is, like it or not, a process necessary to society, then it is left to the men to play the major role in the management of the "other life" (however defined) that is constituted by cultural symbolism and deemed necessary to regulate and channel the dangerous process of sexual reproduction, now associated specifically with women. This other life, free from the taint of decay and death that clings to sexual reproduction and embodied life, is the realm of the spirit, or the "supernatural," and, very often, of death: so, if woman are the sex associated with organic life, then men can trump them by assigning themselves the role of masters of death. This seems paradoxical from our point of view—why should being associated with death be a good thing? But it must be recalled that almost no group in the ethnographic record thinks of death as the mere end of existence; it is rather understood as the entry by some element of the person other than the physical body into another form of immaterial existence.

Male Womb Envy

I am in principle skeptical about deriving a form of social organization directly from an aspect of individual psychology. I will, however, show that the cultural attitudes exhibited in societies with men's societies or any form of all-male association are highly consistent with the idea that men, whether individually or as a group by virtue of culturally constructed symbolism, are envious of the apparently female capacity to create life.[5] This appearance is, as I mentioned, derived from the observable fact that it is women who carry the fetus for nine months, give birth, lactate and nurse, and almost universally bear primary responsibility for the needs of small children. This invidious attitude between the genders, in turn, can be seen as a symbolic representation in another register of the hostility of the cultural channel to the genetic channel from which it tries to free itself but finds itself bound by the necessity of sexual procreation.

Envy is different from jealousy. In jealousy, one begrudges someone the attentions of another person whose love one covets. Thus a child can be jealous of a new sibling who usurps the mother's care that formally was reserved only for it, or a person can be jealous of a spouse or partner's interest in a lover. In envy, however, one hates another person's possession of some perceived good, solely on the basis of the fact that one does not have it oneself. Thus while jealousy requires three people, envy is a dyadic relationship. The key to understanding envy in its bewildering complexity is that the hatred of the other person's possession of the good can and does lead in two opposite directions: on the one hand, one recognizes the value of the good and wants to take it from the other and have it oneself; while on the other, one hates the good, for being out of one's reach, and wants to devalue or spoil it so that the other person will no longer arouse any envy and will be punished for having what the envious person lacks. Moreover, these two apparently incompatible reactions, one idealizing the good and wanting to claim it for oneself, the other devaluing the good and wanting to spoil it for the other, do not cancel each other out or require that a choice be made between them. On the contrary, they can and do coexist, despite the apparent logical fallacy involved (Klein 1975).[6] Envy is thus recognizable by the simultaneous presence of idealization and devaluation of the same thing.

The modern theoretical formulation of men's envy of female reproductive capacity, or of its chief organ, the womb, originates, I believe, with an article by Karen Horney (1926). The idea was taken up in anthropology by Margaret Mead and Ashley Montague among others, and its most well-known, if somewhat idiosyncratic, exposition was that of Bruno Bettelheim (1954), which combined a clinical with an anthropological view. More recently, McElvaine (2006) has published an entire book expounding the thesis of the deleterious role of male envy of women's reproductive capacity in Western culture and history. The idea now turns up fairly regularly in the literature, since once it is pointed out it is hard to ignore or doubt. For example, Tuzin (1995), writing of the men's cult among the Arapesh of the Sepik lowlands, writes: "At the latent level, iconically linked with the penis-cutting, meat-eating and flute-playing, lives the one meaning that men cannot admit to themselves and that women can never be made to understand: it is they (the men) who are excluded from the ultimate condition of all living things, namely, the ability to generate offspring; in our usage, to procreate" (300).

If the ability to procreate, understood as something that women can do and men can't, is defined as a "good," then envious responses on the part of the men would go off in the two opposite different directions implied in the envious response as defined by Klein. On the one hand it would be idealized as a desirable thing, to which the response of the envious man would then be "yes, but it is we men who actually have it, or have something like it only better, or at any rate have ultimate control or the ultimate causal role in it." This response would represent the wish to take or claim the good thing (procreation) for oneself. The other, opposite, response would be to denigrate it and spoil it, decrying it as dirty, polluting, fit only for animals, while at the same claiming that men have the ability to reproduce in a different way that is far superior, being clean, spiritual, and resulting in a "life" that is better than the deeply flawed one that birth from a womb bestows upon us. Concomitant with this view would be a simultaneous idealization of the feminine principle and capacity for childbearing, together with a contempt and disdain for them and an association of women, and of the feminine, with everything that is chaotic, dangerous, or otherwise undesirable. Further, the dependency men have on women, since it is they who alone can allow them to reproduce (not to mention that the women give them birth in the first place), is a scandalous blow to male assertions of autonomous self-sufficiency. An envious response to this dependency is to deny it and claim that men actually reproduce themselves. All this may appear far-fetched in the abstract, yet a great deal of ethnography confirms that all of these ideas take form as real beliefs and practices in a great many societies. I think these ethnographic facts can be elegantly accounted for and understood in all their contradictory complexity on the basis of the reasoning I have put forward. (I probably need to stress that while this appears to be a very widespread pattern, it is neither universal nor necessary.)

Men and "Pseudo-Procreation"

L. R. Hiatt (1971) coined the phrase "pseudo-procreation rites" to describe ceremonies undertaken by men among Australian aboriginal groups in which they ritually enact control over the fertility of game and economic resources as well as of people through mime, dance, song, and the manipulation of ritual objects. These groups likewise convert junior initiands into men by mimicking female reproduction through ritual actions. An ocean away, Jean Jackson (1996) observed this phenomenon among the Tukanoans of the Vaupes region of Colombia, whose ceremonial life features "rituals of male unisexual rebirth":

> Tukanoan myth and ritual speak of a symbolic birth which is superior to physical birth, first because it is cultural and spiritual; second because it is accomplished almost exclusively by men. Men conceive, gestate, and give birth. In the process, boys become true humans, thereby allowing Tukanoan society to continue. Although many of the images and symbols used in this ritual derive from female reproductive processes, they are significantly altered in form, content, and intended effect. (89)

CHAPTER SEVEN

Since Colombia could hardly be farther away from Native Australia, one has to suppose that we are confronted with a cultural ideological formulation which, though not universal, is a compelling one to many people in many different societies around the globe.

Hiatt asserted that we should recognize two different strands of male procreative rituals. (I drop the "pseudo-" part of the phrase, because as I hope should be clear, I see nothing "pseudo-" or any less real about cultural reproduction accomplished with symbols than sexual procreation accomplished with genes.) One the one hand, he describes "phallic" ritual, in which the men claim the reproductive agency of the penis and of semen in the production of new humans; on the other, he defines "uterine" ritual as that in which men re-produce youths as men, ritually (and really) separating them from their mothers. In this case, the symbolism is of female generative processes and organs. As an example of the former, following Róheim (1925), Hiatt describes how many Australian men's societies perform increase rites featuring "a quivering or trembling motion of the actors that shakes off some of the body decoration of white down. The white down represents life stuff that impregnates women, the quivering represents coitus" (1977, 81). Likewise, citing Meggitt (1962), Hiatt observes:

> The Walbiri believe that mythical ancestors decorated themselves with bird- or vegetable- down. As they agitated their bodies, the down was dislodged and transformed into immaterial particles called *guruwari*. When one of these enters a woman's womb, it gives rise to new life. . . . But the Walbiri believe that the *guruwari* will not operate unless men provide the appropriate conditions to initiate their action. Hence men in groups (cult lodges) are responsible for the regular performance of secret songs and rituals connected with their own localities. (1977, 83)

A particularly clear example of a uterine rite is one Hiatt takes from Stanner's (1964) account of the *Karwadi* ceremony among the Murinbata, another Australian group, named after a secret being described as "Mother of All" or "Old Woman." After being circumcised, boys are ritually swallowed by Karwadi and then vomited up. After several more ritual activities, the ceremony ends "when the novices crawl towards their mothers through a tunnel of legs formed by the initiated men. . . . As each youth emerges, he sits momentarily in front of his mother, with his back to her, while all the women wail and lacerate their heads. The youths then return through the tunnel, and all the men rush with loud shouts back to the secret ground" (86). Hiatt interprets this ritual symbolism as a "fabrication by which men, having taken youths from their mothers, symbolically destroy and re-produce them. The passage of the blood-covered novices through the tunnel of legs suggests the notion of male parturition, if not explicit, is at least close to the surface" (86).

Male cults like these are found in many parts of the world, but a great proliferation of them occurs in New Guinea and in indigenous South America as well as in native Australia. While there are or course variations in numerous details, the resemblance between these rituals in the distinct geographic regions is so remarkable that it poses a challenge for ethnological theory: might there have been diffusion or contact, as Lowie (1920) and others suggested? Or if the logic that led to them is so compelling, then why the concentration in these geographical areas? Whatever the cause, the ritual cycles, which generally may be about both the promotion of fertility, human and natural on the one hand, and the initiation of boys into the men's society on the other (the phallic and uterine aspects in Hiatt's terms), include some or all of the following elements: secrecy on the men's part and the threat to woman and children of dire consequences (including death) if they come too close to the men's proceedings; the manipulation of sacred and secret phallic noisemakers such as "flutes" or "trumpets" (long tubular wind instruments) and bullroarers, the sounding of which are presented to the women and children as the voices of supernatural beings conjured during the ritual; the employment of ritual symbols representing female reproductive organs and processes, such as a beeswax gourd representing the womb among the Barasana (Hugh-Jones 1979) in the Amazon, or in New Guinea the Abelam belief that the men's house itself is a symbolic womb (Forge 1966, 1973); the separation of the boys from their mothers and their symbolic rebirth as new "stage-one" initiands; and the belief that the rituals were once in the hands of the women who then were dominant, and that the men stole them and must now guard against a re-usurpation by women, who are consequently feared, denigrated, and often ritually intimidated and attacked. Taken together, it should be clear that these constituents of the rites conform to the envious responses of men to women I proposed above: the assertion of male, phallic procreative power, the claim of male possession of reproductive uterine power, the idea of a symbolic world related to a supernatural realm that is superior to the organic one produced by biological birth, the consequent need for symbolic rebirth into the other symbolic world for the men on the one hand, and the denigration and debasement of women and their reproductive process on the other. The final giveaway is this: the widespread belief that it

CHAPTER SEVEN

is really women who have or once had the "good thing," and that the men had to steal it and put on a spectacular show accompanied by intimidation to convince themselves and the women that they really have it, or have something like it only better.

Gender Antagonism versus Dual Inheritance

It would be absurd to deny that the rituals described above have something to do with antagonism between the sexes and male domination, and indeed it is in these terms that the ideologies underlying them have often been understood in the anthropological literature. It would, however, also be a mistake to limit ourselves to this interpretation. In fact, the rituals, as well as many other practices to which I turn in the next section, seem to arise as a result of a confluence of multiple factors.

The first problem with the "gender antagonism" analysis is the question of why anyone should feel so strongly about this. The men in these cults go to great lengths and expend enormous amounts of energy and resources on their ceremonial life, and guard its secrets with lethal force. What is at stake that leads to such commitment? If one argues that the answer is that it is a way of enforcing male domination, this only shifts the question: why then is *that* so important? Plenty of societies do very well without the marked asymmetry and the antagonism. Furthermore, we learn that in those same societies that have male ceremonials, relations between actual husbands and wives can be quite affectionate and without the ferocity expressed in the rituals. As Gregor and Tuzin (2001) observe of the Mehinaku of Amazonia,

> coexisting with this pattern of misogyny is genuine attachment between many spouses and between male and female kin. The most dramatic cultural manifestations of affection and love between men and women are found in mourning practices, which require a full year in seclusion for the bereaved spouse. . . . More subtle are a host of indications that husbands and wives really care about each other, including anxiety during prolonged separation and overt expressions of concern and affection. (324)

In the same vein, Tuzin (1997) notes the disjunction between the sex antagonism expressed in the men's cult of the Ilahita Arapesh of New Guinea and the actual behavior of men in relation to their wives. In a book written after his return to Ilahita after the men's cult, the Tambaran, had been given up under Western pressure, he writes: "It is tempting to regard the Tambaran as having gotten its just deserts and men like Galan, Mangas, and Tunde as having gotten a taste of the abuse their wives and foremothers have endured for generations. The case for retributive justice is muddied, however, by the fact that the relentless misogyny of the Tambaran belied the generally benign, protective behavior of men toward their wives" (56). And he goes on to cite other ethnographers of the Arapesh who comment favorably on the good relations between actual men and women in these communities. So something else must be going on.

Here is where it will help to shift the focus, as Nadelson's analysis suggests, away from gender duality and sex antagonism and onto the distinction and antagonism between heterosexual and (male) homosocial domains. The actual marital units are in themselves apparently not scenes of domination, control, and violence. Indeed, among the Ilahita Arapesh, it was only after the men's house was demolished that spousal abuse became a problem. Rather, those themes are aspects of a religious ideology enacted in secret among the males by themselves. While women are wise enough to avoid provoking the men, they are often in private quite unimpressed with what their husbands are up to. The real conflict here is between the social scenes of actual biological reproduction, that is, the conjugal pairs, and the wider social group, constituted by cultural symbolism and ideologically associated with men. This conflict, as I have been at pains to demonstrate throughout this book, is inherent in the fact of dual inheritance; it is the conflict between reproduction of society by genetic procreation and the reproduction of society by reproducing its symbolic system.

But why is the latter, that is, culture itself, symbolically associated with men, when in reality culture and its symbolism are possessions of all humans of any age and sex? I have argued that one of the main tasks required of the cultural symbol system is that it somehow or other control the competition among men for mating opportunities. The ferocity of the men's cults, in other words, is not primarily directed from men to real women but takes that form only secondarily: it is really an issue among the men themselves. Without a mechanism for overcoming their rivalry and uniting them into symbolic kin who can live in a social group in amity, the integration required at the group level would be in danger. The danger is seen as coming from the women because the cause of the rivalry is competition for sexual access to women and their reproductive capacity. If there were no women, and men could just reproduce by themselves, the issue would be obviated; there would be nothing to fight over. Since that is not the case, the men must create an alternate symbolic world in which their need for women in order to reproduce is denied or minimized, so they can minimize with it the need to compete relentlessly with one another. At the same time, they must depict women and their genitals, otherwise the sources of sexual attraction for them, as dangerous, repugnant, polluting, or in some other way undesirable, in the interest of maintaining harmony among themselves.

We have seen already that the aggression arising from competition for mating opportunities among men armed with lethal force has to be cast out from among the men. One place in which it is then located is in other groups, seen necessarily as "enemies," since they are the receptacles of the disavowed violence of the men among themselves. The second place into which their aggression can be siphoned off is the realm of the supernatural beings, who thereby take on the (perceived) punitive force necessary to enforce the moral imperatives of the religious ideology. And now we see the third locus for containing the men's warded-off aggression: the women, or rather the symbolic depiction of them in ritual and myth, where they are seen as dangerous and a threat to the solidarity of the men. As long as the reality of sexual reproduction, requiring women, exists, there is grave danger for the men of destructive competition among them. Nor is this merely my supposition; evidence for it comes directly from the ethnography: the Barasana of northwest Amazonia "do not say that if the women saw the instruments [the 'flutes'] they would once again become dominant over the men, but they do say that there would be a period of chaos during which the men would fight amongst themselves and kill each other" (Hugh-Jones 1979, 128).

The assignment of sexual reproduction to the women that is implicit in all this is, of course, made much easier by the fact that women's physiological involvement with reproduction is so much more obvious than men's, including as it does such physical facts as menstruation, pregnancy, birth, and lactation. But these facts alone would not explain sexual antagonism unless it were understood that it is the scandal of biological reproduction combined with the threat of intragroup violence among the males that renders the idea of the female such a potential threat. The domination of them is not really directed at them as people—as I have said, the actual relations between spouses or of men with female close relatives, especially sisters and mothers, does not reflect any of this—but is rather seen as a necessary condition for the control of sexual reproduction according to a set of symbolically established and observed regulations.

The male cult is thus an alternative kin group formed by the fact that, as participants in the symbolism of the men's cult, in which all take a role, they are all informed by shared symbolic instructions, just as genetic kin are related by shared genetic substance. Therefore, it is a further dimension of the men's cults that they must be the scene of symbolic reproduction in the sense that they must produce men who have internalized the same set of symbols and have come to embody and replicate them. So it is no accident both that the men's ceremonies function as increase or fertility rites, and that they also provide for the periodic rededication of men to the religious symbol system that makes their life in society possible. They are also the occasion for the initiation of new recruits, that is, young boys who, having lived as adjuncts of the conjugal social units up until initiation, are now separated from these and made into members of the male group by their immersion in the knowledge of the ritual symbol system.

Initiation is a site of symbolic reproduction par excellence, in that in the course of the rituals the process is begun of teaching the initiands the lore, beliefs, practices, and symbolism that constitute the "other world" that transcends mere biological existence. In often painful and demanding ordeals, the boys are given information by the older initiated men that will enable them to learn to see themselves as identified with the cult and its constituent members. The reproduction of society is "symbolic" in the sense that very often it appropriates imagery, analogies, and metaphors referring to or based on the facts of biological sexual reproduction. But in a more important sense it is "symbolic" in that it is to be understood as making a boy into a man by transmitting to him, though the perceptions and experiences of the men's ritual as well as through explicit verbal instruction, the symbolic cultural information that must parallel, augment, and to a certain extent counteract the genetic information that has brought him into being in the first place.

As I will show in the following chapter, beliefs of this sort are not confined to those societies with elaborate men's cults. I have chosen to use these as exemplars because they make the point so very clearly. But many societies have found a variety of ingenious ways to insist upon the superiority of some form of symbolic reproduction and with great regularity if not with unanimity to associate it with men. It is to a further exploration of some of these that I turn in the next chapter.

EIGHT

Symbolic Reproduction and Reproductive Symbolism

I argued in the last chapter that while the genetic system reproduces itself through sexual procreation, the symbolic system reproduces itself by means of the manipulation of symbols, very often by men, and very often in the context of ritual action. In my discussion of the cases offered by Hiatt in his discussion of what he calls pseudo-procreation, it became evident that this ritual action employs symbols and actions that resemble, mimic, or in other ways recall or represent aspects of the process of sexual reproduction. Thus, in the case in which men ritually cover themselves with white down feathers and then engage in shaking dance movements that give off the down, the dancer is emulating an erect phallus ejaculating semen. In the case in which boys are "reborn" as men by passing, covered with blood, under the legs of a row of men, the channel created by the spread legs of the men is a ritually constructed representation of a female birth canal. Even elements of actual sexual reproduction can function as symbolic elements of ritual: when the Marind-Anim manipulate real semen produced in real copulation, the semen and the copulation are functioning in symbolic terms within a symbolic, ritual system rather than as part of the sexual procreative process: offspring resulting from copulation occurring as a necessary component of the Dema ceremonies are killed.

However, the use of symbolism drawn from the realm of sexual reproduction, while extremely common, is not a necessary characteristic of the reproduction of the symbolic system of a society. To make the matter more complicated, sometimes the ways in which the symbol system is transmitted and acquired across succeeding generations, far from being modeled on sexual procreation, is differentiated as much as possible from the sexual system, the necessity of which for the continuity of society it is part of the role of the symbolic system to deny, minimize, supplement, and/or control. Thus, for example, the Merina descent units are recreated over time by a mystical, spiritual process that is far removed from any hint of anything material, organic, or sexual, while in the circumcision ritual the two systems, sexual and mystical, are contrasted to the detriment of the sexual.

The process of "symbolic reproduction," in the sense of "the reproduction of the symbolic system," is of course a necessary aspect of any society that continues beyond a generation. As a child becomes enculturated and socialized, as it learns language, it is willy-nilly acquiring the symbolic system of the group in which it is reared, without this process being explicitly organized, represented, or conceptualized. It is an implication of the dual inheritance model, and of the antagonism between the cultural and the genetic channels that characterize it, that whether explicitly or not, the symbolic system regularly constructs the reproduction of society, including the reproduction of the people who constitute it, as somehow owing more to the cultural system than to the genetic system—and, very often, to men rather than women. In this chapter I examine examples that illustrate the wide range of ingenious ways in which this dynamic can be enacted in culture beyond the men's cults I discussed in the last chapter.

Baptism and Compadrazgo

Most Westerners are familiar with practices in our own society that supplement biological birth with religious rituals intended to inaugurate a child into the larger social group, such as baptism, confirmation, circumcision, and *bar* and *bat mitzvah*. Here I discuss infant baptism, common to the Roman Catholic and Orthodox Christian traditions, which involves submerging the baby in water that has been ritually sacralized. There are a wide range of interpretations of baptism from within the relevant faith traditions themselves, but one central one is certainly that baptism is a rebirth or second birth that begins the transformation of the biological infant into a member of the church. The ultimate authority for this view is Jesus himself, who tells Nicodemus that "unless a man be born again, he cannot see the kingdom of God," and goes on to differentiate birth of the flesh and birth of the Spirit: the latter requires that a person must "be born again of water and the Holy Spirit" (John 3:3–6).

Jean Daniélou, in an extensive scholarly survey of the patristic exegesis of the symbolism of the sacraments, summarizes the position of the early Fathers thus: "The Church is the mother of the sons of God; it is in baptism that she brings them forth. So the symbolic meaning of the rite is ready at hand: the baptismal bath is the maternal womb in which the children of God are begotten and brought forth" (1956, 48). And Saint Thomas Aquinas wrote: "Just as in carnal generation a person is born of a father and mother, so in spiritual generation a person is born again a son of God as Father, and of the Church as Mother" (cited in Gudeman 1971, 49). The equation of the church—an institution whose priestly hierarchy is male—with the spiritual mother, and of the baptismal font as a symbolic spiritual womb, is consistent with my analysis here: the male society that controls the religious life of the community has a maternal function, and offers a "spiritual" as opposed to "carnal" birth, from which however it borrows its symbolic imagery.

It is of course true that most lay participants in a baptismal rite are very likely unfamiliar with the teachings of the Church Fathers or of Aquinas, but it takes no familiarity with liturgical history to understand that the baptismal font, a container full of water from which a child is brought forth, is a symbolic replica of a womb. The ritual thus negates the sufficiency of the biological womb from which the baby has emerged, while at the same time recreating it at a cultural level as a man-made and sacralized artifact. Otto Rank ([1914] 1952) argued on the basis of many examples from myth, literature, and folklore that emerging from water is a widespread symbol of birth, and baptism seems to be a clear example of his thesis.

That according to at least some exegetes the birth from the font is understood as superior to birth from the womb is clear from this passage from the Zeno, the fourth-century bishop of Verona: "The font, where we are born to eternal life, invites you by its healthful warmth. Our mother is eager to bring you into the world, but she is not in the least subject to the law which ruled over the child-bearing of your mothers. They groaned in the pains of birth, but this heavenly mother brings you forth all joyful" (cited in Daniélou 1956, 48). The pains of childbirth to which Zeno alludes refer to the biblical punishment of

Eve and hence of subsequent womanhood. This in turn, for the patristic writers, refers to the Fall, Original Sin, the loss of innocence and immortality, and the necessity of sexual reproduction. Birth of the Spirit from the font is then regarded as free of all these defects.

For Aquinas, following the earlier traditions, baptism is accompanied by godparenthood, which provides a person with a new set of parents related not by biology but by symbolic means. It was basically Aquinas's formulation of the sacrament of baptism that was adopted at the Council of Trent, in the mid-sixteenth century, and so it was this version that was brought to the New World by the Spaniards during the Conquest. Stephen Gudeman (1971), in an authoritative study of godparenthood, or *compadrazgo*, offers a wide-ranging survey of the variations on the practice now found throughout Catholic Latin America, arguing that they all derive from this one original source. His paper specifically focuses on his ethnographic field site, a peasant community in Veraguas Province in Panama, in which the main constituent units are individual households occupied by a single married couple and their biological children, though other kinfolk may be invited in as co-residents. It is compadrazgo that binds these otherwise autonomous and self-sufficient units into a wider community. The baptized baby thus emerges from the font as both a member of its nuclear family to whom it is related by genetics and also of the wider inter-household community to which it is related by cultural means.

In Veraguas, a baby at about six or eight months acquires three godparents, two godmothers and a godfather, at the rite of baptism. The most essential aspect of the relationship of a person with his or her godparents is thought of as "respect." This means that godparents and godchildren "may not marry, have sexual relations or engage in sexual joking with one another, nor should they be angry or swear in each other's presence" (Gudeman 1971, 56). Needless to say, the household is, by contrast, the scene of sexual reproduction, and a place where relations do not observe the norms of respect behavior. Thus "the parents initiate the child into the physical world and the household; the godparents initiate him into the spiritual world and the community."

An interesting detail in Gudeman's ethnography is the ambivalent meaning of the term "hot hand." A godparent whose godchildren die is said to have *la mano caliente*, but the same term refers in other contexts to a man's sexual potency, thought of as a positive trait. Thus in the context of the family, the hot hand is a creative life-force, but in the context of compadrazgo it is destructive: "in the family 'hot' produces life, while in the *compadrazgo* 'hot' produces death" (58). This usage underscores the fact that the family unit is opposed to that of godparenthood on the basis of the opposite value assigned to sex and sexual reproduction in each one. Gudeman's conclusion is that "the family and *compadrazgo* are in a relation of complementary opposition: one is concerned primarily with intra-household ties, the other pertains to inter-household links. Families are divided into households; *compadrazgo* unites these units. One entails physical and material exchanges, the other consists of spiritual bonds" (60).

Bloch and Guggenheim (1981), in an article building on Gudeman's analysis, argue that what Gudeman has missed is the gender ideology implicit in the system of baptism and compadrazgo. They claim that childbirth and the womb, which are devalued as impure and carnal, are associated with women, while compadrazgo ushers the person into the higher spiritual community dominated by men. Therefore, they conclude, these institutions should be seen as aspects of a political gender ideology that helps create and sustain male domination.

But as I have argued, and as I believe is clear from the ethnographic material I have presented, the institutions of baptism and compadrazgo are better seen to reflect the dual nature of society, divided into reproductive units and the wider community that encompasses them, representing the domains of genetic and cultural relatedness respectively. It is sexual reproduction that is the highest value within the household where biological reproduction takes place, whereas sex is devalued and indeed believed to be lethal (as the "hot hand" illustrates) in the context of the web of inter-household relations created by ritual symbolism rather than by biological birth. Insofar as there is gender asymmetry, which in the ethnographic case of Veraguas is true but not the most salient feature, it should be seen as a common symbolic equation based on the more fundamental distinctions inherent in dual inheritance.

Patriliny

J. S. La Fontaine (1981), one of many anthropologists who joined the backlash against seeing the nature/culture dichotomy as a master key to understanding male/female relations, rejects the argument that the "facts of biology" inevitably associate women with biological reproduction. She points out, quite correctly:

In many societies it is men not women who are credited with reproductive powers. Richards has recorded . . . the pithy comments of an Ngoni discussing the views of the neighboring [matrilineal] Bemba: 'If I have a bag and put money in it, the money belongs to me. But the Bemba say that a man puts semen into a woman and yet the children belong to the woman, not the man.' . . . The Basque shepherds of the French Pyrenees emphasise the act of fertilisation, identifying the process of creating a child with the process of making cheese by the addition of rennet to milk. A child is conceived when the man's semen curdles the woman's blood. . . . Both components are necessary, but the metaphorical identification of semen with rennet assigns causative force to the male contribution, which is seen as the agent of change, the causative element. (336)

This argument, which offers ethnobiological theories of conception that accompany patrilineal descent systems, where men are related to others only through male relatives by the agnatic principle, lends support to my argument about the attempt of men, aligning themselves with culture, to seek to denigrate the female role in reproduction. But it errs, in my view, in taking what needs to be explained as if it were itself an explanation. The Ngoni who attribute conception to the semen alone (thus accounting for the social institution of agnatic descent and relatedness) are no more or less right about how conception works than are the Bemba (or any other group) who attribute reproductive capacity wholly or disproportionately to women—either as a basis for matrilineal descent, or to disparage women as the ones who bring sin, pollution, decay, and death into the world. Both of these theories are ideological symbolic constructions placed upon the natural fact of biological reproduction that have been created and put in place to serve a socio-cultural purpose. Patriliny is an ideology whose purpose it is to override the fact of sexual reproduction by redefining descent as something that bypasses women as much as possible.[1]

Patrilineal kin reckoning creates potentially large groups of men who see themselves as related and, at least in optimal cases, allied or cooperative. These groupings have a collateral dimension, so that members of the same patriline who are contemporaries are united by belief in some shared essence that makes them "kin" with attendant rights and obligations, and they are also descent lines that continue across generations. Men in an agnatic descent line reproduce themselves socially by claiming that it is they who procreate, using women merely as accessories, the necessary means to that end. Women in patriliny do not reproduce themselves, from the cultural point of view. While they certainly give birth and, from a biological point of view, their genes continue as much as do men's, from the point of view of the patrilineal social ideology a woman's descent stops with her: her son is a descen-

CHAPTER EIGHT

dent of his father and continues the patriline. Her daughter marries into another patriline and whether she is incorporated into it, or keeps her original clan or lineage affiliation to some extent, she too will not reproduce herself socially. Men in such an agnatic system are thus the sole carriers of the process of symbolically constructed cultural reproduction over generational time. In other words, the underlying process of genetic reproduction whereby the descent lines of both parents are passed along to offspring is denied at the ideological level, and only the continuity of men across generations is recognized. A good illustrative example of patrilineal ideology is to be found in rural Turkey.

In the agricultural Turkish village studied by Carol Delaney (1991), the hearth, or *ocak*, represents the continuity of the patriline of the man of the household. His fire, like his seed, should ideally continue forever across generations despite the death of individual fathers and sons. The woman of the house uses the hearth to "transform the wheat seed into bread, the mainstay of the village diet, in the same way that she transforms the man's seed into a child, the continuator of the line" (159). Of a man without sons it is said that his hearth is extinguished—his patriline has died out: "The ocak represents the continuity of the line: the flame passes from father to son, as does his seed. 'A boy is the flame of the line, a girl the ashes of the house.' In other words, the boy continues the ocak, the girl extinguishes it" (161). Wives are brought in to feed the fire, and are consumed in the process: it is a woman's task to make fuel out of dried dung, feed the fire, and throw out the ashes, to then be used as fertilizer. A nickname for a wife is "the one who dumps the ashes." The fire, the divine eternal spark, goes on forever passed from father to son, while the woman is represented as a sterile dead end.

The governing metaphor for procreation in this society, as in many others characterized by strong patrilineal ideology, is that of the seed and the soil—indeed, Delaney uses this phrase for the title of her ethnography. The man provides the seed, which requires the nurturing soil to grow in, but "the child comes from the seed." As support for this proposition, she cites what an informant told Michael Meeker, another ethnographer of the Turks: "If you plant wheat, you get wheat. If you plant barley, you get barley. It is the seed which determines the kind . . . of plant which will be grown, while the field nourishes the plant but does not determine the kind. The man gives the seed, and the woman is like the field" (cited in Delaney 1991, 33). Although this metaphor does not deny the necessity of copulation and gestation for procreation, it symbolically frames it by making a distinction between the divine, spiritual essence of male semen and the nurturing, fertilizing role of the female. Women symbolize the womb and the tomb, and thus the physical world and the outcome, organic death. The spiritual substance transmitted by the male seed, by contrast, is in principle eternal, coming as it does originally from God: "The theory of procreation lends itself to a distinction between spiritual/essential and material/supportive, the first male and the second female. However, it is perhaps more accurate to characterize the differences as between a generative principle definitive of God and men and a nurturant principle exhibited by the earth and women" (156). Fathers partake of the divine, women do not. It is thus fathers who give life, which is potentially eternal (since the dead return to their heavenly place of origin), and they thus represent in a finite way the creativity that God exemplifies in an infinite way. Moreover, God, the ideal father from whom all earthly fatherhood springs, and whom earthly fathers represent in relation to their families, created the universe purely from the divine essence, and had no need for a female partner:

The belief that semen is a kind of creative life force that partakes of divinity is a widespread belief. The difference between its occurrence in other traditions and in monotheism is that creativity comes to be *the* characteristic of the one God; it is his creativity alone that brings the world into being. In monotheistic traditions God has no divine partner; instead the feminine element is subordinated and becomes symbolically equated with what is created rather than with the creative power. (37)

Delaney offers an interesting interpretation of the fact that the *ihram,* the garment a man wears on the pilgrimage to Mecca, and which also often serves as a shroud at death, must be unsewn: God's world is one and all of a piece, but anything that is sewn represents the joining together of separate parts. The word for "to sew" also means to "plant," as in planting a flag; in its general sense of "to stick something into" it also implies sexual intromission, the planting of the semen/seed in the womb/soil: "Thus the garment seems to imply not only the unity of God's creation but the fact that it was accomplished without sex and without a partner" (307).

Of course this case is only one particular version of a patrilineal system; and since the majority of societies in the ethnographic record are patrilineal, one could not hope to grasp the entirety of the phenomenon in anything less than a vast comparative monograph. Nonetheless, the bottom line is clearly that the superiority of the nonphysical, spiritual dimension, associated with the religion, is asserted even while the necessity of sexual procreation is acknowledged, by making the paternal line potentially eternal, like fire, for example, which can in principle go on forever as long as there is fuel, and defining the female "line" as one that does not exist but rather begins with the individual lifetime of any particular woman and ends with her death, like the ashes of a particular fire that has gone out.

Recruitment

In societies of a certain degree of complexity, there will be institutions, or sub-societies, to which people are recruited other than by being born into them. Our own very complex society is rife with such institutions. Businesses and corporations, schools and universities, governments, armies, churches, clubs, sports teams, criminal gangs, and a vast variety of other groups besides draw their membership from among people who have been produced within the overarching society by means of procreation, but whose inclusion in a particular subgroup is achieved after they have come into organic existence, by recruitment.[2] Recruitment can range from being voluntary, such as in deciding to join a particular club, to completely involuntary, as in the case of prisons; most institutions are somewhere in between. Anthropology has traditionally differentiated affiliation with such groups as "achieved" as opposed to "ascribed" status, which one acquires simply by being born to a particular set of parents. In less complex societies than ours, such status might include being a member of a certain moiety, clan, lineage, totemic group, sodality, marriage class, or any of a number of such options.

In societies at the more elementary end of the scale of social complexity, there may be a complete absence of segmentation into subunits such as lineages or clans. Since the groups are small enough that everyone knows everyone else, and knows how they are related to each other in terms of biological genealogy, there is no need even for such simple institutions as lineages or other such kin groups. At a slightly higher level of complexity, one means of recruitment rather than birth is the men's group. We have seen among the people of Avatip, for example, that all males belong to the men's society, but while it is thus "universal" among the men, one is not simply born to it; one has to be initiated into it through set of ritual stages. It is a subgroup in that it does not include everybody in society, since females and junior males are not part of it, and it reproduces itself by recruiting (involuntarily) the children of the society who have been born male.

So although, as in this case, a subgroup may be all male, it nonetheless reproduces itself by the simple expedient of recruiting individuals to whom the women of the society have given birth and educating or indoctrinating them with cultural instructions that allow them to participate in the subsociety, in this case, the men's house. However, it must be noted that the men who initiate and indoctrinate the recruits to the men's societies are, at the same time, speaking collectively, the biological fathers of the young boys to whom they will now become cultural mentors. The adult men in such a society thus lead double lives, as we saw in the last chapter: they are both married householders who engage in procreation with their wives, and also members of an all-male group that reproduces itself by recruitment. In such societies it is rarely a boy's biological father who plays a primary role in his induction into the ranks of the initiated, and one of the very simplest forms of social segmentation, that is, moiety organization, often has the function of allowing men of opposite moieties to serve as initiators of the children of men to whom they have a specified ritual tie. This is yet another way in which paternal investment is redirected away from genitor to his biological son and turned into a system of shared and exchanged reciprocity. As such, it appears to be a social strategy for undercutting inclusive fitness considerations and redirecting the impulses associated with it toward the construction of the cooperative tribal society.

Some people, upon discovering that the American religious group known as the Shakers practiced the strictest celibacy, remark that it is no wonder they died out. (Actually there is one very small but still active Shaker community today, at Sabbathday Lake, Maine). The assumption about the Shakers is that since their villages had both male and female inhabitants, but they refrained from sexual relations, then they had no way of reproducing themselves at all. But of course this misses the point that the Shakers reproduced very successfully for close to two hundred years by means of recruitment. Indeed, Charles Nordhoff, who did what amounted to extensive ethnographic research among the Shakers in the late nineteenth century, begins his account of them thus: "The Shakers have the oldest existing communistic societies on this continent. They are also the most thoroughly organized, and in some respects the most successful and flourishing" ([1875] 1993, 117). They flourished by accepting individual converts of both sexes as well as whole families who converted after procreating. Shaker communities also acted, as monasteries and nunneries have often operated, as orphanages and repositories for unwanted or inconvenient babies whom they raised as their own. What misleads contemporary observers, I think, is that the Shakers were quite unique in managing to suc-

CHAPTER EIGHT

cessfully maintain mixed-sex communities in a condition of celibacy. They practiced a separation of the sexes in their dormitories and in the division of labor, and it was thanks to their ability to reproduce themselves without sexual procreation that they were able, quite unusually at that time, to achieve real equality between the sexes, as Sally Kitch (1989) argues. Their ecstatic dancing and "shaking" was, presumably, a largely successful sublimated outlet for eroticism, which was thus rerouted from being a matter of intercourse between couples into a source of communal solidarity.

Sherpa Monasticism

Monasticism represents a more extreme version of the men's society. In this institution, the men in the men's society do not reproduce biologically but act as a separate subsociety that perpetuates itself by recruitment from a lay population of married householders. The Sherpas of eastern Nepal practice the Nyingma tradition of Tibetan Buddhism and have a number of monasteries as well as nunneries that were established in the early twentieth century in association with an influx of wealth into the region. The Sherpas in the Shorung Valley among whom I did ethnographic fieldwork several decades ago practiced field crop agriculture in their fertile fields, supplemented by the herding of dairy animals, principally yak-cow crossbreeds. Unlike the Sherpas living at higher altitudes, they were not much involved in either the mountaineering or tourist trades, but the men did travel extensively outside the region on trading missions and on various labor projects, especially road building, in northern India and elsewhere.

It could be said that while the institution of monasticism was not as robust among the Sherpas as it was in other parts of the Tibetan Buddhist world, it was sufficiently well established to form a sub-community within the wider Sherpa society. Structurally, in fact, it would be possible to say that as in most Tibetan-related communities there are two parallel societies, the lay and the monastic, both of which reproduce themselves over time and over generations, the former by means of biological procreation, the latter by recruitment from the former.

The Sherpas themselves see the monastic world and the lay world in complementary binary opposition: upon coming of age, a young man or woman has the choice either to marry or to enter a monastery or nunnery. (Both exist in the Sherpa region, though the male monastic system is more highly developed and monks but not nuns play the leading role in the public ceremonial life.) In either case, people then inherit their share of the family land, to use to support themselves. The standard way to express the idea "he became a monk" is "he did not have a wedding" or "he did not take a wife," thus implying that these options—marriage or monkhood—are the only two choices available. Unlike the practice in other parts of the Buddhist world, such as in much of Southeast Asia where it is a regular stage in the maturation of every young man, the monasticism of the Sherpas (as for many other Tibetan groups) is supposed to be for life.

One ideal pattern among the Sherpas is for a family with several sons to send one of them, ideally a middle son, to become a monk. An older monk, often a paternal uncle of the new recruit, acts as a sponsor for a fraternal nephew, whom he helps bring into the monastic life. Since the Sherpas have patrilineal clans and lineages, there are thus set up two parallel systems of reproduction: that of fathers and sons in the lay world, and that of paternal uncles and nephews in the monastic world. Both the sons and the nephews are of course begotten in the lay world, but one remains in it, the other enters the monastery. Thus from the perspective of the monastic world, the lay world is a source of the next generation just as it is for the lay world itself. But a boy or man becomes a lay householder by marrying, whereas he is transformed into a monk by more extensive symbolic means: his head is shaven and he receives the red robes that make him outwardly identical with other monks, and he also receives a new monastic name and takes up residence in the monastery, at some remove from his village. While a married person retains membership in his community and descent group, in principle a monk exchanges these affiliations for his new monastic one. As a monk he studies the sacred texts, receives teachings, participates in the round of ritual life, and spends much of his time carving wood blocks from which to print new editions of religious writings. In other words, his monastic identity negates to the degree possible his former identity as a layperson and reconstructs him culturally as a member of a new "brotherhood" united by their incorporation of a shared common doctrine and way of life. The information one needs to become a monk is contained in books, which he himself helps to re-create for the next generation. These may thus be seen as external analogues of the genes that carry the information that created him as a human being in the first place.

The orientation of the monastic world is away from everyday village life and toward death, as it is conceptualized in the Tibetan Buddhist worldview. This is so in a number of ways. First, it is thought that being a monk is the surest way to obtain a better rebirth. Since death for all except certain liberated beings is inevitably followed by rebirth in one of the six realms of existence, one hopes through meritorious action to be reborn at a higher plane of existence, one further along on a path to ultimate liberation. Being a monk is ipso facto a superior rebirth and more meritorious than participating in the round of village life; indeed, the stock answer given by almost any monk to the question "why did you become a monk?" is "to obtain a better rebirth."[3] The ritually prescribed actions performed by the monks accrue merit for them, as well as for the village community around them, and their abstinence from sex and reproduction is itself one of the main markers of the higher status of their way of life.

Beyond this, the main occasions on which monks act in the world of the laity are at funerals. Because the survivors want the consciousness of the deceased person to achieve rebirth in the best possible corporeal form, it is customary for them to hire as many monks as they can afford to recite sacred texts for the benefit of the deceased during the seven week period between death and a new conception in the womb of a lay woman who has just had intercourse. Paying and feeding the monks is itself meritorious action, which redounds to the credit of the reincarnating consciousness of the departed. The reading of the texts is itself meritorious action, but the reading is also thought of as instructions—"showing the way"—to the newly deceased person on how to navigate the passage from death to rebirth so as to achieve the most advantageous outcome. For the layman, the best outcome in strict principle is to be reborn as someone who will be in a position to enter the monastic life, though many would gladly settle for birth to a wealthy family. However, another desired rebirth is in the heavenly paradise of one of the Bodhisattvas. This is not nirvana, complete liberation, but it ensures a long and pleasant existence and a promise of an even more advantageous rebirth the next time.

The most remarkable and unique dimension of the self-reproduction of the celibate society of monks is the institution of reincarnate lamas. The term *bla ma* (superior one) is reserved for highly accomplished religious figures, some of whom, who by virtue of their high spiritual attainments and store of merit, are able to be reborn to the same position in which they died. Thus, for example, the abbot of a monastery may be recognized, a few years after his death, in the form of a recently born little boy who exhibits extraordinary powers, and whose identity as his previous self is ascertained through tests in which he is asked to pick from an array of ritual items the ones that were his in his last life. This little boy is then installed in his former position as head of the monastery and is instructed by a mentor in the religious and ritual knowledge he will need to function as such upon reaching adulthood. He is, quite literally, the monastery's "baby" insofar as he himself has simply guided his own physical rebirth in such a way that he will be discovered and returned to his rightful place in the monastery. His ability to thus transcend death and, in effect, to reproduce himself, using the copulation of a married human couple for the purpose, is a great part of his merit and a source of his charisma and high spiritual status. Thus his biological parents both are and are not his genitors: he receives a body from them, but as the consciousness that survives death and transmigrates, he has given birth to himself.

The chains of transmission of oral teachings and initiations across generations of celibate monks form "lineages" in a way that is explicitly equivalent to the creation of lineages of descent, except that one involves genetic instructions and one involves symbolic instructions: the same word (*brgyud*) is used for both kinds of lineage. This example can stand for many in illustrating the realization of the dual inheritance model in real social formations: the orientation toward death and another world or worlds beyond the present one; the marking of the distinction between them on the basis of the presence or absence of genetic reproduction by means of copulation; the elevation of (some) males to a higher status because of their apparent ability to be relatively free of the impurity of sexual reproduction, replacing it with symbolic reproduction; and the reproduction of the celibate system by methods that do of course require procreation but that also symbolically negate and transcend it: all these features exemplify in ethnographic reality the interaction of the two modes of information transmission constituting the dual inheritance model.[4]

Orientation to Death

As I have now demonstrated with sufficient ethnographic examples, the organic life into which we are born from mothers is in many cultural systems ideologically devalued in favor of another kind of life that takes place on a culturally posited superior level of existence. This higher plane, insofar as it is accessible in the course of organic life, is also in many cases assumed to be a foretaste of another life that follows organic death, one which does not suffer from the self-evident defects of this one. The cultural symbolic system itself, I have suggested, serves as the experiential model for this other life, since it is free of the taint of sexuality and it appears to be immaterial. Writers about the cultural construction of gender have often regularly perceived that men are associated with this other world, in contradistinction to women, who are charged with being the authors of our organic existence and its ills. This means, for practical purposes, that men are often associated with death as women are associated with birth.

Few cultural systems fail to posit realms of existence beyond the present one. Men may be associated with these other realms and the modes of access to them in this life or the next, but they are also in a great many cultures the agents of real death insofar as they are hunters and warriors, who serve the community and meet their own and their fellows' and families' self-preservative needs by providing food and protection through armed violence. Their prowess as warriors may also bring them honor and glory as well as spiritual benefits, as do, for example, the severed heads of the victims of Mundurucu raids. There are some societies in which women bear weapons, hunt, and fight in war, but these are relatively few (I will discuss one of them in the next chapter). For the most part, the possession of lethal weapons is a distinctive feature of men. Indeed, it is this fact, together with the typical male propensity for competition, that I have proposed as one of the main motors of the evolution of the distinctive form of human society united by a shared symbolic cosmos. In the last chapter, I examined the role of ritual life, for example in men's societies, in enabling men to mute their aggression toward each other and redirect it to enemies, to supernatural beings, and often enough toward women—in the abstract or in the flesh—in the various forms of misogyny that are so prevalent in the ethnographic record, as Gilmore (2001) has documented in detail. It is the disavowed aggression of each that constitutes and puts compelling force into the collectively constructed cosmic order and set of moral imperatives that serve to regulate society and prevent chaos. But the orientation to death itself, which seeks to differentiate men from women and elevate them in status, nonetheless frequently borrows its symbolic meaning from the realm of biological reproduction it seeks to disavow, as the next examples illustrate.

Kwaio

The Kwaio, a society of swidden taro farmers who inhabit a central region of the island of Malaita in the Solomon Islands, are just one of many examples of a society governed by the oversight of the ancestral spirits, that is, the dead who, in the other world, are now capable of observing human affairs and either rewarding or punishing people for their behavior insofar as it conforms to or violates the traditional *abu* ("taboos," prohibitions) (Keesing 1982). These ancestors are called *adalo*, and the evidence of their existence "is on every side. If a snake bites, or a tree falls on someone, it is because an *adalo* has willed it" (42). Retrospectively, it is understood that if these things happen there must have been some earlier infraction. On the other hand, the adalo can also imbue humans with power or mana, which protects them from harm and makes for good outcomes.

Inhabiting a mountainous region, the Kwaio have a cosmos that privileges the uphill direction as symbolically as well as physically "higher" and associated with males, and the downhill direction as female and symbolically inferior. The men's house is at the upper reaches of the village clearing, with households in the middle region, and the women's menstrual hut downhill from there. If there is a priest in the settlement who conducts rituals dedicated to the adalo, he occupies a special house at the uppermost end of the clearing. Thus the layout of the village is itself a map of the value system of the cosmos, which any inhabitant automatically experiences in the course of everyday life just by walking around the settlement.

The menstrual hut is conceptually associated with reproductive powers, gendered as female, and with childbirth; this is in direct contrast with the men's house, associated with death and the realm of the adalo. This distinction can be clearly seen in the contrast, and parallel, between menstruation and sacrifice. When a woman is menstruating, she must sleep in the menstrual hut, with is abu to men. When she finishes, she must wash herself and spend a day in the clearing before she can reenter her house. Likewise, when men and boys sacrifice pigs for the adalo, they must sleep in the men's house, which is abu to women, wash ritually in the morning, and only then return to their homes. Women are in an abu state when they are stained with menstrual blood, as men are when stained with the grease of a sacrificed pig. The ritual washing returns them to the morally neutral state called *mola*, the opposite of abu.

More striking still is the symbolic mirroring of birth and death when the death is that of a priest. When a women gives birth, she must build a childbirth hut out of sight in the forest below the women's downhill area. She gives birth there and remains there out of the sight of men for fifteen days. She is in isolation except for the company of an unmarried girl who brings her water and food from a special "menstrual garden." Then her head is shaved and she undergoes ritual purification, after which she may enter the women's domain, where she can be seen by men, but she cannot yet return to her own domestic household. Only after further seclusion will she rejoin her husband in the neutral space of their home and recommence normal everyday life.

When a priest dies, or after a crematory sacrifice or "holocaust," an equivalent process occurs but now involving men. A man who is called the "taboo-keeper"

> retires into a sacred men's house and remains in a partitioned off area, entering and leaving by a special door, and staying out of sight of women. When he is not participating in rites, he is supposed to stay on his bed. He is attended by a young man . . . bringing water and food from a special sacred plot. Eventually, before being seen by women at the upper margins of the clearing, he undergoes purificatory rites (including washing and head-shaving). These sequences not only dramatize the symbolic mirroring of WOMEN'S POWER and ANCESTRAL POWER; they express the parallel mirror imaging of CHILDBIRTH and DEATH: the 'taboo-keeper' is symbolically giving birth to an ancestor. (Keesing 1982, 68–69)

While of course the men's ritual is a rite for a death, it is conceptualized not as destruction but as its opposite: birth into the realm of another form of life that is eternal, immaterial and invisible, and morally perfect in a way that the organic world is not. One exits the world as one came into it, only in a purely symbolic birth carried out by men rather than through a physical birth from a woman. The same dichotomy is observed when a man is wounded in battle: he too retires to a hut that is the male equivalent of a childbirth hut; so by a similar logic, birth is both contrasted with and equated with another form of male association with death, that is, warfare.

Bara

Like the Kwaio, the Bara, a sedentary pastoral people of southern Madagascar (the Malagasy Republic), employ the distinction between male and female to order a set of cosmic oppositions (Huntington and Metcalf 1979). According to their cultural view, the master polarity is between order, associated with men, and vitality, associated with women; human social life represents an ideal balance between them, that is, vitality controlled and kept within proper boundaries by order. Too much vitality is chaotic, but too much order equals sterility and death. Conception, in this scheme, is the molding into ordered form of the fertile blood of women by the male semen, and life is a journey from mother's womb to father's tomb. Like many other cultural symbol systems, that of the Bara divides the human personality into the female blood and the male bone; the ritual handling of death brings to the fore the decay of the flesh and blood and the survival of the bones, which, in the third and final stage of the funeral cycle, are reburied, after they have had time to become dry, in the paternal ancestral tomb complex. The first stage of the funeral cycle is the actual first burial of the corpse, while the second, the "gathering," is a communal feast sponsored by the family of the deceased.

While the gathering is a relatively staid affair, the first burial juxtaposes solemnity and mourning during the daytime with unusual license and obscenity during the night. Right after the death, the corpse is placed for three days in a house designated as the female house, where the women gather to cry and mourn. Meanwhile, in a parallel male house, the male relatives of the deceased receive condolence calls and plan the funeral. After dark, the men and women leave their respective houses and go into the public space where they engage in sexually provocative dancing and the singing of obscene songs. This display of vitality, equated with sexual energy, offsets the excessive order entailed in the death.

Then on the third day, in a symbolic reframing of marriage by capture, the men raid the women's house and seize the corpse over the protests of the women and place it in a coffin. There follows a procession up the mountain side to a burial cave, led by young but sexually experienced men; while the girls, in disheveled attire, try to distract and stop them. Finally the men use the coffin as a battering ram to get past the girls, an act symbolizing copulation according to the ethnographers, and insert the coffin into the cave head first:

> But the symbolism shifts as attention focuses on the arrival of the deceased among the ancestors. The dominant theme becomes that of birth, with the deceased entering the world of the ancestors head first like a fetus. When asked to comment on the meaning of burial, the Bara invariably use the metaphor of birth. This is evident . . . in the tomb-side address to the ancestors: "Here is your grandchild, born here. Do not push him away from here." Just as one must be born into the world of the living, so must one also be born out of it and into the world of the dead. (Huntington and Metcalf 1979, 116)

Here once again, as in the case of the Kwaio, death, associated with men and contrasted with fertility, is both likened to birth and contrasted with it: death is birth into the world of the ancestors, that other world from which life comes in organic form via mothers and to which it returns in the ancestral tomb of the fathers.

The cultural symbolic system itself is also like death insofar as it is not alive; it is only brought to life and vitalized by humans generated by means of actual organic reproduction. But unregulated reproduction, pursuing the aims of the genetic program, as the Bara see it, brings on the chaos of incest, adultery, witchcraft, and violence, and must be given order by the cultural symbols. Themselves without life, the symbols are of a higher order than mere organic life because they are not subject to ordinary death but rather, by their creative and transformative capacity, raise death itself to something more sublime than embodied life.

Male Parturition

As I have just discussed, one way men representing culture have of trumping women's ability to give birth is to use it as the basis for a symbolic construction of death as a birth into an immaterial and better life. Another way is to claim that men can actually produce children by themselves. The case of reincarnate lamas in the Tibetan Buddhist system, for example, comes close to realizing this claim by reducing the role of the biological parents to that of transmitters of the consciousness of the deceased lama into a new body in which it can begin its former life afresh.

Another example is the belief in some societies that men can become pregnant. The Hua of highland Papua New Guinea hold this belief, and their ethnographer, Anna Meigs (1984), claims that similar beliefs are to be found among the Tauna Awa, also of the Eastern Highlands, as well as among the nearby Gimi, and the Keraki of the Trans-Fly region. For the Hua, both sexes can get pregnant, but men have no way to give birth. Instead, the abdomen of a "victim" of male pregnancy becomes more and more distended as a blood clot grows in him and develops into a fetus-like being. Unless the man is treated by means of eating a small amount of his eldest brother's wife's feces, the clot will eventually burst inside him with fatal results. The causes of male pregnancy, for the Hua, include eating food contaminated by a menstruating woman, eating meat of an opossum (an animal with female connotations), and sorcery. Meigs observes that while male pregnancy is officially abhorred, this may be only part of the full picture: "Yet one cannot resist suggesting that it is also desired. All the facts deny the premise that males can become pregnant. I submit that the reason males believe they can become pregnant, and believe in the fake fetuses provided by the curers, is that they have strong psychological reasons to do so. They have a will to believe they are fertile" (55). Such a will would of course be entirely consistent with the argument I have been developing here about the envy of men, on behalf of the cultural system, of the women representing the biological system, both desiring and devaluing fecundity simultaneously.

Meanwhile among the nearby Keraki, who practice anal intercourse between men on ritual occasions, it is explicitly assumed that such intercourse may result in male pregnancy. Eating lime is supposed to counteract this outcome, which is feared not in its own right but because it might betray the secret of male ritual intercourse in the men's house to the women:

> Cases of what appears to be *prolapsis ani* have been described to me in awed breath and put down to the unavailing effort of the male mother at delivery. The native indeed fears that such unduly corpulent men may actually succeed in delivering their children and thus betray the secret of sodomy to their womenfolk—a revelation which, they say, would cause extreme shame to every man. (Williams 1969, 201–2, cited in Meigs 1984, 47)

Wana swidden farmers inhabiting the hills of the eastern arm of the island of Sulawesi in Indonesia have a profound ambivalence about sex and reproduction: procreation is the most powerful mystery of Wana thought, and not to procreate is a sin, but magical pains are also taken to avoid childbearing, and having too many children is a source of dismay: "A variety of stock reasons to avoid pregnancy are given. Having children interrupts work. Cleaning up children's excrement is a nuisance. And for women's part, the threat of death in childbirth is of paramount concern" (Atkinson 1990, 74). Furthermore, pregnancy and childbirth are not exclusively female processes for the Wana. Jane Atkinson argues that one reason for this inclusiveness is that Wana thought in general emphasizes sameness, including between the sexes, but also that "in terms of Wana sexual politics, men cannot leave anything as emotionally powerful as childbearing to women" (75): "As Wana tell it, men used to be the ones to give birth. Men suffered terribly in childbed. They'd stamp their feet in rage, tear down the rafters, and after all their suffering, they'd deliver measly, unviable little infants—smaller than tadpoles, and rather like ants. Then a woman said to a man . . . let me take over, I have a 'pocket' for it [i.e., a womb]" (75–76). Since then, women have given birth, but before they do, males actually become "pregnant" first: "A man carries the child for seven days then puts it in a woman. Sterility is attributed to men, not to women. . . . Some say a woman is fertile all the time, whereas a man is fertile only seven days a month" (76). Similar beliefs are reported for the people of Negeri Sembilan in Malaysia (Peletz 1996).

This belief about the mechanism of conception is also essentially the same as the "one-seed theory" among the Greeks of classical times. For them, the male semen already contained a complete homunculus, which only grew to full size in the woman's uterus (Leitao 2012). This belief persisted in European thought through the seventeenth century. Greek myth also, as is generally known, attributed the ability to give birth to the male high god, Zeus, who delivered his daughter Athena from his forehead and his son Bacchus from his thigh. Indeed, creation myths from around the world, including the biblical one, often credit the origin of humans to a primal male being. One might also mention in the present context the widespread phenomena covered by the term *couvade*, which, while it varies greatly in form and meaning from society to society, nonetheless by definition clearly implicates the father in the experience of gestation and giving birth.

Selk'nam (Ona)

As a final ethnographic example of males symbolically procreating by producing a "baby," let me cite the ritual called *Klokoten* or *Hain* among the Selk'nam (Ona) of Tierra del Fuego, at the southernmost tip of South America. The Selk'nam were hunters and gatherers who exploited the meat, hide, grease, and other products of the guanaco, a large mammal related to the llama and the vicuña. Unlike many other foraging peoples, the Selk'nam, who no longer exist as a society, had an elaborate men's house and men's cult tradition, supported by a version of the usual myth whereby hegemony had once been in the hands of women under the leadership of the cruel divinity Moon, until men managed to kill the women and seize the instruments of power.[5] The Hain ritual was the highpoint of the ceremonial cycle and was the occasion for the initiation of young males into the men's cult; the initiates were called the klokoten, whence the alternate name for the ritual.[6]

A large circular ceremonial house was constructed for the Hain, symbolizing the cosmos with its four wombs, sources of mystical shamanic power. Men wore masks and were painted in ways that identified them with the mythical beings whom they were to impersonate in the ritual. Every young male had to be transformed into an adult man by going through the Hain ritual, in the course of which he endured various tests and rigors and learned the important ceremonial lore and the moral code governing adult life. The ceremony itself entailed a long series of dances and dramatic enactments for which the women and children were the audience. The sequence of ritual events is far too long to be described here; I will only concentrate on one particular scene, that involving the (symbolic) birth of a child.

The most dreaded of all the deities present at the Hain was Xalpen, a man-eating female whose aroused rage and fury against the men and the initiates expressed the collective wrath of the women from whom dominance had been stolen in primordial times. She was accountable to no one and unless the women appeased her by feeding and singing she was likely to kill one or more of the men or initiates. Xalpen was also dominated by an insatiable sexual urge that caused her to copulate with all the young klokoten. (She was the only figure in the ceremony represented by an effigy rather than by a masked and painted man.) With all this love-making, she inevitably would become pregnant; her birth pains drove her into an uncontrollable frenzy, in the course of which she threw a bow out of the Hain hut. (The interior of the hut was naturally invisible to the spectators, who only learned what was happening by the sounds and cries coming from within.) This gesture meant that she was going to kill someone to appease her wrath, and the hut quaked and shrieks were heard as Xalpen disemboweled the klokoten one by one with her sharp fingernails. The mothers of the initiates wailed in despair and sang in a vain attempt to persuade Xalpen to be merciful to their sons.

Suddenly there was silence and the women knew that a baby had been born and that Xalpen had returned with it to the underworld. Everyone was overwhelmed with grief for the death of all the klokoten, who were now brought out of the hut as if dead, bleeding and with the disembowelment scar running from neck to genitals, while the women chanted a lament. A day or two after the massacre of the initiates, a beloved figure named Olum, the restorer of life, went to work inside the hut and brought the corpses of the klokoten back to life. Everyone was very relieved, and a shaman announced, "Soon you will see something beautiful! Get ready!"

> The new arrival is Xalpen's baby K'terrnen, who is impersonated by a slim *klokoten*. The baby could be of either sex, and if it is a girl the actor presses his genitals between his legs and binds them there with a string made of guanaco nerve or tendon. . . . The infant is decorated with parallel rows of down, from the tip of its conical head to its fingers and toes. The tiny feathers are glued to the body paint. . . . The bright paint colours as seen through the soft white down produce a glimmering effect, making the baby seem all the more supernatural. . . . When they hear the soft hand-clapping from the Hain, the women begin their welcoming . . . chant, stretching out their bent arms to draw the baby out of the Hain so that they can rejoice in admiration of 'Xalpen's gift of life.' (Chapman 1982, 141–42)

The baby was presented to the women, taking small stylized steps and keeping its body stiff and unmoving, and then returned to the Hain hut; shortly thereafter a shaman emerged to announce "It went below," that is, to the underworld.

Although the Hain ceremony was much more complex and the symbolism vastly richer than this short synopsis of one section of it is able to convey, one can clearly see the key themes of the male assertion that they can control the production of life symbolically: once again we find that the deepest secret source of life as well as the greatest threat to society is unbridled female sexual power, which the men alone can control through symbolic action; and we encounter the claim by the men to produce a baby whom they put on display—albeit by conjuring up the female arch-enemy of men and of society. But of course, it is the men themselves who are enacting this drama, and who seem simultaneously to know that what they are doing is a hoax and also to take it with deadly seriousness; when in the grip of the dancing, they experienced the spirits they are calling forth and embodying with costumes and masks as terribly real.[7]

Child Purchase: The Mamluks

The Marind-Anim, once they stopped, or were prevented from, taking children from other groups in headhunting raids, continued to obtain children from neighboring societies, only now by buying instead of stealing them. Another interesting example of a social system that reproduced itself consistently over time through recruitment by purchase is that of the Mamluks of Egypt. I have used them briefly in chapter 4 as a metaphoric example of a master being displaced by his slave—a process I attributed to the ascendency of the cultural channel of inheritance over the genetic one. In this section I want to examine in more detail the Mamluks' system of asexual reproduction and its relation to sexual reproduction.

The ambitions of the Islamic movement from early on required that the military strength of the Arabian forces be augmented if they were to prevail. It was recognized early on that this enhancement could be achieved by recruiting non-Muslim men into the Islamic armies by offering them a number of perquisites. Caliph al-Mansur seems to have been the first to prefer non-Arab warriors as his military aristocracy. Once the Caliphate had conquered Transoxania and moved into the lands occupied by Turkish groups who were not yet Muslim, this region became the preferred source for recruited soldiers. These recruits were regarded as the slaves (Arabic *mamluk*) of Arabic masters, to whom they owed complete loyalty and obedience. David Ayalon (1977) observes:

The main reason for the success and durability of the *mamluk* system in the Muslim countries was undoubtedly the superior military quality of the peoples from whom the *mamluk* slaves were taken (mainly the peoples of Central Asia, of the Caucasus and Transcaucasus, and of southern Russia). By recruiting these slaves as youngsters (it would appear that the age of puberty was considered to be the most suitable one), they could be trained, molded, and imbued with ideas which always served the interests of their masters. . . . By making the *mamluk* institution an aristocracy of only one generation, the life of the institution could be prolonged (at least theoretically) almost indefinitely, without its losing its military sharpness. (chap. 9, 55–56)

Puberty was considered the best time to recruit new Mamluks, because they were then old enough to have absorbed the fierce qualities of their native groups but young enough to be tractable and amenable to formation according to the wishes of their new masters. The system was a "one generation" aristocracy because the loyalty of a Mamluk was only to his master, and the position of Mamluk could not be inherited by the biological children of a Mamluk. When the master died, his successor brought with him his own Mamluks, to whom the Mamluks of the previous master were usually hostile and vice versa. Therefore there was no relaxing of discipline and hence loss of military prowess in succeeding generations, as was often the case in hereditary dynasties descended from successful warriors. In those cases, the patriarchs' very success allowed their offspring to live in a more luxurious manner than the one that had nurtured military zeal and skill in their fathers, and

CHAPTER EIGHT

left them less fit for war (this pattern was of course originally observed by Ibn Khaldun).

The Mamluk system possessed another advantage, as Nizam al-Mulk wrote in a poem in his treatise on government:

One obedient slave is better
than three hundred sons;
for the latter desire their father's death
the former long life for his master.
(Cited in Ayalon 1977, chap. 10, 216)

Since each generation of Mamluks lost power with the passing of their master, as the successor brought in his own Mamluks, however, the question arises as to the means of replication of the system. The answer was child purchase:

For the success and perpetuity of the *mamluk* system one condition had to be safeguarded: an uninterrupted supply of *mamluks* from their countries of origin. The main reason why this condition could be so often fulfilled was the readiness and willingness with which the people of those areas sold their kinsmen (to say nothing of the readiness of the heads of tribes, etc., to sell their subjects). This was the backbone of the whole system. Raids and kidnapping were not as important a factor as that willingness. (Ayalon 1979, chap. 9, 56)

The sultans of Egypt, like other potentates of the medieval Arabic/Islamic world, created a corps of royal Mamluks to act as their militia. In the early days these Egyptian Mamluks were purchased mainly from Turks of the Qipchak horde; in the late thirteenth century, recruitment came mainly from among the Circassians (Cherkess) of Transcaucasia. In 1250, these Mamluks overthrew an incompetent sultan and placed one of their number, Baybars, on the sultan's throne. From that date until 1517, when they were defeated by the Ottomans, the Mamluk dynasty ruled Egypt. Their rule for most of three centuries bears witness to the durability of their system of recruitment.

When a new Mamluk was purchased by a royal slave merchant, he was brought to a slave market in Cairo and sold to a sponsor or master, often enough the sultan himself or one of his top *amirs* (military commanders)—all of whom at this point in history were themselves Mamluks. This master, called the *ustadh*, commanded the absolute loyalty of his slave, whom he sent to a special academy for training. The first part of the curriculum consisted of conversion to Islam, instruction in literacy, and study of the Qur'an; the teachers were generally eunuchs. Then when he grew older, the young man was taught military skills, including riding, the handling of various weapons, and so on. Progression through the various stages was slow, and graduation came at around age thirty-five. Upon graduation, however, the Mamluk was liberated from slavery, along with the others of his cohort, by his ustadh, to whom he now owed even greater loyalty out of gratitude for his having freed him. He became a full-fledged soldier, equipped with his own horses and weapons, and could begin a military career that had the potential of landing him in the position of an amir or even that of the sultan himself.

Mamluks were allowed to marry and have families, but they could not be succeeded in their positions in the military leadership or government by their sons. On the contrary, it seems that the sons often were seduced by the luxury of civilized life in Cairo (much as Ibn Khaldun had observed) and did not distinguish themselves. The true line of descent in the dynasty was in any case from ustadh to 'atiq (liberated slave). All the Mamluks who had had the same master had unswerving loyalty to him and to each other, which made for great esprit de corps among the cohort. Competition among cohorts kept a keen edge on the fighting skills of each. The rules laid down for inclusion in the society of Mamluks were that one had to have been born a non-Muslim outside of the realm, purchased and trained in a Mamluk academy as a slave, and then liberated.

An individual Mamluk thus produced two different hereditary descent lines, his biological one and his cultural one created by his reeducation as a Mamluk and his subsequent becoming an ustadh to newly purchased recruits. In the former case, his biological issue simply entered mainstream Egyptian society, where they remain as a recognized ethnicity to this day. Many of the Circassian Mamluks also liked to bring as many of their biological relatives as they could to Egypt, where they often installed them as *aghas* and amirs to lend themselves military support. In doing so, they augmented the strong loyalty of the fellow former slaves of the same master who supported them with the tribal loyalty of blood relatives. Indeed, at some points in history, it was said that the reign of certain Mamluk sultans was the "rule of brothers-in-law" who commanded their troops.

As for the line of inheritance among the Mamluks themselves, the situation was very intriguing. The ustadh was addressed by the same term as father, and various textual sources confirm the equivalence of the two statuses. Furthermore, an older Mamluk while still in the mili-

CHAPTER EIGHT

tary academy could take on a younger one in the relation of mentor and apprentice. The older one was called *agha*, the younger one *ini*: "The relations among Mamluks united by a common servitude and a common liberation, and those of an older Mamluk with his cadet, present a striking analogy with family ties. This analogy becomes even more evident thanks to the following facts: according to the dictionary one of the meanings of ustadh is father; of agha, older brother; of ini younger brother" (Ayalon 1979, chap. 1, 35, my trans.). Ayalon cites several textual sources that illustrate the interchangeable usage of the terms for father and for ustadh. This equivalence also applies to "fellow slaves of the same master" (*khushdashiya*):

The fact that in the mamluk sources the word akh (brother) is often a synonym of khushdash, and the noun ikhwa (brothers) a synonym of khushashiya, illustrates again the analogy of relations among mamluks liberated by the same master with those of the members of the same family. The more so since the plural designates *brothers by birth* and not brothers in an idea or a religion (ikhwan). (chap. 1, 36, my trans.)

Thus it appears that the relations among cohorts of Mamluks were thought of as identical, and not just analogous, to the relations produced by procreation in the biological family, whom they conceptually outranked.

At the same time, the potential conflict between the two forms of descent, the biological and the cultural, appeared in the historical situation whereby the Circassian sultans often attempted to place their biological sons on the throne as their successors, in hopes of establishing a hereditary dynasty. The opposition of the other Mamluks to this was so strong, however, that these sultans by right of biological heredity never lasted on the throne for more than a few months before being deposed and replaced by a Mamluk chosen from among those who had successfully rebelled. Ayalon writes that

a surprising feature about the Circassian sultanates is that even the later rulers, who were well aware of the fate that lay in store for the sons of sultans appointed by their fathers to succeed them, did not draw the obvious conclusion, and continued the practice in the certain knowledge that their sons would be deposed. This fact arouses the astonishment of Ibn Taghri Birdi, who can find no explanation for it. In one place he remarks: 'We have seen the same reward meted out time and time again, from the day that Barquq deposed al-Mansur Hajji down to our own day. They all drink from the same cup handed to them by their *atabeks*, and the contents of the cup are prepared by their fathers' Mamluks.' (1977, chap. 4, 139n.)

As Ibn Taghri's observation suggests, the Mamluks of the former sultan regularly dealt particularly harshly with the biological sons of former sultans so as to prevent them from becoming rallying points for usurpation.

Enough examples have been given, I think, to demonstrate the prevalence in the ethnography of symbolic reproduction, in which a society or a subgroup in society reproduces itself through symbolic means, all the while appropriating an array of symbols borrowed from those same processes of biological procreation they claim to be rejecting in favor of their own methods of reproduction. The various cases I have given all involve men, who cannot reproduce by themselves, claiming the capacity to control or replace the process of birth to their own advantage, or even to be able to do it themselves.

Nothing could be more misleading, however, than to assume that this book is primarily intended as an examination of gender relations and techniques of male domination. On the contrary, my point is that these latter phenomena are themselves the frequent, though not necessary or universal, cultural expressions of the problem that I am proposing *is* universal and must be addressed somehow, namely the problem of how to integrate into a functioning human society capable of reproducing itself two different systems of information transmission with different characteristics and different agendas. To further offset any impression I may have given that the domain of the cultural must be embodied by men in opposition to women, however, in the next chapter I devote my attention to socio-cultural systems in which women exercise prerogatives often associated with men.

NINE

Beyond Gender Asymmetry and Male Privilege

As I have made clear, it has not been my intention to write a treatise on gender and gender inequality. That subject has nonetheless emerged willy-nilly because any survey of world ethnography cannot avoid it, and because the actual subject of my book—the consequences of dual inheritance for human social life—is closely intertwined with the frequent symbolic assignment of the cultural mode of inheritance mainly or exclusively to males and the genetic mode and the problems it poses for society to women. Furthermore, the issues entailed in the relations between the two modes of inheritance find clearest expression in many instances in elaborations of male ritual symbolism and ideology, as I have demonstrated in the last two chapters. The intrinsic ethnographic interest of these facts, as well as their widespread occurrence in different and unrelated parts of the world, justifies my having focused attention on them. Taken together they emphatically show that the dynamics I have been identifying are quite real, and that while the solutions cultural systems find to the problems posed by dual inheritance vary widely and often exhibit great creativity, upon analysis they also reveal surprising similarity and a common origin in the attempt to deal with the difficulties human society has in its efforts to manage the genetic system and its inherently antisocial aspects.

To support my contention that it is the distinction between the two modes of inheritance, not that between genders, that is the key determinant of the myriad social forms encountered in the ethnographic record, I have thought it essential to devote this chapter to an examination of social systems and cultural formations in which male dominance and privilege are either absent or offset by institutions in which females are not confined by gender ideology to the sphere of biological procreation but are active participants in the tribal society beyond the marital pairs to which they may also belong. In the course of doing so, I must also address the question of how such societies deal with the challenge of overcoming, redirecting, or otherwise containing potentially disruptive male competition without resorting to the methods involving gender asymmetry that I have examined in previous chapters.

Sherbro

Many West African societies feature a more equal or complementary relationship between the sexes than is the case in much of New Guinea, Native America, Australia, or Eurasia (including much of the contemporary West). Of course, a feature such as this is somewhat subjective and hard to measure, but certainly one can cite the fact that in contrast to the case in many societies with men's cults in those other areas, some of which I have described, the well-known men's secret society called Poro found among various West African groups including the Mende, Kpelle, Temne, and others is complemented by a parallel female secret society, Sande, of equal prominence and status. Like Poro, Sande features initiation rituals and ranked hierarchies, and prepares girls ritually to become wives and mothers just as Poro converts boys as initiands into men ready to become future heads of households who may legitimately procreate. In these societies, then, important ritual institutions display a complementarity in which both sexes have parallel spheres, rather than having a ritual cult for men only, consigning women to a lesser valued domestic sphere.

The Sherbro people go a step further in the direction of gender complementarity and equality (MacCormack 1980). They inhabit a stretch of the coastal region of Sierra Leone, where they subsist on fish, rice, and other crops, and, like other neighboring groups, they also have both Poro and Sande societies. In addition, however, they also have a third secret society, called Thoma, which unlike Poro and Sande accepts both males and females. Each Thoma chapter is headed by a man and a woman, referred to as "husband" and "wife" in the ritual context, though they are not in fact married. Thoma is thus explicitly constructed on the model of a marital domestic unit designed for procreation while actually achieving its goal of social reproduction by symbolic, not sexual, means:

> Sherbros hold a strong cultural assumption that society must be perpetuated. Adult husband and wife farm, rear children, and have other explicit responsibilities for maintaining healthy, orderly, rule-bound society. The cooperation of adult men and women in these activities which perpetuate society are conceived as metaphors for the cooperation of man and woman in procreative sexuality. . . . The theme of gender complementarity within cultural unity is restated in a different modality in Sherbro ideology of cognatic descent. (97–98)

The symbolism of Thoma, in contrast to that of some of the other ethnographic examples of secret societies I have discussed so far, does not turn on the separation of and conceptual asymmetry between the sexes. On the contrary, while recognizing their differences, it incorporates them both, symbolically and in reality. Instead of being based on a male/female contrast, the key idea of the initiation rituals is to transform "wild," unsocialized children of both sexes into cultured adults, who will however retain the vigor associated with wild animals and the forest. Therefore, dancers in the ceremony wear the masks of animals, specifically the duiker and the hippopotamus. In the course of the initiation, the initiands are secluded in a hut in the forest, which is understood as the maw of the forest spirit *min*, a bisexual being with both a scrotum and a womb/vagina. In their ritual enclosure the initiands are stripped naked and their heads are shaved, so that they are "like a foetus," in preparation for a new birth. This birth takes the same form as their final birth as ancestors at their death, when they are "laid naked on the ground, on forest leaves, covered with more leaves, then with earth" (105).

During their period of marginalized symbolic death in the forest, the initiands receive instructions regarding proper ritual and social behavior, including secret knowledge, and then they are "born":

> Members of Thoma beat on the 'belly' of the forest spirit (beat on buttress roots of the cotton tree or an up-turned canoe), announcing its labour has begun. Young trees in the grove are shaken, and people on the outside 'hear' the pains; and 'see' the *min* thrashing about in labour. The sounds of the long double-headed Thoma drum and women's chanting voices are also heard. The chant ends in a drawn "uh wheeee' of labour pain, followed by silence. (106).

The parallel with the Hain ritual of the Selk'nam is striking; except that while among the latter it is men who stage the death and rebirth of the initiates and the birth of Xalpen's child for an audience of the women, among the Sherbro, Thoma members of both sexes participate in the ritual and the initiands too are of both sexes. Symbolic reproduction takes place on the model of procreation, but in this case it is not the men claiming sole authorship of it. Rather, adults of both sexes act together as representatives of society as a whole to transform the community's children into grown-ups. Gender complementarity is maintained as a factor necessary in all aspects of social life, including both procreation and its parallel in the ritual enculturation of youths.

To further stress the fact that this initiation system is based not on gender antagonism but on the participation of both sexes, a figure called the Gbana Bom, the master of ceremonies, stands in the clearing to welcome the reborn initiates back into the village. This figure is a Thoma man wearing a woman's wrap-around skirt in a symbolic union of the sexes. Thus aspects of the sexual complementarity involved in sexual procreation are brought into the ritual context through the role of the "husband and wife" as leaders, the miming of the death and rebirth from a forest spirit of the initiands, and the bisexuality of both the forest spirit *min* and the master of ceremonies; but no actual procreative sexuality is involved. However, the desired end result is that the initiates will now be socially recognized as full-fledged persons able to marry and themselves procreate and re-create the cycle in which genetic and cultural reproduction alternate in the production of fully social beings, with both men and women cooperating in both processes.

I have given this example to show that the close interweaving of symbolic and sexual reproduction in a cultural ritual system, explicitly designed to turn what are considered to be unsocial children with animal natures into culturally regulated adult humans, does not necessitate a battle of the sexes but may on the contrary involve their cooperation, complementarity, and union. This emphasizes my point that the opposition that is being expressed and overcome is that between the genetic channel of reproduction that has produced the children, understood as wild, unregulated creatures who only follow their own appetites and drives, and the cultural channel by which society transforms the children into adults who play by the social rules and regulate their self-centered behavior—in particular their sexuality—in appropriate ways that perpetuate the society as a whole. There is no intrinsic need to assign one of side of this duality to one sex and the other to the other, if the sexes can be understood as relatively equal and mutually necessary for life and for society.

This case also illustrates another important point, one made by others many times before me, namely, that the opposite of male-female asymmetry in the form of male domination, hegemony, or privilege is usually not a reverse asymmetry in which women dominate men. Such social systems are very rare in reality, though they may flourish in the imagination of men. The ancient Greek legendry about the Amazons is one such ideological construct based on an imaginary reversal of the actual state of affairs in that male-dominated and sexually asymmetric society.[1]

Instead, the more usual "opposite" of gender imbalance favoring males is some form of complementarity, such as the one to be found in the Sherbro case, which may or may not pass muster as equality but which emphasizes cooperation rather than conflict between the sexes and views the totality of the society as encompassing both sexes, rather than defining one sex's society as the main legitimate part of the community as a whole and relegating the other sex to a symbolically devalued, subordinate, and limited domestic sphere.

The two different systems may be found together, of course. In an earlier chapter I described the Merina descent system which, at the level of the descent group, does not differentiate at all between men and women, since membership in the descent group stemming from an ancestral couple is thought to cross generations by mystical rather than genetic means. Yet during the circumcision ritual for young boys, it is men outside the house who represent the descent group while women, within the house and associated directly with biological reproduction, are devalued. In the Merina case, both gender differentiation and gender equality are present, but in different social contexts.

My point is that the association of women with sexual procreation, and of men with a transcendent form of life not subject to organic processes, expressed in cultural symbolic form, may be one very common outcome of the attempt to construct a society out of humans informed by both genetic and cultural inheritance. But this construction is not socially necessary, much less given in the nature of things; rather, it is a conceptual arrangement that is readily suggested to the cultural imagination by certain facts about life. But these facts are themselves subject to varying cultural interpretations, so that, for example, the man's semen can be seen as the only true source of the resulting child, on the one hand, or conversely as merely the equivalent of rain that waters the fetus envisioned as a seed that the woman herself produces, as it is said among the matrilineal Na, on the other.

For the Sherbro, it is children, not females, who embody the animal-like appetites that culture, in the form of ritual action, must transform and constrain. And it is no mere metaphor to say that these appetites are "animal-like." As I have shown, animals, dominated as they are by the genetic program and without a countervailing symbolic program, pursue strategies designed by natural selection to enhance inclusive genetic reproductive fitness; these strategies preclude effective cooperative social life without some adaptation or other such as reciprocal altruism or dominance hierarchies. Children are as good a category of humans to whom to attribute these same qualities as any, indeed a better one than women: while their enculturation has already begun even in the womb, children really are not yet fully enculturated and socialized, as adults of both sexes are. It is only their small size and relative ineffectiveness in causing serious harm that prevent them from becoming sources of social disruption; instead they are granted license that must be taken away through socialization, and often enough through rigorous ritual action and education, and replaced with the culturally constructed social code.[2]

Having argued that gender asymmetry to the advantage of men is not a necessary state of affairs in human social life, though a very common one, I will now devote the rest of this chapter to presenting some examples of societies in which women occupy domains more typically associated with men. I will begin with one of the few societies I could identify in which the entire system, and not just one aspect of it, exhibits a gender asymmetry in which the whole overall social advantage goes to the women.[3] Interestingly enough, it is in Europe, in a country that, while not technically on the Mediterranean, is often thought of as a part of the circum-Mediterranean culture area, and thus often presumed to be characterized by machismo and patriarchy, namely Portugal.

Nazaré

Nazaré is a maritime community on the Atlantic coast of Portugal about a hundred miles north of Lisbon, famous for its fine surfing beaches (Brøgger 1989). Some men work for large fishing industries located elsewhere, but many are self-employed fishermen, while others work in the merchant marine. Tourism in this picturesque town is second to seafaring as a source of income. Women predominate in the tourist industry, whether as labor or as entrepreneurs operating real estate properties for the tourists. Women also manage the marketing of the fish that the men catch; indeed, they hold a monopoly on this activity. Men of Nazaré do not like to work in any environment but the sea, and the men who work in the tourist industry are mainly from other parts of the country. Unlike the situation that prevails in much of the rest of Europe, in Nazaré "even the casual observer would be struck by the dominance of women in both private and public life. The streets of the village are indeed teeming with women who are obviously following a tight schedule" (21–22). The women are the true heads of households and are not only responsible for such traditional female tasks as food preparation, housework, and child care; they also represent the household in public as well as private matters:

The house is the domain of the women, and outside the bed and the dinner-table, the husband does not really fit into the average household. As soon as he gets up in the morning, the husband is expected to leave the house after a lonely and usually perfunctory breakfast. . . . More often than not he will go to his favourite *taberna* where he usually has his first drink. Unless he is busy with his own fishing gear or that of his patron, he will divide his time between his *taberna* and leisurely strolls along the beach. (22)

Marriage in Nazaré does not foster bonds between families; quite the contrary. In-laws take great care to avoid each other—a result of female dominance and the tendency toward matrilocal marriages. The new couple are closely affiliated with the wife's mother and preferably live close to them. The wife's mother and her other female relatives make relentless efforts to draw the new husband into their family circle and to separate him from his own mother. Various psychological strategies are employed to put the husband on the defensive and to incorporate him into the wife's family's household as a relatively dependent and junior partner. His own parents are not welcome as guests in his home and often will not even attend the baptism of his first child. All of this on the ground reality, meanwhile, is at odds with the stated ideology of patriliny and male dominance so common in other parts of the Mediterranean world. This discrepancy creates considerable strain felt throughout community life.

Brøgger provides a vignette to show that the publicly maintained facade of patripotestal superiority is offset by an opposite ideological construction: a woman owned a gold chain in which her husband's family claimed a share. Her sister's son however, contested this claim by asserting that he and his sister, not the chain owner's husband's kin, were her true relatives. He put forward this argument: "Because a child develops in the body of a woman, from where will it get blood and flesh if not from the mother" (35). This ethnobiological idea is, of course, in marked contrast to the general Mediterranean idea that the child grows from a seed planted by the father in the mother's womb/field, or is the bread which the father places to be baked in the womb/oven: "The dominant position of women thus appears to be a firmly entrenched cultural trait in Nazaré. Behind a superficially Mediterranean ideology, matrifocality . . . exists together with an almost full fledged matrilineal family organization" (35).

But Nazaré social structure also differs from that of other more overtly matrilineal societies in that the woman's brothers do not exercise any authority over her, as generally happens elsewhere where matriliny is the regnant system. On the contrary, the male members of the matrilineage leave the management of the family compound to the women. Brøgger therefore has no hesitation in describing this social system as not only matrifocal and latently matrilineal but as being matriarchal in the true sense of the word. A man has little connection to his own grandchildren through his son—indeed a man may not even know the son's new married address—but he also has no influence over his sister's or daughter's children. Since there is no land to inherit, lineages on the male side that might unify male relatives and thus grant them some leverage do not form. The three-generational mini-lineage of a dominant grandmother and her daughters and granddaughters with their families dissolves upon the death of the old woman. It is true that a few men who have become very successful financially as fishermen are able to assert dominance within their own households, but the typical pattern is one in which men are out of place not only at home but also in the marketplace and even the public square, finding their true home only at sea.

The occasions on which men spend time with other men occur either aboard the fishing boats or in the taverns. Crew members on fishing boats regard each other as *camaradas* and *companheiros*, implying a degree of esprit de corps among them. But they do not form lasting corporate groups, nor does the composition of crews remain stable from one outing to the next. Crew members are entitled to a fair portion of the catch, but are not wage workers or contracted in any way other than perhaps by kin affiliation. When a crew comes ashore after

CHAPTER NINE

a successful catch, they share a collective meal; but that is the end of their association until another trip, which may be with a different captain and crew:

> Although fishing is the *raison d'être* of Nazaré, the organization of fishing enterprises has failed to create groups with any measure of corporate strength. The crews are really *ad hoc* groups of men, each of whom is pursuing his own, separate career. If a better opportunity occurs with a different crew, no loyalty is to be expected. . . . This lack of corporateness is clearly demonstrated at sea. Each *companheiro* brings his own food. . . . There is no sharing of food, not even a fixed time for a meal. . . . There is a conspicuous lack of conviviality, each fisherman keeps to himself. Sometimes the *mestre* [captain] passes a keg of wine around as one of the few expressions of togetherness. On reaching the fishing-grounds fishing operations require an input of labour and attention which excludes social intercourse. (111)

Friendships of the moment, struck up over a bottle in the taberna and as quickly dissolved, seem to be more prominent than any more lasting ties between men. The typical Nazaré male is an individual whose reputation is pitched against a multitude of others in the community, and self-interest in alliances of the moment preempts strong loyalties to anyone. All in all,

> the behaviour of the mature male is, within a wide range of variation, less self-assured than that of women. One might expect daring fishermen to behave with a certain measure of machismo, but this trait, which is so characteristic of the Latin male, is strangely absent in Nazaré. The dominant personality trait of grown men may even be described as gentle, sometimes resigned, and somewhat introverted. (42)

Brøgger speculates that the absence of macho self-assertion on land may be the result of the Nazaré man's ability to express and experience the "tough" aspects of his masculinity at sea: "Fishing in the waters of Nazaré is dangerous, and the pursuit of fishing is to a certain degree a heroic enterprise which requires strength, persistence and daring. The cultural expectations of maleness are therefore fulfilled and do not require any measure of vicarious display" (42).

The gratification of the Nazaré man's urge to test himself and exercise his skill and fortitude is, however, very largely a solitary one. In very few societies are the opportunities for male camaraderie so few and the subordination of men in everyday social interaction on dry land to the will of the woman so striking as is the case in this community. What then of male competition, which I have identified as an issue with which every society has to deal in some way? With regard to fishing, each man is in competition with others for positions in the crew of a respected owner, and each owner is in competition with others for good crewmen as well as for desirable fishing grounds. While this is not a wage economy, it is what can be described as a competitive labor market. As Brøgger observes, in this market "there is little scope for corporate action. The common fishermen compete with each other for the most lucrative positions and the *mestres,* skippers, for the most able men." The only enduring and reliable relations are among close kin, but these cannot coalesce into effective groups either, because of the dominance of the matrilineage. Thus there is an atmosphere of what appears to be rather sullen stand-offishness among men, who live in many ways as solitaries.

One of the outlets for this somewhat stifling emotional atmosphere is the tavern, where men, their tongues loosened by alcohol, are able to unburden themselves to "true friends" of the moment or to close kin, because what one says while drunk is not taken seriously or taken out of the tavern. Every man is, indeed, expected to get drunk, because this is the only opportunity for authentic connection in this otherwise rather tense environment. The other place where the suppressed competitiveness goes is into widespread and deeply rooted beliefs in sorcery and evil spirits. Fishing being an enterprise in which luck plays a critical role, efforts to ensure good luck for oneself or to bring bad luck for rivals by ritual means are undertaken by employing *bruxas,* or witches, who know how to cast spells that can affect the catch.

In this society, in short, various circumstances prevent men from collaborating in groups that might trump the effective domination of both the household and the public arena by cohorts of related women. The household is only very weakly a "family"; women socialize with other women, while men are either marginalized in the taverns alone or with a few companions or are out to sea. Perhaps, indeed, it is implicit in the world of maritime economies that since many men spend so much of their time away for relatively long periods, they leave a power vacuum on terra firma in the public space that women can and do fill.

Dahomey

The making and wielding of lethal weapons that enable a person to kill a large animal or another human have of course played a crucial role in human evolution. Evolving humanity spent the vast majority of its time on earth living by foraging which almost always included a significant proportion of hunting, though the quarry varied from large herds of huge animals to individual small varmints and everything in between. Hunting as an activity requiring the skilled handling of a lethal weapon has been predominantly an activity reserved for males. Women and even children have been employed to support the hunt in various ways, for example by driving game into waiting nets where the men kill the trapped animals; and women, perhaps armed with sticks for digging rather than with weapons designed for the kill, have in many societies provided small game for the larder. But in many societies women are prohibited from even touching much less using hunting weapons, and where this is not so they are not usually trained in the effective deployment of blowpipes, bows and arrows, spears and spear throwers, and so on.

The Agta, hunter-gatherers of the Philippines, are widely known in the literature as one society in which women hunt for large game such as wild pig and deer using bows and arrows, which they do with more success than the men. They tend to hunt in groups and can leave their children behind with female relatives while out on the hunt, though they do not range as far afield as do the men. If breastfeeding, they can hunt with an infant carried in a shoulder sling. This ethnographic fact certainly demonstrates that women are capable of the strength, agility, and patience required for hunting with a bow and arrow, but in most societies the duties of childcare are thought of as incompatible with hunting. As Robert Kelly (1995) observes, "the Agta may be the proverbial exception that proves the rule" (268) that women do not typically hunt because combining childcare with gathering is logistically much easier than combining it with hunting. Furthermore, it can be argued that throwing a spear or powering a blowpipe with sufficient force to kill a large animal require chest and shoulder strength that is more usual in human male anatomy than in that of females (though of course this fact could itself be the evolutionary result of a genetic adaptation to the cultural association of hunting with men, rather than the other way around).

If women rarely use weapons to hunt, it is likewise equally uncommon to find them armed for acts of violence against other people, as men so frequently are. There are many societies in which a man is not found in public without his weapon—I picture the Nepali hill man with his inevitable long curved knife in his belt—but this is hardly so for women. Although there are numerous cases on record of women engaged in some aspect of intergroup violence (Jones 1997; Edgerton 2000), it is very rare in premodern societies for significant numbers of women to follow military careers and form fighting groups composed entirely of women, in contrast to the case with men. One such well-known case, however, is that of the women's troops that served the kings of the West African kingdom of Dahomey for the better part of two centuries when that political entity, in what is now Benin, was at its zenith (Alpern 1998; Bay 1998; Edgerton 2000).

The origins of the institution of units of female warriors in Dahomey are unclear, and the reasons for it are in some dispute. Some have said they may have evolved out of an elite group of female elephant hunters, though why women would have been employed hunting elephants in the first place is also unclear at least to me. Another reason advanced for the use of women in the army is that the population of Dahomey was quite small compared with that of its main adversaries, such as the kingdom of Asante, and therefore required all the person-power that could be marshaled to succeed in combat regardless of the sex of the soldiers; this is at least plausible. Yet another factor sometimes offered to explain the origin of women's legions is that the king of Dahomey would not allow males to guard his person after dark, for security reasons, and therefore established a personal cadre of armed women to protect him from plots and attempted coups. The women warriors, their own careers dependent on the king's favor, would have little motivation to depose him and every reason to keep him on the throne as long as possible.

What is not in dispute is that beginning as early as 1708 there are reports from European observers testifying not only to the existence of all-female fighting forces in Dahomey but to their extraordinary zeal, ferocity, and success in battle, often surpassing that of the male troops. These women warriors occupied positions of great prestige at court and were given special favor that differentiated them markedly from ordinary village women. In the villages, women were subordinate to men, but in the royal palace the situation was reversed, and women ranked higher than men. Women villagers were socialized to prepare for marriage, motherhood, and household work, and were also prepared, such as by means of elongation of the labia, for a rewarding sexual life.

All single women, however, were in principle wives of the king, and could on that basis be called to his service at any time. Those whom he summoned who were young and attractive became his sexual consorts, but many who were distinguished by their size or muscular physique would be inducted into the ranks of the women's army. Besides local

CHAPTER NINE

women of Fon origin (Fon being the ethnonym of the indigenous citizens of the kingdom of Dahomey), another source of recruitment consisted of foreign women captured in war or in slave raids. All women warriors were sworn to a vow of absolute celibacy, which could only be overridden by the king himself if he were to choose them as consorts. The barracks of the women warriors were guarded by eunuchs to ensure their chastity, though inevitably there were occasions on which the rule was broken; if the culprits were caught the punishment was death.

Once they had killed an enemy, the women now regarded themselves as men:

> In an impressive testimonial to gender stereotyping, the Amazons also chanted, again and again, that they had become men: "As the blacksmith takes an iron bar and by fire changes its fashion so have we changed our nature. We are no longer women, we are men.." . . . After an Amazon killed and disemboweled her first enemy, she was proud to be told by other women soldiers that she was a man. (Edgerton 2000, 26)

This is rather ironic, because, as Edgerton notes, male Dahomean soldiers sometimes fled from battle, something the women never did, thus earning the women's scorn. In fact, when this happened, the women warriors derided the men by calling them "women." This reflects both the reversal of the traditional village gender roles that took place in the palace, as well as the force of gender stereotypes even in the face of the self-evident failure of empirical facts to conform to them.

The loyalty of these women to their king was unshakable, since they were separated from their kin and had neither husbands nor children to support them. Since they were not threats to the king's actual wives, as men would be, and since it was in their interest to protect their protector, the mutual bond between the king and his women troops was unassailable. One can compare this situation with that of the Mamluks: like the Mamluks, the Dahomean women warriors were often recruited from foreign sources, though by capture rather than by purchase. They were removed from the process of biological reproduction, and thereby formed a group in which their culturally endowed status was their source of group spirit among themselves and of their subservience to their patron. And like the Mamluks, their loyalty to their ruler was profound but extended no further than their own lifetime. Dahomean women warriors outranked ordinary village women whose domain was sexuality and reproduction, just as the Mamluks, slaves though they were, formed an elite in Egyptian society.

The fact that one group was male and the other female actually makes no difference, since these factors are important and determinative only in the realm of biological reproduction. In other spheres of life, such as where celibacy prevails, gender characteristics can be assigned to people of any sex according to cultural, symbolic agendas: thus the Dahomean cultural ideal of a man was a brave warrior and that of a women was one of weakness and subordination. When these were reversed in actuality, as in the case of the women warriors, the women were redefined culturally as men. In reality, once they were removed from the process of biological reproduction, both Mamluks and Dahomean women warriors achieved a status in which their sex was no longer relevant. Once again, these facts serve to reveal that the key opposition in the construction of society is not that between men and women, or male and female gender, but between sexual genetic reproduction and asexual symbolic reproduction.

Okinawa

Most organized religions are dominated by male specialists or a male hierarchy, and in some instances the male cult is in effect the religion of the entire community, as we have already seen. Susan Sered (1994), who has surveyed the world's religions and identified those in which women play leading roles, asserts that there is only one instance in which the performance of official religious ceremonial and worship on behalf of an entire social group is carried out by a female priesthood. Other examples of religions in which women play leading roles, she argues, are "cults" within the context of a larger male-dominated religious domain, such as the *zar* possession cult of East Africa, Candomble and Macumba in Brazil, or *nat* worship in Myanmar (Burma); in these cases, the nation or surrounding region is officially associated with one of the "great" male-centered world religions—Islam, Catholicism, or Buddhism—while the cult represents a separate encompassed institution, unauthorized by the putative official religion. The all-female Sande secret society of West Africa is complemented by the male Poro society, while shamanism in Korea, heavily dominated by women, exists alongside Confucianism, Buddhism, or Christianity, each with a predominantly male hierarchy. But in Okinawa, the largest island of Ryukyu Prefecture in Japan, the main and only and "official" religion of the whole society features women as priestesses, with men playing only auxiliary roles (Sered 1999).

CHAPTER NINE

Actually the word "priestess" is misleading, as Sered points out, since the term for these women who lead the community's religious life, *kami-sama*, is also the word for a divinity. (The word for divinity is *kami*; *sama* is an honorific suffix.) According to Sered's account, the people of Henza, a small island community off the coast of Okinawa in which she did fieldwork, do not make strong distinctions either between human and non- or super-human actors, or between the sexes. Thus, older women who are "called," often after an illness, to be kami-sama are actually themselves embodiments of divinities, so that their ritual actions, usually consisting largely of summoning non-human kami-sama and feeding them, are bound to be efficacious.

Officially, Henza, like the rest of Okinawa, features patrilineal clan organization, and the government bureaucracy, including the office of headman, such as it is, is in the hands of men. But in most other respects, Henza is matrifocal and to some degree matripotestal. Sered insists, however, that they are not "matriarchal," since the ethos and ideology of Henza is strongly egalitarian with regard to individuals and to the sexes. The economy combines horticulture and shore collecting, both usually done by women, with fishing at sea as the province of the men. Women are responsible for all food preparation, and the major part of the diet is provided by women's work, which is, however, less strenuous and certainly less dangerous than fishing on the ocean. The principle ceremonial activities involve feeding the non-embodied deities, who take in the essence of the food offerings and in turn imbue them with spiritual power that strengthens those who then partake of their material substance.

In addition to doing most of the land-based work, women also manage the households and are responsible for the majority of social interactions in the community. They form strong matrilineal bonds with close kin and tend to be the responsible parties in managing the household (often a necessity if the men are out to sea). At a more formal level, there are numerous all-female organizations that bring groups of women together on a regular basis. These include twice monthly meetings to practice the dances for *obon*, the annual festival in which the ancestors are invited back to the village; free-loan clubs; a women's association for younger women (meaning under sixty-five in this community remarkable for its average longevity); and senior groups. Even in the context of the nominally patrilineal clans, the main clan rituals involve groups of women performing together, and a woman kami-sama serves as the clan priestess. Women also attend functions involving their husband's clans and have dense networks of relations with relatives in both consanguineal and affinal groups:

> Taken together, kin associations, neighborhood associations, clan gatherings, ritual groups, and age group associations ensure women an extraordinarily high level of social integration. . . . Women's gatherings are institutionalized, public, visible, and (often) sacralized, and therefore an important and acknowledged part of the social and power fabric of the village. (Sered 1999, 104)

Men constitute almost all the membership of the Henza village council, and the rotating headman position is held by a man. Men who have fished together feel solidarity with each other, and there is also a fisherman's association. Men also meet to talk and drink in bars in the evening, just as women socialize at beauty parlors, and there is a young men's association. However,

> at least in the eyes of village women, men's groups are not really comparable to women's groups—a distinction that, I suspect, reflects the fact that the men's groups are only [mutual aid groups] whereas women's groups also have a strong ritual component. . . . The women's groups are neighborhood-focused and permanent, whereas the men's groups are composed of colleagues, and therefore are more temporary. (106–7)

Because of the nature of the seafaring life, including not only fishing trips but also time served as sailors and in traveling to see the world and to try to find paying employment beyond the confines of the little island of Henza, men are very often absent, and it is partly to this imbalance that Sered attributes the important roles taken by women in the economic, social, and religious life of the community, just as is the case in Nazaré. Meanwhile, women are linked by shared participation in ritual in a way that more usually is the case with men in men's societies in other societies, as I explained in the last two chapters.

Sered stresses that the priestesses of Henza are not enforcers of a moral code nor are they teachers of a doctrine; their religious status arises from the fact that they have at a certain point in life, usually reluctantly, been entered by a non-embodied kami-sama and therefore are themselves now divinities who do not so much *do* religion as *be* it. They do know the songs and dances and how to perform the rituals, but their efficacy comes from their own sacred status. This status does not, however, carry over into ordinary life; once they doff their ceremonial robes, they work in the gardens, forage along the beach, and chat with their neighbors just like everybody else. Each clan has its kami-sama, and the whole community is represented by a head kami-sama called the *noro*. In the days of the Kingdom of the Ryukyus, before the islands were incorporated into Japan, the noros of different communities were appointed by the king and served as key liaisons between the royal capital of Shuri on the main island of Okinawa and the various outlying localities such as Henza.

The dominant ethos of the island is summed up in the word *yasashii*, which means "easygoing" or "easy." Harmony, good feelings, and the absence of conflict are supposed to characterize not only the social life of the community but also its relationship with nature and with the divine realm. The interviews with informants that Sered quotes extensively make clear that it is "not done" to bring up or emphasize areas of difference or conflict. The rituals of the women are designed to emphasize smooth and cooperative social relations between the people and the non-embodied kami-sama, mainly through food sharing. The non-physical or "divine" kami-sama are thought of as immanent in the world, and thus are in some sense neighbors with whom the priestesses encourage friendly rapport. Women's religious performance thus underscores the key cultural ethos and, as Sered also observes, emphasizes presence, as in the presence of the kami-sama, both embodied and non-embodied.

Men, by contrast, play a larger role in those particular elements of the ritual life that involve some element of absence, separation, conflict, or danger. For example, men are more involved with funerary rituals, and any ritual involving a blood sacrifice of an animal is performed by men. Sacrificial rituals are typically performed to ward off some impending threat to the community. Men also play a leading role in obon, the annual ceremonial occasion on which the ancestors are invited back to the village:

Men's rituals [are] carried out at times of communal danger. All involve killing an animal and using parts for ritual purposes. Although women would cook the meat from the animals, killing was the job of men. . . . In a very broad sense, the ritual division of labor in Henza places men in the sphere of death-related rituals and women in the sphere of life-related (and especially food-related) rituals. We have seen that men have the key roles in burial rituals. . . . Traditionally (and to this day), pregnant women do not attend funerals. In the past, men were not usually present at birth (although there was no prohibition involved). The *noro* emphasizes that she only does "the good things" and not "the bad things," and priestesses do not attend funerals. (123)

I have argued in an earlier section that men are often oriented toward death, or a realm that is associated with the dead or the ancestors or life after death, one that in any event is different from and transcends the material world that one enters by being born to a mother. In many such societies, women, associated as they often are with birth, are also equated with decay and death understood as bad things, in contrast to men who have access to a realm beyond mortal life that is associated with death understood as a transition to that other life. The Okinawan case represents an interesting transformation of this theme. Here there is definitely an association of men with death, killing, absence, and danger, and an association of women with life in its biological dimension, that is, its sustenance with food and its procreation through pregnancy and birth. However, whereas their association with death and killing is often understood in other social contexts to elevate men's status above mere biological existence, in the Okinawan case this association is not valorized as leading to higher ritual status for the men. On the contrary, men are given the difficult but "bad" work of sacrificing, burying the dead, and dealing with supernatural threats as well as with the ancestors, who are also problematic figures.

The women, who are associated with food, birth, and "good" things are able to be divine beings themselves and to interact with the divinities, and in this sense they take up the role of providers of access to the "transcendent" more often aligned with men in other societies. However, as Sered makes clear, this "transcendent" realm is really quite "immanent" in that the deities are not far distant but live among us as neighbors—literally so in the case of the embodied kami-sama. In this conceptual system, biological life is marked as "good," and death as "bad." One notes with some surprise that this correlation might seem on the face of it to be the obvious one, yet it is striking how rarely this conceptual equation is actually encountered in the ethnographic record.

A further observation to be made is the parallel between the Okinawan case and that of the people of Nazaré described above. While they could hardly be farther apart geographically and culturally, they do share some very important common features: both are fishing societies in which the absence of men from the social life on dry land seems to have the corollary of marginalizing and isolating them, while women take up the role of creating social integration often put in the hands of men in other societies. Men in both societies have outlets for their tendency to violence, not only in ritual in the Okinawan case but in fishing itself, which after all involves the wholesale killing of large numbers of living creatures. But their role in ordinary social life when on dry land is to be somewhat solitary and withdrawn, present in body but absent in spirit, as Sered puts it. And in both societies, the men find solace in the bottle, turning whatever conflicts they have either into drunken monologues and confessions, as in Nazaré, or inwards in the quiet, introverted, and sullen style of intoxicated comportment, as in Okinawa.

On Henza, equality between the sexes, while sometimes expressed as complementarity, is more often evinced through the simple lack of emphasis on sexual difference or on gendered roles and ideologies. These clearly exist, but their public expression is muted. And in Okinawan society, the state of things in the world in which things grow is thought to be a benevolent one, so that birth is contrasted with funerals and death ritual in a way completely the opposite of, for example, that to be found among the Merina, where birth is considered polluting while the cult of the tombs of the ancestors is the high point of the religious and social life. One could certainly argue, as does Sanday (1981), that horticultural societies in which women contribute heavily to the economy are more likely to develop such an ideological system. And it appears that perhaps fishing added to horticulture reinforces this tendency by reducing the men's role in society still further. It should be recalled that the Sherbro, too, divided their time between fishing and hoe agriculture of rice and other crops.

Minangkabau

The Minangkabau of Western Sumatra are proud to refer to themselves as a *matriarchaat*, a matriarchy, having adopted that Western concept from the Dutch as a positive description of their distinctive social system. I have already reiterated the by-now well-worn observation that there are very few or perhaps no "matriarchies," if that term is taken to mean a system that exhibits the same gender asymmetry in terms of domination and subordination as does "patriarchy," except with the gender roles reversed. What does appear to be the case is that in many societies, some of which I have discussed above, women and men appear to be relatively equal in status, either in a way that establishes an essential difference between the sexes but sees them as complementary, or in a way that mutes or erases gender differences (or some combination of the two).

The Minangkabau are matrilineal and matrilocal rice-cultivators, and title to land passes through the matrilineage. This does not mean, however, that men exercise no power; on the contrary, as in most matrilineal societies, they play an important role as brothers and as maternal uncles to their sisters' children. Nonetheless, there are very real senses in which the women in this society are key decision makers, hold real social and economic power, and do most of the agricultural work as well as household work. They are also the ones who perform the elaborate ritual cycles of life. The key affective ties between mother and children as the central social bond, and the relatively ephemeral role of in-married men compared with the ties of matrilineal kinship qualify this society as "matrifocal" (Tanner 1974). Indeed, the Minangkabau are an instance in which it is corporate women's groups that exchange men as husbands, rather than the more usual situation in which male lineages exchange their sisters and daughters to be wives in another such group.

As is the case in Okinawa, Minangkabau men are often absent; this is a centrifugal society for males. Young men are extruded from their natal homes and villages and are expected to seek their fortunes in the "migration area." There they will attempt to establish themselves in trade or business, or in some activity that generates income and thus renders them welcome as suitable husbands to be incorporated into an affinal matrilineage. This feature of their society has had the result that Minangkabau men have played prominent roles in Indonesian national life and society as businessmen, intellectuals, professionals, and so on, but it has also weakened their tie with their own home region and its core social units. In other words, the same role that fishing plays in Okinawan society is played among the Minangkabau by the imperative for men to leave home and, upon their return, never to fully occupy a central place in their affinal group.

Peggy Sanday (2002), in a recent ethnographic monograph on the Minangkabau, has argued for accepting their self-definition as a matriarchy by proposing a new and more positive meaning for that term. For her, the key idea is that in some societies, such as the Minangkabau, there is an understanding of power that is very different from our own. In a matriarchy as she defines it, "power relations are inscribed in webs of significance, taking growth in nature as a model for the ordering of culture" (235). Insofar as the symbolism of natural growth in turn implies maternal nurture, as it does among the Minangkabau, a social system is matriarchal when its core symbolic values are built around a respect for the benign attributes of the processes of development found in the growth and ripening of a rice seedling or the maturation of sapling into a tree (by this definition, Okinawan society would also seem to qualify as a matriarchy):

> I reserve the term *matriarchy* for structures highlighting maternal symbols and meaning such as I have described for the Minangkabau. I have argued that the dominance of maternal meanings in the Minangkabau case is due to the derivation of social meaning from the perceived sources of life and death in the whole of Nature according to the proverb "Growth in nature is our teacher." . . . Such considerations lead me to propose that matriarchy be redefined in terms of *cultural symbols and practices associating the maternal with the origin and center of the growth processes necessary for social and individual life.* (236–37)

The guiding ethos of Minangkabau social life is made up of the interweaving of three overarching strands, namely Islam, the laws of the national state, and *adat,* the customary law or tradition. These do not actually go together particularly well, and often contradict each other; but while everyone is devoted to some degree to Islam, and usually obeys the rules of the nation state, it is the traditional adat that the Minangkabau themselves identify as the most fundamental basis of their social order. Adat long predates the arrival of either Islam or the idea of the Indonesian nation, and originally it was the traditional religious system found throughout the Malaysian culture area. Over time, and partly to distinguish themselves from would-be conquerors, the Minangkabau associated their adat with their matriliny and created an ideological system that valorizes processes in nature and in the cosmos as attributes of motherhood. They ascribe their origin as a people to a divine queen, and each matrilineage recognizes an apical ancestress. Core concepts such as "home," "origin," and "center" are associated with motherhood as well. The achievement of social order and harmony, it is believed, entails the inculcating in each generation the adat imperative to encourage nurture, to care for the weak so that they will grow strong, and to exemplify smooth and polite comportment as opposed to rough or overly direct speech and interactions. These benevolent qualities, it is held, characterize nature itself viewed as a realm in which everything grows and matures according to orderly processes which are inherent in the cosmos. Thus adat is not just a feature of human society but is the way in which human society puts itself in tune with the larger natural and cosmic order.[4]

At the same time it is recognized that nature is not only benign but that it has another, destructive aspect as well. This other, evil side of nature is what the world is like without adat, without order: "I was told that without adat human beings would be like wild animals in the jungle: 'the strong will conquer the weak, the tallest will defeat the shortest, and the strongest will hold down the smallest.' *Jungle* here refers to nature in the raw, the world where adat does not rule" (28). A covert fourth strand of Minangkabau customary life, then, is the magical control of evil forces in nature, "punishing and ameliorating the evil that produces death and decay" (28). The ceremony at birth, for example, is intended to protect the newborn, who does not yet know adat, from "the forces of evil flowing through the village." The power that gives magic efficacy also can curse transgressors and thus acts as the "teeth" that enforce the following of adat precepts.

The picture of the world without adat as being one of unconstrained competition serves as a good description of what the world would be like if it were only the genetic program of the pursuit of reproductive advantage that dominated social life. And the supposition that in fact that is how most animals live is not far off the mark from an evolutionary perspective (though as we have seen many have devised more or less effective social forms nonetheless). Adat, which protects the weak from the strong, is part of a cultural strategy of countering competition by customs and regulations that encourage a sense of equality. It will be obvious, in fact, that this worldview holds two contradictory ideas simultaneously: that the natural world, and presumably humanity as part of it, is governed by the cruel law of the jungle; and that at the same time the natural world as well as human society is governed by an adat that is intrinsic in nature but that must be taught in humans, which counteracts and negates the disruptive force of selfish competition prevalent in "bad" nature. In this system, the opposition is not ideologically gendered; rather, equality and complementarity between the sexes are emphasized, as is the key role of women in inculcating adat in the course of rearing children.

The complementarity of the sexes in maintaining adat is exemplified in the distinction between *adat ibu* and *adat limbago*. The former, adat ibu, refers to the extensive menu of women's life cycle ceremonies that structures the lives of individuals and of the community. The latter, adat limbago, designates customary laws, norms, and rules that regulate material concerns, primarily the inheritance of and title to land, the primary resource in this agricultural society. Reference to these formalized rules guides the judicial councils composed

CHAPTER NINE

of men bearing hereditary titles that settle disputes regarding land ownership:

> It is in the interaction and mutual support of women's ceremonies (*adat ibu*) and male dispute settlement (*adat limbago*) that I locate the Minangkabau matriarchaat as practiced and experienced in village life. Adat ibu gives local adat its aesthetic, emotional, social, and sacred center; adat limbago makes adat intelligible as a body of rules to new generations and affirms its legitimacy in a world that pulls people in many directions. Adat ibu designs a path to guide and protect individuals by weaving them into the tapestry of village social relationships; adat limbago applies the rules that lead individuals down this path and punishes transgressors. (21)

Sanday contrasts women's and men's role in adat as the difference between practice and theory. The performance of the ceremonies, and especially the contributions of food, themselves constitute women's adat: it is realized not in verbalized explanations, which the women do not provide, but in the actual doing of the work. Thus, in a way somewhat related to the situation in Henza where priestesses *are* the divinity, among the Minangkabau women *are* adat in their enactment of the customary rituals surrounding birth, marriage, housebuilding, death, and other occasions as well:

> Without properly conducted ceremonies, I learned, a human being is not considered fully Minangkabau, cannot get married, and, if male, cannot take on a hereditary title. Likewise, a house cannot be properly built, and errant husbands may never return to their wives. People who move from one village to another cannot celebrate a birth or a wedding in the adat way without first undergoing a ceremony inducting them into the local clan. Ceremonies are thus part and parcel of the reproduction of adat social structure. (81)

Thus, for the Minangkabau, women are the agents of the reproduction of the cultural system of adat, through their performance of life cycle ceremonies and through their teaching of it to their children. Furthermore, adat itself is conceptualized in maternal terms as the benign force of the universe that leads to growth in an ordered way, in human society and in non-human nature both. To use the language of nature and culture (which, it is true, I myself have abjured) as a shorthand of convenience, Minangkabau women are, in a sense, both combined, while men, though they exercise political power and hold the clan land in title as hereditary heads of the matrilineage, are executors of a (cultural) law that is thought of as female (natural) in origin.

Again, as in Okinawa, it is the women, not the men, who are able to create solidarity through ritual, while men are dispersed, not on the high seas, but in the "migration area," and have less chance to form bonds among themselves and in their matrilocal domestic units.

It must be stressed, however, that adat is only one thread of the ideological system regulating life among the Minangkabau. Leaving to one side the national political and legal system, there is also the strong influence of Islam, a religion that is traditionally much more male-oriented in its value system and in its institutionalization. Thus, there is another dimension of relations between the sexes among the Minangkabau that to some extent contradicts the ethos enshrined in adat. How the tension between these two is resolved is an interesting question, but it appears to be accomplished thanks to a rather tolerant and forgiving understanding of Islam:

The easy accommodation between adat and Islam was apparent to me in the early weeks in Padang and later traveling throughout the province. Daily life is marked by the call to prayer from the loudspeaker in the village mosque, which can be heard from most houses five times a day beginning before dawn. . . . Despite the seeming ubiquity of Islam, most people I knew were not Islamic fundamentalists, nor did Islam rule all aspects of their daily life. . . . As I saw it practiced in Belubus, I came to understand Islam as a spiritual state of mind and a basic cultural premise translated into ways of relating to others. (19–20)

But Islam and the national institutions of the Indonesian state do provide men with an alternative set of values and with venues for the exercise of both solidarity and competition, as do the adat councils in which senior titled men sit with each other to reach consensus about community disputes. Men in Minangkabau society have ample opportunity for assertion in a male sphere, but ideologically, and to a large extent in actuality, it is the female based solidarity system that is fundamental to the socio-cultural system as a whole.

The ethnographic cases I have discussed in this chapter provide ample evidence that human social systems can be arranged without gender asymmetry, or even with an asymmetry that favors women in some way. In these various cases, women may join with men as partners in initiating the youth, dominate the public space, fight wars, represent the community in its interaction with the divine, and provide the source for and maintain the traditional order that makes society function. As these examples show, there is no essential reason why culture must be associated with men and sexual reproduction with women; societies which have complementary and relatively symmetrical and equal parts for men and women to play are quite possible to achieve. They also show that even where there is no strong gender asymmetry, there is nonetheless still the influence of the struggle between the genetic and the cultural programs that inform human existence, however the resulting social formation may be configured. Having established these various propositions, I now return to the not yet fully answered question of just exactly how sociality in humans is achieved at all in spite of the obstacles to its existence posed by the effects of their being organisms that reproduce sexually as well as culturally.

TEN

Sources of Human Sociality

Humans form ties of friendship, amity, affection, and long-lasting solidarity with others far beyond the range of those to whom they are closely genetically related. These ties involve taking actions that benefit others in some respect, taking delight or pride in the accomplishments or good fortune of others, and, most fundamentally, enjoying others' company for its own sake. People feel closely connected to the point of identification with others, they empathize with others, and for the most part and most of the time they prefer to be in the company of others, at least known and friendly ones, rather than to be alone. Not only that, but under most circumstances they do not do each other serious harm, though quarreling, grumbling, and envious gossip may certainly be rife in a society of any size.[1] All this is so obvious as to need no defense, yet it violates the most fundamental principle of evolutionary process, which holds that the pursuit of genetic reproductive inclusive fitness entails competition with all others for these same ends, so that cooperation can be managed only among close kin, or under conditions of reciprocal altruism, or, perhaps, as means to set up the conditions under which competition can take place. There still remains only incompletely answered the question of how we humans manage to do this.

Part of the answer, as I have discussed in chapter 4, is that there are two different evolved forms of sociality inherent in humans today, and one of these is the "tribal" sociality identified by Boyd and Richerson. However, my

CHAPTER TEN

view of such evolved and now innate propensities is that for the most part they are only incomplete dispositions and tendencies, the specific forms of which must be provided by the symbolic system of the socio-cultural group into which children have been born or in which they have been socialized. It is this combined form of self-creation, employing innate predispositions completed by culturally provided specificity, that gives humans such enormous adaptive advantage as they have had, since it moves the particular form of expression of any innate tendency to a medium of information that is much more capable of change and quick evolution than is the genetic channel. It follows that there must be ways in which the presumed evolved prosociality of humans is culturally realized in the course of development, especially given that this prosocial tendency must contend with a powerful counter-current in the competitive agenda furthering the aims of genetic inheritance. In this chapter I will examine some of the ways this is achieved.

Donald Campbell (1975) argued that the moral systems of the ancient civilizations promulgated the idea that right behavior required the negation and overcoming of the egoistic urges that, as he understood, arise from the genetic program. When it came to providing an answer to the question of where resistance to the promptings of such powerful biological forces could be derived from, he proposed selection and evolution at the societal level. Boyd and Richerson (1985) reiterate Campbell's view approvingly. They propose that different imperatives from the cultural and genetic channels can reach a stable equilibrium, and they argue that

> most of these mechanisms will share the qualitative behavior . . . analyzed here; namely, an intermediate equilibrium phenotype may often be achieved through a balance of strong opposing forces—genetically inherited propensities pulling the individual in one direction and culturally inherited beliefs pulling in the other. . . . This model evokes a familiar picture of the human psyche. Many authors (see Campbell 1975 for a review) have portrayed humans as torn between the conflicting demands of an animal id and a socially acquired superego. One is tempted to conclude that because so many observers have found this picture of the human psyche plausible there must be something to it. It is intriguing that this view of human behavior is a natural outcome of a dual inheritance model of human evolution. (197)

As should be obvious by now, I concur with them in this. My own analysis of how the situation can have arisen in which the selfish urges of the genetic program are balanced or even overridden by the cultural program among humans rests on the observation that the cultural system, composed of symbols that convey meanings just as genes can convey instructions, is a public reality that can be perceived, and participated in, by any number of people, without diminishment and usually with enhancement, whereas genetic success, by definition, comes at the expense of rival alleles. People can have positive relations with lots of other people—their extended family, their colleagues at work, their neighbors, members of all sorts of groups and clubs to which they may belong—and care about them deeply; but one can only copulate and reproduce with one other person at a time. Likewise one can convey information by symbolic means such as via language to many people at once, but one can only share genetic information necessary to procreate with one other person at a time. Those united by shared symbolic formation—and this includes people who are classified as "kin" insofar as kinship itself is a symbolic construct—are thus in a better position to realize a social relationship based on good will than those who are in competition over reproductive mating opportunities. Furthermore, since the beneficiaries of inclusive genetic fitness, those sharing the most genetic information, namely the first degree relatives—parents, children, and siblings—are barred from legitimate procreation with each other (except in a very few cases) by an incest taboo and/or avoidance, it seems to follow that sexual reproduction is by its nature to some extent at odds with amicable society, as the ethnographic examples I have presented in the course of this book have demonstrated.

    I have shown that the elimination of sexual reproduction from social life can be accomplished either in reality, by some means of recruitment other than by procreation, or in imaginative symbolic constructions, by various methods I have illustrated. If it is not eliminated completely, then it is managed by symbolically cordoning off the social time and space for sexual reproduction from the rest of social life so that it does not introduce unwelcome uncontrolled expression of the competition intrinsic to the genetic program into the realm of the amicable and social. "Marriage" in the sense I have used the term here, as the socially recognized and respected right of a heterosexual couple to copulate so as to procreate, is a widely found solution that solves the problem of mating competition by allowing each member of society a regular long-term procreative partner of the opposite sex (or sometimes two or more) and conceptually separating the social entities thus created from other wider aspects of tribal society. In the absence of universally available and reliable birth control methods, the intermingling of men and women without some such social constraints would create the problem of how the care and social assignment of the resulting offspring would be resolved. Societies that do allow such unrestricted sexuality usually do so only for a certain designated period of "carnival," or during a certain age period, often one at an early stage of adulthood when fertility seems to be reduced. And in such instances, the acceptance of resulting children as social actors and their placement in satisfactory care-taking arrangements has to be facilitated.

In many societies including our own, the marital relation that legitimizes sexual reproduction is also supposed to be accompanied by long-lasting "diffuse and enduring solidarity," to use Schneider's memorable phrase, and the affection that is also supposed to characterize kin. Often the supposed (and sometimes real) warmth between siblings may be conceptualized as opposed to and better than the distance, danger, and distrust thought to be intrinsic in married relations with strangers or even enemies. One cultural response is to downplay the sexual dimension of marriage and emphasize its companionate qualities, often using the sibling model. We have seen that the Culina, for instance, are eager to define their entire society as one of universal siblinghood, confining the space of marital reproduction to a specific demarcated social moment in the life career.

Among the Kamea of the southern highlands of Papua New Guinea, to take another example, the Sun and the Moon are described in myth as being both a brother and sister pair and also a married couple (Bamford 2007). This apparently does not simply imply an incestuous union (which would in any event certainly not be unusual in a myth of origin) but rather a general tendency to reinscribe marital relations as sibling relations:

> The implicit sameness that underlies siblingship and affinity in myth has terminological correlates in the lives of actual people. . . . Although a husband may refer to his wife as *nka apaka* ("my woman") and a woman to her husband as *nka oka* ("my man"), an equally common practice is for spouses to refer to one another using terms for siblings. . . . This same practice applies to third-party interactions. A man meeting a married couple in their garden may greet them, "Afternoon, brother and sister." As Oates and Oates . . . note of the Kamea language more generally: "When speaking of one's own or one's hearer's relationships, no distinction is made between the spouse and sibling relationship." (74)

In the light of the considerations I have been discussing here, the implication would seem to be that a sexual relationship is, for public purposes, better thought of as if it were a nonsexual and therefore a less disruptive, shameful, or conflictual one.

One is reminded here of Lévi-Strauss's depiction of the patrilineal Cherkess (Circassians) and other peoples of the Caucasus region:

> There, it is the brother-sister relationship which is tender—to such an extent that among the Pschav an only daughter "adopts" a "brother" who will play the brother's customary role as her chaste bed companion. But the relation between spouses is entirely different. A Cherkess will not appear in public with his wife and visits her only in secret. According to Malinowski, there is no greater insult in the Trobriands than to tell a man that he resembles his sister. In the Caucasus there is an analogous prohibition: it is forbidden to ask a man about his wife's health. (1967b, 40)

For the matrilineal Trobrianders, to whom Lévi-Strauss alludes, it is famously the other way round, with the sibling relationship being fraught and dangerous and the marital tie comparatively relaxed (Malinowski 1929). But further reflection shows that far from violating the pattern, the Trobriand data is in fact consistent with it: sexual reproduction does not form an aspect of the husband-wife relationship in Trobriand thought. In that society, any relationship between sex and reproduction is disavowed, so that copulation in marriage is undertaken for pleasure but is not linked with reproduction (Malinowski 1929). The latter process, in this cultural system, flows through the consanguineal matrilineal relation: a woman is impregnated by a *baloma*, a spirit of her matrilineage which enters her body when she bathes in the sea; marital copulation only serves to open the channel in the woman's body for the baloma to enter. Therefore, shame and avoidance are required between siblings, who, as members of the same matrilineage and close consanguineal relatives, in contrast to the husband and wife, represent the reproductive couple in this instance. It might be observed that this fact seems to indicate that at least in many cases, what is "shameful" is not sex in itself but rather its connection with the scandal of biological reproduction. Where the latter is assigned to the sibling pair at a conceptual level, the sexual relationship between husband and wife may be far less stigmatized, indeed hardly stigmatized at all, while the tie between brother and sister through which the line of descent actually runs is viewed as highly charged and in need of strong containment via strict taboos, as is the case in the Trobriand Islands.

Affect Hunger

Given that marital relations are often (though not always) to be kept apart from other aspects of public social life, and its central act performed in relative privacy or secrecy, the question still remains of why it is that humans pursue and value social relations in the generalized public arena beyond those involved with mating. The evolutionary advantage is great, of course; cooperative action in production, defense, and other enterprises provide adaptive rewards aplenty for the individual who cooperates with others in a society. But how should it come about that people feel things like affection and long-term identification with one another? What could be the source of culture's motivational resources to overcome and control the powerful imperatives of the biological code? Walter Goldschmidt (2006) addressed this question in his last book, *The Bridge to Humanity*, written when he was well into his 90s. In that work, Goldschmidt, a cultural anthropologist who had been explicitly trying for decades to integrate cultural and biological ways of thinking, addresses himself to the precise question that also interests me here, namely, how did humanity manage the feat of transcending the evolutionary law of (inclusive) self-interest to create its distinctive and viable social life. His answer is summarized in the subtitle of his book: "How Affect Hunger Trumps the Selfish Gene."

Goldschmidt believes that learned values and behavior, that is, the symbolic world of culture, are indeed capable of overriding the imperative of the genes if followed with energy and strict adherence. But such commitment of energy requires very strong motivation on the part of actual actors to pursue cultural goals when these conflict with strong biological urges. He looks for the motivation to do so in a distinctive aspect of human life, that is, the greatly extended period of infantile immaturity, requiring the very young child to depend crucially on an adult caretaker (or plural caretakers) for years in order to survive. All mammals have such a need to a degree but in humans it reaches a great extent. The result is that human infants in order to survive must come equipped by natural selection with a robust need to seek and retain the nurturant care of an adult, who in the usual case is the mother. This in turn requires that the mother herself have an evolved powerful motivation to provide such nurture, despite the fact that this is so costly in terms of energy, effort, and attention. The resultant mother-infant dyad united in an affectionate bond then creates the ideal situation in which both socialization and enculturation can take place, the former implying the development of the ability to care for and about other individuals, the latter implying the learning of the basic cultural symbolic system of language and the wide array of instructions and information it encodes and transmits supporting prosocial behavior.

"Affect hunger," for Goldschmidt, is the residue of this earliest experience of life, in which the infant is absolutely dependent on the benevolence of the mother, and—very importantly—will do whatever is necessary to retain her nurturant care. This includes imbibing her cultural instructions and social values and, in a word, accepting what she says and obeying her wishes so as to protect the relationship. The long-term impact of this initial experience is a wish to receive approval from benevolent and meaningful others, which now becomes a motive for a goal sought for its own sake. A key developmental move in the individuals' maturational process is that the social group itself becomes the new locus of the relationship of nurture and dependence one once had with the mother, and the greatest anxiety becomes fear of the loss of approbation of the group, of falling into a state of public dishonor, disgrace, shame, guilt, or general disrepute. It is this distinctive feature of human emotional life that, for Goldschmidt, makes adherence to cultural values, even ones that demand limitations on one's own self-interest, genetic or otherwise, so compelling as to achieve the difficult task of "trumping the selfish gene":

Affect hunger is rooted in biology and emerges with culture. It ties the two together. Affect hunger does not leave the realm of biology, for its very existence plays a role in survival, first by contributing to the development of the central nervous system [since holding and other forms of maternal contact enhance neuron growth in the infant] and second by motivating us to entice the maternal care that is needed to live in a human world. Affect hunger thus brings the infant to culture by inducing it to attend to the lessons set forth on what the culture expects and what must be done to satisfy its demands. It is truly a bridge—a bridging concept that leaves both biology and culture intact. (141)

I find this formulation to be eminently plausible and consistent with what has gone before in this book. Some source of affectively generated adherence to social norms needs to be posited to account for the effectiveness of cultural symbols which, unless they mobilize such affect in actual people, will remain lifeless and inert. It is a persuasive proposal to derive the necessary affect from the bond of attachment that links mothers and infants, since it is at the same time precisely the delayed postnatal maturation of the infant that provides the matrix within which social and cultural information and values are effectively taught, experienced, and learned. And since without some effective prompting of mothers to care for their infants there would be no reproduction, it must be that humans have evolved to a high degree the mammalian pattern of nurturance, and that infants are primed by evolved processes to seek out and return the affection they are given to the extent that they can. Recognition of the mother's voice, odors, and face, mutual gazing, the smile response, distress at separation, and stranger anxiety are some of the well-known features of this bond.[2]

Goldschmidt is therefore right to point to the idea of "attachment" as it was originally formulated by John Bowlby (1983) and has been developed by any number of subsequent researchers. For Bowlby, who was deeply influenced by the field of ethology just coming into prominence when he was developing his own theories, attachment would have to be an evolved response on the part of infants to seek out and "attach" to a protective adult. This need to be near such an adult, and in time of potential threat to cling to it for protection and reassurance, would, Bowlby suggested—following Harlow's famous experiments with artificial terry-cloth rhesus monkey mothers who provided no milk versus uncomfortable wire mothers who provided milk—trump even the need for nourishment. To account for this evolutionarily one may reason that protection from predators in what he termed the "environment of evolutionary adaptation" would be an even higher priority than feeding, since it poses an immediate threat to survival whereas the latter can be to a certain extent postponed without fatal results.[3]

Given these various considerations, then, it makes sense to suppose that the developmental pathway marked out in many societies for the maturing child entails the reallocation of the original affectionate feelings away from the mother and toward new objects, including group mates and at a more abstract level to the society and its subcomponents (clan, lineage, totemic group, secret society, and so on). What this means is that the symbolic system, in order to allow this reallocation to take place, must draw its power from a source that is strong enough to overcome hostility rooted in rivalry for selfish goals, of which the most salient is the pursuit of mating opportunities, and replace it with prosocial affection. That source would appear to be emotions intrinsic to the human mother-infant bond. Does ethnography support this view? Without being able to give a definitive answer, I can offer some examples that would tend to support it.

Kaingang

Since divining inner states, including affective experiences, is not easily accomplished, I will illustrate the developmental process I have suggested with examples that lend themselves more easily to direct observation. This observation builds on the assumption that extensive holding and direct body contact is a key dimension of early mother-child interaction and the creation of an attachment bond. It was apparently the irresistible soft warmth of terry-cloth in Harlow's experiments, to which the infant monkeys clung despite receiving no milk, that brought home the point to Bowlby in the first place. In some societies children are expected to forego the pleasure of regular body contact at an early age, but the Kaingang, foragers of the southern Brazil highlands, follow a different path (Henry 1941). Once weaned from being held by their mothers, children in this society become the constant source of warm affection of a very direct physical kind from any adult in the group, and vice versa: "They [little children] are at the beck and call of anyone who wants a warm little body to caress. As Monya, aged two, wobbles on fat, uncertain legs Kanyahe calls to him. He slowly overcomes the momentum of his walk, turns about, smiles and wobbles obediently over to Kanyahe. Children lie like cats absorbing the delicious stroking of adults. . . . They like to cuddle next to an uncle, aunt, or step-mother" (18). Sooner or later, this sensuous cuddling takes a more overtly sexual turn, and most children are at some point initiated into sexual life by someone considerably older. Jules Henry observes that despite the ubiquity of public talk about sex, and the relative lack of shame about it, he never observed directly sexual play between children in this society, a fact he derives from the satiation they receive via regular affectionate body contact.

This pattern of interpersonal relations does not cease with childhood once sexual life has begun. On the contrary, it comes to play a critical role in the cohesion of Kaingang social life:

Kaingang young men love to sleep together. At night they call to one another, "Come and lie down here with me, with *me!*" . . . In camp one sees young men caressing. Married and unmarried young men lie cheek by jowl, arms around one another, legs slung across bodies, for all the world like lovers in our own society. . . . But I never saw them make an overt sexual gesture. (18)

From this phenomenon Henry derives an important conclusion:

The basis for man's loyalty to man has roots in the many warm contacts between them. The violent, annihilating conflicts among men in Kaingang society were all among those who had never shared the languid exchange of caresses on a hot afternoon under the green arched shelter of a house nor lain together night after night under a blanket against the cold. The very transient, unfixated nature of these contacts leaves no ground for jealousy. The relationships built up on these hours of lying together with anyone at all bear fruit in the softening of conflicts that are so characteristic of the Kaingang. . . . Men who have hunted together day after day, raided the Brazilians together, slept together beside the same fire, under the same blanket, wrapped in each other's arms, hold this relationship above their kinship with their brothers. (19)

Here then, in a small foraging society with minimal organization and without any formalized men's society or male initiation, one can see quite plainly the forging of solidarity among unrelated, or relatively distantly related, males through the employment beginning in childhood and into adult life of the experience of close affectionate body contact that begins in the mother-child relation and then refinds itself in the complex "body" of the group, bringing to it the feelings of unity, identification, safety, and generalized connection that are required to overcome within the group the violent aggression that the men are so very capable of enacting on others outside of it.

Ache

The Ache (Guayaki) of Paraguay were, before settlement on a reservation, hunters and gatherers who at some earlier point in history, like the Kaingang, had been horticulturalists, but were presumably driven off their traditional lands by rival groups and took to living in the forest in small bands as foragers (Clastres 1998). Being nomadic, it was essential to their style of life that small children had to be regularly carried often and for long distances. Weaning was relatively late and body contact with a caregiver thus a constant feature of everyday life. But unlike the Kaingang, adult Ache, especially men, did not tolerate physical contact. If one wished to hand an object to another person, one put it on the ground in front of him and let him pick it up, rather than risk touching him in the act of handing it over. These were, as we say, "touchy" people, quick to reach for their arrows, and personal distance was a virtue and a wise policy.

There were, however, exceptions to this pattern. One was the application of a complete body rub to assuage dangerous male rage. Clastres (1998) gives a good example of this stemming from the increasing encroachment of Paraguayan society ever deeper into traditional Ache hunting territory. The worst aspect of this was that the *beeru* (whites) made a practice of stealing Ache children, who fetched a price on the market for sale as servants and slaves. This, needless to say, saddened the women and infuriated the men, who were unable to put a stop to the process:

> When the men suffered too much, they became *yma chi ja*, violent forces—then they would start fighting, they would want to strike out at everyone. The children would run away and the women would yell: "They want to shoot their arrows! Don't shoot! Do-on't sh-oot!" And they would bravely rush at the men, who were getting excited and clashing arrows; the women would put their bodies in the way of the men's arms as they were about to release the string of the bow. The men allowed themselves to be calmed fairly easily; they had only been half angry, and a good *piy,* a caressing massage over their whole bodies, would be enough to soothe them, though sometimes a woman would be sent rolling over the ground by an enraged punch to the head. (65–66)

The men who, when adequately provoked, have a tendency to enter a state in which they are capable of indiscriminate violence within the group itself, can be restored to relative calm by the application of direct and well-timed complete bodily stimulation and contact—in effect returning them to the state of being comforted by being picked up and held by mother as they once were when throwing a tantrum at being forced to set out on their own.

This interpretation is supported by an observation by Hill and Hurtado (1996), who also worked among the Ache. When children who are between the ages of three and five and still cannot walk well enough to keep up with the adults protest being put down on the forest floor by their mothers, they are carried by other adults. But at five or so they are definitively weaned from being carried. When this occurs, "children scream, cry, hit their parents, and try everything they can think of to get the adults to continue carrying them. Often, they simply sit and refuse to walk. . . . This leads to a dangerous game of

CHAPTER TEN

'chicken' in which parents and children both hope the other will give in before the child is too far behind and may become lost" (222). I think it requires no great leap of imagination to see in the women's placating the enraged men with a full body massage an adult version of the childhood experience of being held after throwing a tantrum, which the men accept, albeit with the possibility of an accompanying resistant final expression of protest in the form of a punch to the head—a grown-up version of the child's hitting its parents in protest at being made to walk instead of being held.

Another instance in which the usual avoidance of physical touch between men is abrogated occurs during the annual great festival, the *kybairu*, in which, during the season when honey is gathered, scattered bands reunite and enjoy seeing each other after a long separation:

But they especially enjoy seeing the women, the daughters and sisters of the *cheygi* [people of other bands]. They miss women; and the game they play not only celebrates the festival of honey, but also the festival of love. It is a chance to abandon themselves to the joys of *pravo*—of seducing women. . . . Festival of honey and court of love. (Clastres 1998, 221)

The men of the different bands first enact a mock combat and subsequent reconciliation: they are now at peace and it is not necessary for the men to try to steal the other group's women (the usual purpose of raiding between bands under other circumstances), since these will be exchanged freely during the festival. The men from different bands then pair up and sit down together. "This is the only time the Atchei [Ache] allow physical contact, which is strictly forbidden in everyday relation. At this moment, however, it is even encouraged" (226). What then ensues among the men is a remarkable negotiation of the relative aggression and/or harmony that is to prevail during the festival. Normally, physical contact carries with it the implication of an act of aggression. But,

if you want to deny what seems to be an aggressive appearance, what better way is there than to accept what would ordinarily be interpreted as an act of hostility—bodily contact? If men welcome this now, it means they are not really enemies. So they enter into a ceremonial game, the necessary prelude to all phases of the ritual, the *kyvai*: the tickling. (226)

The aim of this game is to get the other man to burst out laughing by tickling his armpits and sides; he inevitably does start laughing hysterically after holding out stoically as long as he can. The tickler cries out "We are doing it so that we will laugh, so that we will be happy. . . . We are friends" (227). As Clastres says, by allowing men to touch each other, "the true function of the *kyvai* becomes clear: to establish or strengthen friendship between the two men."

The climax of the festival is a game called *proaa*, the object of which is to grab a bean pod held by someone under the armpit or in the fist. This is done by everyone—adult men and women both—tickling the one with the pod until that person gives it up to someone else, at which point everyone tries to get it away from the new holder of the pod, and so on until everyone has held the bean pod. As each one tries to hold out,

he is tickled by dozens of hands at once. He rolls over the ground, people fall on him, he tries to escape, they catch him. Piles of people collapse on top of one another. It is a wild scuffle, the women crying out sharply and the men grunting softly. . . . This is when the men and women choose each other. If they are married they can find an extramarital partner; if they are single, they can declare their love and eventually get married, if the passion lasts. (230)

The *to kybairu* fulfills everyone's secret expectations and answers the sacred call of the joy of living—it is the Festival. (231)

As Clastres's description suggests, tickling is midway between an attack and a caress: it is aggressive and "defeats" the other, but in a spirit of play and without any real harm done. In this way, it acts as a perfect mediation of aggression and affectionate sociality, leading to the wider indulgence of deregulated eroticism that the festival aims to accomplish: aggression is expressed in the mock battles and in the tickling itself, but at the same time is overcome by laughter that begins between pairs of men and eventually encompasses the entire group. There is a temporary armistice in which touching and pleasurable body contact not experienced regularly after childhood is allowed and exploited to reestablish the Durkheimian good feelings of collective effervescence for each within the whole group that such gatherings are intended to engender.

As this example illustrates, when unconstrained expression of affection through bodily contact is carried on in a mixed-sex group, it will as often as not lead to more overtly sexual relationships, since the stimulation of the skin can be either sensuously soothing or erotically arousing, or some combination of both. Thus the Kaingang practice of close physical content is confined to males among themselves, and apparently does not lead to sex involving the genitals, whereas the Ache tickling ceremony takes place in a mixed-sex group with the express purpose of leading to extramarital liaisons as part of a temporary suspension of normal social time in favor of the general enjoyment of the permitted promiscuity of the Festival.

Initiation and Childbirth Symbolism

It follows from the argument thus far that groups would have ways to mobilize the motivations arising from positive experiences originating in the infantile dependency of the mother/child nurturing dyad and, through symbolic chains of association, redirecting them toward the group itself, which now can take on the role of the protecting and loving encompassing "container" that was once taken by the mother. The symbolism of rebirth achieves this by putting a developing group member in the position of infantile dependency and repeating the process of cultural instruction that occurred in the original maternal surround, when the infant absorbed language and early social and cultural understandings in the context of that protective environment and in a state of dependency. "Initiation" is the term of art given to those formalized rituals or procedures that do this work to create socialized persons, and it does in fact recreate, with the wider group now replacing the mother, the earliest scenes of attachment and instruction.

Not every society marks one or more of the stages of development through life with a ritual of passage or initiation, of course, but quite a remarkable number of them do so. We may understand this in broad terms as the overlaying or reinterpretation of the biological processes of birth, growth, and maturation within the realm of the symbolic, like the ritual of baptism and compadrazgo which I described earlier. In her survey of initiation rituals in a range of ethnographic contexts, J. S. La Fontaine (1986) notes that almost every instance she presents makes use of symbolism that alludes more or less explicitly to childbirth. Indeed it has been a truism of anthropology since van Gennep ([1909] 1960) that rites of passage take the form of a symbolic death and rebirth, in which the initiand moves from one social status to a new one by "dying" to the former and being "reborn" in the new one after passing through a liminal period of transition. This is true for society-wide rituals in which children advance to new age grades or statuses but also for individual initiation into secret societies, religious cults, healing sodalities, and other such partial groups within a wider social system. In the context of our present discussion, emphasis should be placed on the fact not just that the initiand is reborn but that he or she becomes the equivalent of an infant in a dependent relation to the group as mother, and responds by attaching and by absorbing the lessons that the group seeks to impart.

Thus, to cite examples discussed by La Fontaine from primary texts, in initiation into a Chinese Triad society, the slang term for which is "to be born," the candidate has a sponsor called the Mother's Brother, while the senior officiant is called the Father; in the Cantonese version, this individual is called the Mother (48). The initiand must crawl through the legs of this person to reach the inner sanctum in an explicit miming of birth. In Masonic initiation, the novice is blindfolded and must be pushed or led; he is helpless like a baby and is made to feel so; in this and the Triad case, "both rituals represent the initiate as a new-born child" (50, 51). These are examples of "secret societies" within more complex urbanized social groups, but the same symbolism happens in non-modern contexts as well.

The Nyoro of Uganda have elaborate spirit possession cults in which women afflicted by spirits are initiated into the cult and themselves become mediums. In the ritual the senior medium is called "great mother" or "grandmother," and others are addressed as "mother." "There seem to have been local variations, but in every case the initiand is placed in the dependent position of a young child, which is demonstrated by the medium's feeding or even suckling the 'child'" (La Fontaine 1986, 64). Later, one of the mediums, an older woman,

> stands with her legs wide apart while the initiand lies on the ground behind her. Another medium, the 'midwife,' seizes the initiate through the legs of her colleague and, helped by others, pulls her through the legs while the 'mother' groans like a woman in labor. The initiate must then cry like a new-born baby, is placed on her 'mother's' lap and must put her mouth to the breast. As in the situation of birth, the 'mother' then ritually curses the 'baby' and immediately retracts the curse, a magical action which was believed to ward off evil. (66–67)

In the course of the Hopi Indian Tribal male initiation,

> the symbolism of birth is perhaps more pronounced [than in the preliminary initiation] in the Tribal initiation, for not only do the initiands have to adopt the same cramped posture they took up in the preliminary initiation, but they must maintain it for four days, during which time they act like fledgling birds, are fed by their sponsors and must use bird-like calls to communicate their wants. Moreover, as a

CHAPTER TEN

preliminary to the joint rites they are carried out of the *kiva* like babies, on the backs of their sponsors. (90)

In the course of the ritual they grow from infants to maturity, and in the final rites they are ritually "killed," so as to be brought back to life as men.

The West African Poro initiation, like the Thoma of the Sherbro which is a variation on it, revolves centrally around the visit of the main Poro spirit, Gbende, to the world of men. The symbolism of the ritual involves the initiates' being swallowed by Gbende, who then gives birth to them; the scarification they have received is supposed to be a sign of their having been eaten and reborn. At the high point of the ceremony, after a prolonged and severe period of seclusion, the initiates return to the village:

On the third visit, Bende and the initiands enter the house in the village where Poro meetings take place and stay there. The women and children emerge and join in the dancing which lasts till dawn when they retire again. The 'birth' of the initiates is then mimed and the ritual party leaves. (98)

Having completed the ritual the initiates join the ranks of the *hinga*, "those who procreate," that is, future heads of families.

La Fontaine concludes that

a common theme can be found in all these rituals: a reference to sexuality and birth. It appears in the rites of secret societies as it does in those of other transitions, such as funerals, as well as in initiations where it might seem most appropriate, since they are overtly concerned with parenthood. Where a direct reference to sexuality seems out of keeping with the occasion, it is common to find interpretations which assert the Van Gennepian ideas of rebirth. . . . Yet this hardly explains why sexual symbols are so ubiquitous; are they just 'natural symbols,' material available? (189)

Good question! For La Fontaine, procreative symbolism invokes the fundamental ability to generate novelty and to bring about change. Since rituals are about changing people from one status to a new one, social processes therefore "harness" sexual ones to accomplish the desired transformation. This I find to be a rather vanilla-flavored interpretation. More convincing would be an interpretation along the lines of Victor Turner's great contribution (1967) to the interpretation of ritual symbolism. For him, ritual symbols have both an affective or "orectic" pole, rooted in reproductive processes, as well as a social pole, and thus they can act as transformers, able to endow social concepts such as "matriliny" with the affective force borrowed from reactions to biological phenomena such as milk, menstrual blood, and so on; while at the same time the social pole of symbolism channels and gives socially constructive meaning to otherwise inchoate responses to innate biological processes and experiences. Thus for example the white sap of the *mudyi* tree used in rituals celebrated among the matrilineal Ndembu, conceptually linked with mother's milk, serves to give experiential force to the ideas of matrilineal clan unity by generating in those experiencing the ritual powerful (unconscious) memories of being nursed by a real mother, now symbolically extended to the matrilineal kin group.

This example ties the discussion back to the considerations I raised earlier: childhood symbolism evokes the idea of profound affective ties, of which the first is that between mother and child, entailing a mix of protection, safety, sensuous and erotic pleasure, and pleasurable interaction with and the benign approval of meaningful others. The key, from the point of view of dual inheritance, is that what is first experienced individually, in a dyad, and on the basis of mainly embodied experiences is, in ritual, re-created now as something symbolic instead of physical, that can be experienced by the whole of the relevant community rather than just in a dyad. The experience becomes public and available to simultaneous perception by all participants, so that the space first of the womb, and then of the scene of nurture on the mother's lap and in the heart of the domestic space, is further symbolically transformed into the public arena into which the initiand enters as if being born into it, not by sexual means but through cultural processes that ritually transform the initiand and then instruct him or her with the rules that apply in the public arena. Yet in order to accomplish this, the cultural system is required to evoke the very sexual and reproductive images, experiences, and affects it is attempting to transcend, because only in this way can it command the level of motivation and investment of attention, affect, and interest necessary to render the symbolic system experientially effective.[4]

Back to Dual Inheritance

We can elevate this to a more general observation. Symbols, of which we may take linguistic signs as exemplars, are themselves without any force or agency; they are, in the simplest way of putting it, patterns inscribed in a medium such as sound or light waves or some more solid material. By themselves they can do nothing. Humans, like other living organisms, by contrast, are endowed with the ability to generate action and are motivated to do so by powerful evolved physiological processes experienced as wishes, urges, compulsions, fears, anxieties, and so on, which have directive force. The genetic information system, inscribed in the DNA, is, like language, a registration of sameness and difference in a molecular medium that, unless it is put into action in the realm of the phenotype, is, again like language, inert. The organism alone can act, but it does so first of all under the guidance of the instructions in the genetic code that enable it to construct itself as an organism and that then endow it with neuronal, chemical, psychological, and other systems that provide motivations and directions for its actions.

DNA, without the cell in which it is housed, is manifestly incapable of doing anything by itself without conveying its instructions to the cells and thus to the whole organism which alone has the capacity for spontaneous action. (The organism can also, to a degree greater than was previously appreciated, influence the expression of the DNA depending on information from the environment, just as DNA can issue instructions to it.) For most organisms, this dialectic between gene and phenotype-in-its-environment can result in an adequate account of the organism's behavior. It is the premise of a dual inheritance model, however, that in humans there is a second source of instructions, exemplified by the external symbolic system, which is in many striking ways analogous to the genetic system except that it is carried outside the organism in differences inscribed in the perceptual field rather than inside the organism in differences inscribed in the "text" of a long molecule.

In order to shape the motivations and actions of the organism to its own ends, what resources does the symbolic system have at its disposal to influence, manipulate, direct, and otherwise control the actions of the organism? If the symbolic system is just patterns inscribed in matter that act as vehicles for meaning, where exactly is the "meaning" located and how does it get there? Stated in this way, the answer is, I believe, quite obvious. The symbol system can only make use of what is already there in the organism, provided by its self-construction under the aegis of the DNA, but in order to do so it must redirect the actions of the organism from those provided by the imperatives of the genetic program and orient them toward the performance of its own priorities which are in many key areas at odds with or even opposed to those of the genetic program.[5]

To effect this redirection, one of the most effective strategies the symbolic system has at its disposal is to make use of the strong imperatives of the genetic system and, by "symbolizing" them, that is, by putting them in a public, extrinsic, recognizable form available to collective perception, draw to themselves the energies that would otherwise go into the actual acts that the symbols "symbolize." It therefore follows that, to successfully perform its role, the symbol system must often and to a considerable extent clothe itself in symbolic evocations of the very sexual and procreative program that it is in the business of counteracting or overcoming.

Genital Surgery

So far I have been discussing the ways in which a society and its symbol system elicit adherence to values in opposition to urges rooted in fundamental biological processes by masquerading in the trappings of these very processes themselves and thus inducing interest in and (positive) affect toward the symbols, which thus enlivens them and allows them to do their work of informing and shaping the society and the actors who compose it. But of course in addition to such "carrots" inspiring positive motivation, society also has at its disposal the "stick" of negative sanctions for failure to behave in accord with the norms, values, and ethics inscribed in the symbols that constitute the social system. Given the inevitably superior power of the collective over that of any given individual, societies can threaten miscreant members with physical punishments such as the infliction of pain or death; the payment of money or valuables; banishment or abandonment by the group; and perhaps most pervasive and ubiquitous, some form of disrepute in the public eye, manifested as shame, dishonor, loss of face, and in general being made the object of negative gossip and disapproving gazes. The aboriginal Tasmanians, it is said, punished a miscreant by making him stand still while everyone else in the group pointed their fingers at him. In this section I want to concentrate on one such threat, that of an attack on the genital organs.

The reason for singling out this particular sanction, which may not on the face of it seem to all readers like such a real threat, is that it is represented, in quite stark symbolic form, in one of the most widely practiced of ritual actions, namely genital surgery, or, as it is sometimes referred to, "circumcision." La Fontaine (1986) writes concerning this designation:

> Circumcision is so widely distributed as a feature of rituals of maturity for boys that the term 'circumcision rituals' may be used as a synonym for them. The comparable operation for girls is excision of the clitoris, clitoridectomy, which is sometimes called 'female circumcision.' The use of the terms is somewhat misleading, though. Not all operations on the genitals are circumcisions or clitoridectomies; various operations, some more severe and others less so, are reported from different parts of the world. (111)

Why, we may ask ourselves, do people in so many different societies around the world think it is important not only to routinely inflict pain and leave permanent alterations on the bodies of their children, but to do so in the most sensitive and psychologically fraught parts of the body, precisely those which serve the purposes of biological procreation?

Societies that perform these ritual surgeries give various reasons for doing so. Sometimes, for example, it is claimed that it needs to be done to remove a residue of the opposite sex from the sex organs— the female-like foreskin from the boy, the male-like clitoris from the female. It is sometimes claimed that it is done to enhance later sexual pleasure, but perhaps more often it is said to be intended to inhibit what would otherwise be a dangerously disruptive craving for sexual gratification. This would be consistent because I have argued throughout this book that the imposition of control, constraint, and inhibition on sexual and reproductive action is a key prerequisite for long-term social stability in humans. A surgical "attack" on the genitals, and the resultant scars and supposed reduced capacity for sexual sensation, seem well calculated to drive home to the developing person the idea that the entry into socialized adulthood requires such control of the genitals and their urges, with an additional implied threat of what the punishment might be for disobedience to this dictum.

A lively argument for the positive effects of the inhibiting and socializing effect of circumcision is to be found in a quotation attributed to a late-thirteenth-century rabbi from southern France, Isaac ben Yedaiah. According to the sage, circumcision of the man reduces the pleasure that a woman takes in copulation. Therefore the uncircumcised man, thanks to the unmitigated pleasure he gives to his partner, finds himself utterly distracted by her demands for constant lovemaking:

And so he acts night after night. The sexual activity emaciates him of his bodily fat, and afflicts his flesh, and he devotes his brain entirely to women, an evil thing. . . . He is unable to see the King's [God's] face, because the eyes of the intellect are plastered over by women so that they cannot now see light. (Saperstein 1980, 97)

The circumcised man, by contrast, try as he might to satisfy his partner, cannot do so because, thanks to his lack of a foreskin, "he will find himself performing his task quickly, emitting his seed as soon as he inserts his crown . . . as soon as he begins intercourse, he immediately comes to a climax." This is however, a good thing, despite the frustration of the circumcised man's wife (or lover), because "he who says 'I am the Lord's' will not empty his brain because of his wife or the wife of his friend. He will find grace and good favor; his heart will be strong to seek out God" (97–98). Reduced sexual capacity and enjoyment thanks to genital surgery is thus seen to turn a man's attention away from sex and toward the higher values of religion. By a similar logic, clitoridectomy in some Middle Eastern societies is sometimes explained on the grounds that it lessens the likelihood that a woman, once married, will be sexually promiscuous and thus bring disharmony and shame into the community by setting men against each other.

Circumcision of men may be thought of literally as a covenant, as it is in Judaism, perhaps with the Almighty as it is said in Jewish thought, but surely with other men: it is a sign of sexual restraint that proclaims to the community that one may be trusted as a neighbor not to covet one's neighbor's wife. It is the mark that the man has survived a symbolically castrating attack and emerged as a mature person capable of managing his sexual (procreative) desires in a way consonant with the dictates of the Law and of social harmony. In women, circumcision likewise signals modesty and sexual constraint consistent with the social pact. Circumcision thus represents in stark observable and experienced form the supersession of the cultural system over the sexual one, and so serves the creation of adherence to the tribal society.

Nuer Ox Sacrifice

The Nuer, cattle pastoralists of the Sudan made immortal for anthropology by E. E. Evans-Pritchard (1940), do not circumcise young men during initiation, but instead they scarify their foreheads—also a painful test of endurance though to be sure not a direct attack on the genitals. In addition, however, they do something else very closely related

CHAPTER TEN

symbolically to genital surgery, that is, they give a young man his first ox, castrated with his personal spear. This ox becomes an alter ego for the young man, who lavishes care and decoration on it, and from then on this ox's name is his own personal ox-name, which he calls out, for example, when he spears an enemy or kills a game animal or fish. If circumcision amounts to a partial castration from which one emerges strengthened in moral fortitude, the real castration of a recognized double of the self can surely be understood as having very similar symbolic meaning.

An insightful analysis of Nuer ox symbolism by T. O. Beidelman (1966) augments Evans-Pritchard's own by directly confronting the sexual dimensions of Nuer culture that Evans-Pritchard discusses only indirectly. Beidelman addresses himself to the question of why it is specifically an ox, rather than an intact bull, or a cow, that is the preferred animal for Nuer sacrificial ritual. He argues that since the object of sacrifice is to mediate between the material and the spiritual realms, the ox, being itself "betwixt and between" as a neutered, asexual animal, serves the purpose well. But his analysis goes well beyond this initial reasonable but rather generic observation.

Following Evans-Pritchard, Beidelman identifies a moral dualism in Nuer thought, understood as a contrast between the auspicious and moral right side and the left side associated with weakness and misfortune. Adult men embody both of these attributes, having both an ideal moral self and a flawed personal one. The contrast is particularly sharp between the agnatic lineage relations, within which ritual and political action occurs, and the ordinary kin relations of daily life within the immediate family. The goal of a man is to transform his "hectic machinations in the field of domestic kinship" into permanent agnatic statuses (Beidelman 1966, 456). Thus, individual men struggle to increase their domestic groups via polygyny, to acquire cattle and sons, to found settlements, and so, ultimately, to become ancestors of structural lineage segments. This struggle clearly requires the maximization of reproductive fitness (though Evans-Pritchard does not put it that way): "A man's memorial is not in some monument but in his sons. . . . This is the only form of immortality Nuer are interested in. They are not interested in the survival of the individual as a ghost, but in the survival of the social personality in the name [immortalized by descendents]" (Evans-Pritchard, cited in Beidelman 1966, 456). From the adult man's point of view, women occupy an ambivalent position in this social system. They are necessary for the production of the desired sons who will provide advantage in the skirmishes of day-to-day political life, and so ultimately establish one as a lineage ancestor; but, because of polygyny, they are divisive, insofar as sons of different mothers form interest blocs in opposition to their half-brothers and threaten the unity of the agnatic group. Thus, it is "by marriage and sexual union that the most powerful divisions in Nuer society threaten to occur" (459). In other words, the pursuit of reproductive advantage and the requirement that this be achieved through phallic sexual prowess (represented symbolically, as Beidelman makes clear, by the spear) in the impregnation of multiple women, stands in opposition to the ideal moral and political unity of the agnatic male group, whose cohesion it threatens to disrupt. All this should by now seem quite familiar and understandable to the reader in light of the theses propounded in this book.

The bull in Nuer thought embodies maleness and vitality; the word for bull (*tut*) is related to the word *tute*, meaning "to impregnate." The bull is a metaphorical stand-in for the dimension of a man's character that is aggressive, self-serving, and self-aggrandizing, and as such ambiguously both admirable and problematic. The violence of bulls is equated with a man's bravery, and fighting among bulls is prized. The ox, by contrast, is a bull that, by means of castration, has been made to sacrifice its reproductive self-interest in favor of the asexual and uncontentious symbolic maintenance of the enduring moral order of society: "In initiation of cattle, a bull's sexuality and associated bellicosity have been subordinated in terms defined by society; in the initiation of men, the immature, non-jural youth is transformed into a jural man whose sexuality and desires are ideally subordinated to the ranking and discrimination of the moral order" (462). Beidelman's formulation could not be improved upon as a description of the requirements of dual inheritance *avant la lettre:*

> Sexuality embodies the ambiguity of human actions, their vitality, but also their tendency toward morbid conflict. This is not important in the case of most women, who are outside the jural, moral sphere, but it is crucial in defining the moral conduct of men. Removal or isolation of sexuality achieves . . . a temporary ordering and clear expression of the moral dichotomy by which social ideals are framed. Desexualization may be seen as providing medial or synthetic objects between the ideal and actual spheres of existence. Such objects serve as basic models of the ideal ordering of a world subordinated to the moral concepts envisioned within a society. (462)

The ox, then, as a male being whose phallic sexuality and its attendant propensity for violence and conflict have been removed to make it trac-

CHAPTER TEN

table, is thus the ideal animal for Nuer sacrifice, a practice that serves the cohesion of society as a whole above and beyond the interests of individual men. At this point in his analysis, Beidelman cites a passage in Freud's essay *Group Psychology and the Analysis of the Ego* (1921) to underscore his own theoretical conclusion:

Those sexual instincts which are inhibited in their aims have a great functional advantage over those which are uninhibited. Since they are not capable of full satisfaction, they are especially adapted to create permanent ties; while those instincts which are directly sexual incur a loss of energy each time they are satisfied, and must be renewed by a fresh accumulation of sexual libido, so that meanwhile the objects may have changed. (Cited in Beidelman 1966, 263–64)

The opposition of the bull and the ox, then, in Nuer symbolism, as Beidelman analyses it, implies the opposition of uninhibited sexuality versus aim-inhibited sexuality; greed and uterine divisions versus social constraint and sharing; real domestic groups and attachments versus ideal solidary (male) groups such as age-sets or agnatic society in general; divisive, capricious, and transient relations versus corporate, ordered, enduring relations; and bull reputations and reference to offspring versus ox-names and spear-names referring to clans—that is, ideal corporate groups as opposed to actual biological sons. All these oppositions reflect the distinctions implicit in dual inheritance.

There is a further important point in Beidelman's analysis that links the Nuer example directly to my analysis of the consequences of dual inheritance, namely the association of the overarching ideal social order, in this case identified with men, with a world beyond the mundane one of ordinary social relations, tainted as these are by the correlates of sexual and genetic competition. This other world, in being contrasted with the world of actual organic life, is an immaterial world associated with death:

The spear, as a socialised penis, represents the aggressive, superordinating aspects of the penis, the subjection of women to the demands of the agnatic group. Furthermore, it is the vehicle of death and it is only through the transformation of death that ideal aspects of agnatic values are attained. This appears both as in the common solidarities expressed in terms of warfare and feud . . . and in the transformation of cognatic (*mar*) to agnatic (*buth*) relations, memorialized in ox-names and spear-names of oneself as a pivotal ancestor in a segmentary political framework. (464)

In other words, not only is male solidarity associated with the dealing out of death in battle, but also the culmination of a man's career, if he is fortunate, is to become an ancestor, which requires that he himself be dead. Indeed the entire ritual system resting in the hands of men, which maintains the social order, revolves around the sacrifice to and divination concerning an immaterial spiritual realm linked with death that is opposed to ordinary embodied life.

Beidelman concludes:

> In the spearing of an ox a Nuer expresses a kind of transfiguration, through immolation, of his sexual self and an anticipation of his own living transformation, through death, into the agnatic ideal person which his own living, domestic, self cannot wholly be and, indeed, cannot wholly accept. . . . It is their recognition of their deep commitment to domestic values which conflict with these ideals that leads them to focus ritual upon the ambiguous tensions these cross-cutting values produce. The values of agnation are more inclusive and enduring than domesticity, but Nuer seem to recognize that these are less compelling in everyday life. (465)

This formulation of the conflict between egoistic goals and strategies favoring reproductive advantage, on the one hand, and a value system that tries to transcend these in favor of a unified but asexual male social bond oriented toward a spiritual, immaterial realm associated with death and violence as opposed to birth into organic life, is precisely what is implied by dual inheritance.

From Animal to Human

In seeking the "origins" of human sociality it has not been my intention to tell an evolutionary story about how it happened in hominin history; what I have tried to do instead is to show how, building on an innate propensity for prosocial behavior, a biological factor might, through transmutations worked upon it by a cultural symbol system, contribute to the means by which prosocial amity and cooperation come to prevail in spite of the threat posed to social life by the competition inherent in the quest of the genetic program for reproductive success. Following Bowlby and Goldschmidt, I have proposed the attachment system as such a factor: the premature birth of the human infant, and its prolonged period of dependency on its mother, necessitate bonds of need and affection between mother and child that can

CHAPTER TEN

in the course of development become a source for later affiliation to larger groups of non-relatives. I have shown how through the recreation of this situation in initiation, in which an initiand symbolically becomes like an infant, while the group itself takes on the role of nurturing parent who also demands obedience to social regulations that limit the expression of sexuality and aggression, the feelings originally evolved to ensure the care and survival of a particularly helpless mammalian infant can be mobilized on the part of both the initiand and the initiating group to produce attachment to and identification with the group itself and its individual members.

I have also shown how painful procedures, and especially genital surgery, can reinforce the social cohesion built out of early attachment, by indicating through ritual and the resulting fear of punishment that overt sexuality and the aggressive competition that goes with it must be subordinated to the regulation and control of the asexual tribal society.

The power of cultural symbols resides not in the material vehicles of the symbols themselves but in their ability to evoke emotions and motivations originally evolved for biological purposes and to reframe and redirect them in such a way that, diverted from their original aims, these emotions and motivations can now support the social system. To do so, the symbols have to simultaneously represent and call upon affects of attachment, sexual desire, aggression, and fear, while at the same time rearranging and ordering these in such a way that they serve social rather than purely individual biological purposes. As I mentioned briefly earlier, Victor Turner's analysis of ritual symbolism captures this process well, so let me quote him here:

> It is possible to further conceptualize the energetic meaning of dominant symbols in polar terms. At one pole cluster a set of referents of a grossly physiological character. At the other pole cluster a set of referents to moral norms and principles governing the social structure. . . . An exchange of qualities may take place in the psyches of the participants under the stimulating circumstances of the ritual performance, between orectic [sensory, desiring] and normative poles; the former, through its association with the latter, becomes purged of its infantile and regressive character, while the normative pole becomes charged with the pleasurable affect associated with the breast-feeding situation. (1967, 54–55)

Those symbols, within the vast system of symbols that composes a socio-cultural system, that are able to effect this transformation are regarded as sacred; they are the means by which a society is created from a group of individuals and sustained and reproduced over time by a process as apparently mysterious as it is critical. It is thus comprehensible that the symbolism surrounding the men's society found in so many forms and variants in the ethnographic record, as I have illustrated in chapters 7 and 8, so often invokes the maternal function of the cult itself and of its inner symbolic mysteries. It is not just that men want to have the ability to procreate, but that the cultural system, so often identified with the men, recognizes the crucial role that their cult and it ritual symbolism plays in making them able to unite in bonds of mutual identification and friendship, recreating for them the experience of being held by a protective mother. In societies in which women occupy statuses with influence and prestige, such as Nazaré, Okinawa, or Minangkabau, on the other hand, the transfer of allegiance from mothers to a male group need not be made.

Finally, I have argued that the commitment of socialized adults to the community and its members can be traced back to the infantile dependent bond with the mother, as it goes through cultural transformations in the course of growth and development. But I would also suggest that by another route, the same early experience of affection, protection, and bonding is what makes possible the institution of marriage, the relatively enduring establishment of a coupled relationship that includes sexual reproduction. I have been stressing the fact that marriage is the usual social institution whereby sexual procreation is cordoned off from the wider society. But marriage is not just a sexual relationship: we know from our friends the chimpanzees and bonobos that sexual consortships do not necessitate long-term self-perpetuating bonded social units. And we just saw that Beidelman argued that it is aim-inhibited (affectionate) rather than uninhibited (overtly sexual) ties that make for long-lasting connections between and among people. So, just as society beyond the mating pair channels and controls sexuality by confining it (more or less rigidly) to marriage in the vast majority of human societies, so too marriage itself manages sexuality within itself by embedding legitimized copulation within a relationship that also draws upon and recreates in relation to the spouse the original relationship to the mother, in which affection, dependency, and close mutual bonding form the normative "good-enough" experience. In those societies in which the marital bond does not have such an affective dimension, it is often assigned instead to the sibling pair, which again recreates the first dyadic relationship of identification, affection, and loyal commitment.

Thus there is an evolutionarily plausible story about the origins of

CHAPTER TEN

both sociality at the tribal level as well as marriage (as opposed to mere mating) at the procreative level, through the symbolic exploitation of the pre-adaptation of the general mammalian pattern of maternal nurture. In the human case, this period of immature dependency is used to provide both the affective source of later sociality and the opportunity to inculcate the distinctively human second channel of coded information, the symbol system that has the power to use the affects acquired in the first period of life to create the social glue, already prepared for by innate prosociality, that makes possible both married couples and the larger tribal world in which they are embedded.

Conclusion: The Giant Yams of Pohnpei

This book is about some of the consequences of the fact of dual inheritance for the distinctive forms of human society that are to be encountered in the ethnographic record. In contrast to those approaching the issue from the evolutionary side, however, my approach has come from the socio-cultural side, in an effort to advance and enrich socio-cultural anthropology by recognizing the two kinds of inheritance that constitute human beings, highlighting some of their main features and implications, and then showing that the evidence from the ethnographic record illustrated in the samplings of it I have discussed in these pages shows that a great many aspects of any human social system, no matter how different each is from the others, can be fruitfully understood as solutions to the problems posed by the fact of dual inheritance.

Contemporary "dual inheritance theory" is an attempt to extend Darwinian theories and methods from the realm of evolutionary biology and population genetics, where they were developed, to that of human social and cultural life. In order to do this, "society" has often been conceptualized as if it coincided with the biological concept of a "population," that is, a collection of individuals, and "culture" has, as we have seen, been defined as the sum of the ideas in the brains of each of those individuals in a particular society. This formulation makes it possible to argue that since there are bound to be differing variants of these ideas in different brains within a given population, it is the task of the scholar to create models of adaptation that can be captured in mathematical terms, showing the evolving trends of how these ideas either are copied by other individuals in social learning and thus are perpetuated and spread, or fall by the wayside. That process constitutes ongoing adaptation involving the selection of variant ideas in the brains of individuals in a population pool, and selection is expected to be guided by principles that can be formulated and their implications tested in mathematical models. This approach entails predictions that individuals will take in information stored in the brains of others whom they admire, or who represent the mean or normative form of behavior, or who have the greatest prestige or success. These others do not have to be genetically related to the people who learn from them, but it is nonetheless an assumption of the evolutionist program that most of the time the behavior that is guided by the information transmitted by social learning, from brain to brain, will serve to support the inclusive genetic fitness of the individual doing the learning.

As I have noted, it is to the credit of such thinkers as Cavalli-Sforza and Feldman, Campbell, Durham, Boyd and Richerson, and others that they have recognized that cultural learning need not and could not be directly deduced from or tied to genetic inheritance and might indeed be quite independent of it. Boyd and Richerson have perhaps gone the farthest within this paradigm in asserting that while cultural adaptation could be conceptualized in the same theoretical way as genetic adaptation, in principle "culture," as the information transmitted through social learning rather than via sexual reproduction, can and does diverge considerably from the goal of enhancing reproductive fitness, since that was the reason why it evolved in the first place. They suggest the historical demographic transition as a prime example of this process.

When they turn to examine the role of symbolism in evolution, upon which I have erected so much of my present theoretical edifice, if they can find no adaptive function for it, they see what they call a "runaway" process, much like the runaway processes, produced by sexual selection rather than natural selection, that gave us the peacock's tail and the huge antlers of the extinct giant elk in the realm of biological evolution.

My approach to the application of dual inheritance theory to sociocultural systems, while in many respects indebted to theirs, is different in this respect, since I come at it with a preexisting interest in the study, interpretation, and explanation of symbolic systems. I make the assumption that they are real entities in the world that have their own properties separate from those of the individuals who constitute and have been formed by them. This means first and foremost that they are systems that have their own organization and structure that can be described in their own terms and cannot be reduced to the summed behavior of the constituent members. These structures and organizations can be reproduced across generations, just as is genetic information, by means of their instantiation in the individuals comprising succeeding generations. The reality of socio-cultural systems is further to be understood, in my view, as arising from their dependence on cultural symbols, that is, forms and patterns inscribed in the perceptual world that serve as vehicles conveying the meanings of the things, people, institutions, practices, and relationships in terms of which life is to be lived in a society. These symbols are real, material objects, that, by conveying instructions in the form of information, have the capacity to inform and direct the actions of people who interact with them—just as does DNA. Symbols are made to do a lot of work in my view of dual inheritance theory, and I resist viewing them as epiphenomenal, decorative, or merely "aesthetic" in value.

My view of the substantiality of human societies and of their reliance on symbolism is not foreign to some dual inheritance theorists. D. S. Wilson, as I have discussed, is willing to treat socio-cultural systems as wholes, or "organisms"; and E. O. Wilson in his most recent work has granted them the status of "superorganisms." Richerson and Boyd themselves account for social forms using the current concept of "institutions" that has emerged in some economic circles (Aoki 2001; Greif 2006).

A crucial step in my argument has been that while it is certainly true that symbols have counterparts inscribed in the brains of the people who participate in forming a society, the effective life of symbols occurs in their transmission into and through the medium of the sensory world, in the realm of things seen, heard, smelled, tasted, felt, and experienced. By virtue of being transmitted this way, rather than via copulation, symbols, unlike genes, can be perceived by and can inform many people at once, and thereby produce a sense of kinship among groups that is real in the same sense that genetic kinship is real: that is, it describes the relationship of people whose behavior is informed by the same instructions. The kin groups cultural symbol systems create may or may not correspond to groups of biological kin, but whether they do or not, their potential reach is far greater. Because of this, the symbolic kinship group can create the "tribal" society that, in human

CONCLUSION

societies, transcends and unites kin groups based on actual inclusive genetic fitness, much as Richerson and Boyd have argued. Their approach does, of course, recognize the important fact that shared symbolism serves as an identity marker for who is within the group, but the analysis or interpretation of symbolic systems themselves does not normally form a part of their research paradigm.

I have lived in the community of anthropology long enough to recognize the difficulty either of convincing some evolutionist scholars, and even some social and cultural anthropologists, of the position that society and culture are real in the sense I have discussed, or of convincing cultural anthropologists that taking the theory of evolution seriously and incorporating it into one's theorizing about social and cultural formations is anything other than dangerous heresy. But, in one final effort to make my case plain and persuasive, in this conclusion I want to submit one more ethnographic case to an extensive analysis.

As I said in the introduction, the case of the giant yams of Pohnpei, and the ethnography of Pohnpei in general, discussed by Boyd and Richerson (1985) in their consideration of the role of symbols in dual inheritance theory, was unknown to me before I read their analysis of it. It seemed to me that, by familiarizing myself with the ethnography of Pohnpei, I might be able to augment Boyd and Richerson's understanding, and at the same time provide for myself a "test case" to see how well the formulations presented in this book would be borne out in an examination of an ethnographic case that was new to me. It was important to me to see if the various consequences of dual inheritance I had proposed in the course of this book held true in this randomly selected instance, and if so, how. In laying forth this case here, then, I provide a model illustrating what many of those consequences are as I understand them, as well an account of how they play out in this one particular socio-cultural system.

What I will conclude is that Boyd and Richerson's own theoretical ideas, implicit in their formulation of dual inheritance—and expanded in the ways I have undertaken here—provide the basis for a much richer understanding of why men of Pohnpei grow outsized yams than they themselves presumed.

The Giant Yams of Pohnpei

Boyd and Richerson, in trying to understand the place of "symbols" in the dual evolutionary process, offer the huge yams grown by some men of the island of Pohnpei as a possible example of the "runaway" process in cultural adaptation.¹ The rudimentary ethnographic facts are these: the island now known as Pohnpei, located in the Eastern Carolines in the Western Pacific, is the capital of the Federated States of Micronesia. It is only about thirteen miles across at its widest point, but very steep, with mountain peaks rising to 2,600 feet above sea level; only some flat coastal areas are inhabitable and arable. It began to be in contact with whalers and explorers from the West in the early 1800s and in 1886 was incorporated by Spain into its Pacific holdings. Following Spain's defeat in the Spanish-American War, Germany acquired Pohnpei along with other parts of Spain's Pacific empire in 1899, but lost them again after their own defeat in World War I. At that point, the League of Nations awarded Germany's Pacific territories north of the equator to Japan, which administered Pohnpei along with most of the rest of Micronesia until it too suffered military defeat in World War II. The United States then took over the mandate in 1945; and finally in 1986 the Federated States of Micronesia became an independent nation, with its capital in the old Spanish city of Kolonia on Pohnpei.

The Spanish period had little impact on the traditional way of life, but the German administration, brief as it was, imposed important innovations, for example ending the traditional ownership of all the land by chiefs and instituting individual titles to land instead. The Germans also prohibited the chiefly feasting that was central to the political system that sustained the feudal land tenure system. However, the Japanese reinstated feasting, though the change in the system of land tenure became permanent. Although the chiefs are now no longer feudal owners of the land, they retained and retain much of their ritual and real political authority.

A striking feature of the cultural system of Pohnpei is the great emphasis on the hosting of feasts as a central element of social, economic, and political life. At feasts sponsored by chiefs, the attendees are expected to bring quantities of ordinary food such as fresh breadfruit, coconuts, fish, and taro. The chiefs are expected to provide kava, the mildly intoxicating drink widely found in the Pacific and prized for its quality of inducing peaceful sentiments and harmony among those who share it, and domesticated pigs. (In traditional times only chiefs were supposed to have access to kava.) Two other important items offered to the chiefs at these feasts are pit breadfruits and giant yams. Pit breadfruits are, as the name suggests, breadfruits stored in pits where they are added to and allowed to ferment over the course of many years before being brought to a feast. They are regarded as a prestige item, their value increasing with their age—rather like an aged whiskey, as several ethnographers observe.

The most remarkable prestige items, however, at least to Western observers, are gigantic yams that are grown by men in carefully hidden garden plots (Bascom 1948). These yams can attain a length of ten to fifteen feet and can weigh between one and two hundred pounds; the largest may require as many as six men to carry them in a sling supported between two poles. These yams are accorded prestige value based entirely on their size; the man who brings the biggest yam to a chiefly feast gains favor in the eyes of the chief and wins the admiration (and envy) of his fellows. It is the growing of these yams that Boyd and Richerson suggest can be analyzed as a case of a cultural "runaway process."

The runaway process in cultural evolution, as Boyd and Richerson define it, involves traits that have the following properties:

1. More exaggerated variants are associated with greater prestige;
2. The values of the trait observed should not make sense from an adaptive point of view;
3. The observed variant should be plausibly interpreted as an exaggerated example of a sensible indicator trait. (Boyd and Richerson 1985, 268–69)

In the biological realm, female choice, or "sexual selection," would substitute as the equivalent of the prestige value of a trait in the cultural realm. Thus, the large antlers of the giant elk once indicated an adaptive trait that led females to choose males with big antlers to mate with: the antlers signaled that the male was a good fighter, and therefore that a female's male offspring would have a good chance of themselves reproducing by virtue of probable future success in mating combat. Thus the females' choice at first "made sense" insofar as it enhanced their reproductive fitness. But as the males competed for the attention of females, the antlers at some point evolved to become so large that they were in fact unwieldy; that is, they no longer "made sense" from an adaptive point of view. This was because sexual selection had encouraged the evolution of antlers that were an exaggerated (and maladaptive) example of what had begun as a "sensible" indicator trait. "Sensible" here means that the trait in question can be seen by an outside observer to have adaptive value.

Boyd and Richerson propose the giant yams of Pohnpei as a possible analogue of the giant antlers, but in the realm of symbolism. The assumption underlying this analogy is that growing ordinary yams is adaptive, since they provide nutrition. Although prize yams are an exaggerated version of regular yams, they do not "make sense" from an adaptive point of view: "It is likely that farmers could devote the time and energy necessary to grow gigantic yams to better purposes" (277). The value of the prize yams is a matter of relative prestige, and, so it would appear, for Boyd and Richerson markers of prestige, being "symbols," could have no real adaptive value. (Actually, the giant yams do get eaten, either at the feast or following a redistribution of the leftover food by the chief to the attendees. But it is also quite true that this is a relatively inefficient use of effort on the farmer's part in terms of the ratio of energy expended to calories produced.)

At some earlier time, these authors propose, no one on Pohnpei devoted any effort to growing large yams. But "it seems reasonable . . . that more skillful or industrious farmers might have tended to bring larger yams to feasts, and thus the size of a man's yams would provide a useful indicator trait for all kinds of skills and beliefs associated with farming" (269). Naive individuals, that is, junior members of society still undergoing essential aspects of social learning, would then imitate such successful farmers, including their growing of large yams, acting on a bias to imitate successful individuals, thereby increasing their own chances that the cultural variants they acquired would lead them to be successful in farming. This would be an adaptive trait in itself, and might possibly be a way to attract a highly valued female mate who would enhance the fitness of their offspring. The next step in the argument is that at some point what was once a sensible indicator of an adaptive skill, namely growing big yams, turned into a symbol independent of its original adaptive purpose, and became an icon of value in and of itself. This trait, the growing of huge yams, would in turn co-vary with a propensity on the part of at least some people to admire large yams, and this in turn would encourage people to devote effort to producing them:

How does the runaway process produce meaningful symbols? . . . All other things being equal, it is plausible that the best farmers would tend to grow the largest yams. Thus at the outset the indicator trait is an index of farming skill, not yet a symbol of prestige. If the runaway process ensues, the most admired variant of the indicator trait will not be the most useful index of an individual's adaptive traits—it will confer prestige, but only because the rest of the population believes that it is prestigious. (277)

Thus a cultural symbol, in this view, is produced when something that originally either had use value, or was an indicator of the ability to produce use value, was transformed into a relatively "arbitrary" symbol conferring prestige on an individual without enhancing his adaptive potential—or indeed reducing it, since according to the basic assumption the time and calories he expends tending his secret giant yam plot could have been expended by the farmer in some more adaptive way.[2]

Boyd and Richerson are quite aware that the case they are making is hypothetical: no one has observed any of this happen. Rather, its plausibility arises from the fact that it parallels a well-established process in biological evolution and can be successfully modeled in mathematical form. Nonetheless, it stands as an example of what they believe can be accomplished by way of understanding an otherwise puzzling ethnographic phenomenon in Darwinian terms. The trait is "puzzling" because the authors cannot imagine how getting up at two in the morning to work on one's giant yam plot under cover of night is in any way enhancing one's adaptive fitness. Hence the raising of prize yams—or, we have to assume by extension, any activity undertaken for prestige rather than for nutrition or reproduction—is presumed to be an example of the cultural symbol system cutting itself off from its function of supporting genetically adaptive behavior. So while these authors grant that cultural symbol symbolism can indeed operate independently of the genetic program, when it does so they are likely to see this as a malfunction or deviation from what would be "sensible."

The ethnography supports Boyd and Richerson's conclusion that the raising of giant yams is a maladaptive trait insofar as one can easily imagine less costly ways of signaling prestige as well as of producing nutrition. Bascom, the ethnographer on whose account Boyd and Richerson base their own analysis, writes: "Not infrequently families go hungry at home when they have large yams in their farms ready for harvest" (1948, 212). And further, "the labor expended in growing prize yams is far greater than would be necessary to produce the same quantity of foodstuff from a larger number of smaller yams of the same variety" (217). Such observations support Boyd and Richerson's argument that the pursuit of prestige is strictly speaking maladaptive and so from this perspective does not "make sense."

Whether, on the other hand, the socio-cultural system or the population of Pohnpei ever actually suffered in any way from this practice is not obvious. One does not get the impression from the ethnography that food shortages were ever a major problem on Pohnpei in premodern times. As an island with plentiful rainfall set in the tropical zone, it is remarkably fertile, and until recent decades it sustained a relatively small but stable population. Bascom writes that "in a favorable environment, Ponapeans have seldom experienced serious shortages of food. Pit breadfruit and yams, neither of which are destroyed by the tropical storms of the Pacific, constitute important food reserves against periods of shortage" (221). As a result, as Riesenberg (1968) writes: "Nature is generous enough to provide ample leisure for non-economic activities, and the Ponapeans have chosen to use this leisure time for the direction of a great part of their energies into political channels" (110). Riesenberg cites census data that show that, after the island's population was reduced by about half in the mid-nineteenth century due to a smallpox epidemic, the population remained quite stable at between 3,500 and 5,000, only soaring to over 34,500 in modern times (partly due to immigration) (6).

To better understand this system, then, in which a benevolent environment enables leisure time and an opportunity to put surplus energy into "political channels," it is to these political channels that I now turn my attention in order to comprehend why men might overlook their own and their families' own immediate nutritional needs for a time, and neglect their regular garden, while working overtime to ensure that the household's prize yams were as big as possible.

Saul Riesenberg (1968), whose comprehensive ethnography is the authoritative source on the subject of premodern socio-political life on Pohnpei, draws a contrast between what he understands to be the typically "Melanesian" social pattern of relative status equality and low political complexity and hierarchy, on the one hand, and the "Polynesian" pattern of graded hierarchies and a high degree of social and political complexity on the other. The Micronesian islands vary along the continuum between these two poles, but Pohnpei, among these, has one of the most complicated and hierarchical political systems of all.

Pohnpei, small though it is, is divided into five independent states, each of which is further subdivided into "sections" of which there number about fifty on the island. Each of these units—the states and the sections within them—are presided over by a dual chieftainship; two chiefs always rule together. However, one of them, the Nahmwarki (or Nahnmwarki), is the absolute ruler and in former times had sacred and taboo status; the other chief, the Nahnken, acts somewhat in the manner of the talking chiefs of Polynesia.[3] While the former makes final decisions on matters that concern the polity, the latter is the one who announces them to the people. In former times, when the Nahmwarki sponsored a feast he himself sat in a private room out of sight,

CONCLUSION

while the Nahnken actually managed the affair. It was taboo to touch or even look at the Nahmwarki, though these stringent rules no longer apply. The chiefs of the states in former times were the owners of all land in their domains and had absolute authority in disputes and matters of state policy. They could order the death penalty, make war, and in general exercise final authority on all political matters. They also oversaw economic production in their states and had as a duty the obligation to equalize resources, partly through the hosting of great feasts to which surpluses of food were brought as offerings to them by the people and then redistributed. When one of the two chiefs died it was the duty of the other one to appoint his successor. Often the two chiefs came from clans or subclans that exchanged spouses, and were thus often closely related.

Each chief presides over a hierarchy comprising a number of ranked titles, each one ideally occupied by the hereditarily highest ranking individuals of matrilineal clans having inherited "noble" status. These titles carry great prestige. When a holder of a title dies, everyone behind him in the ranking moves up one step, and the chief then offers the vacant title to a new occupant. The chiefs also have an array of other titles that they can create and distribute in a less prescribed fashion, and even in the case of the established ranked titles, there is some flexibility: chiefs could dismiss a title-holder whom they judged to be not sufficiently generous and industrious in his duties, or elevate a favorite who might "leapfrog" into a higher titular role than would have been his lot according to the ordinary rule of succession. The structure of dual chiefs and associated vertical lines of title holders is repeated at the section level, with the result that most adult men hold titles at some point in their lives.

There are in addition to the system of chiefs, nobles, and titles about twenty matrilineal clans, which own clan land and provide mutual assistance in feasting, house building, canoe building, and so on. (Peterson [1982] gives the number of clans as eighteen; Riesenberg sets the number at twenty-three.) Some of these clans have one of two recognized degrees of noble status, and it is from them ideally that most high title-holders and chiefs are drawn. The clans are not localized but are scattered among the different states and sections. The clans are also ranked, as are the subclans that constitute them, and the senior man of a subclan is its chief. It is thus plain that a vast and complex system of gradations of status by clan heredity, chiefly office, and title blankets the island. The titles are coveted for the prestige they carry, but they also require extensive service by the title-holders to the chiefs.

The prestige ranking comes to life in the many feasts that mark life on Pohnpei. Some of the feasts are set according to the calendrical cycle, such as for example the feast at the start of the yam season; there are said to be six such seasonal feasts in all. (Riesenberg 1968, 85). There are also multiple feasts for the various stages of construction of new fishing seines, dwelling houses, and community houses. There are feasts before a war, and feasts when chiefs visit each other on friendlier terms, as well as funeral feasts, marriage feasts, and so on. The political and kinship divisions determine who should bring food to the feasts, and the rank system dictates the order in which people are seated and served.

Feasting is the site where the social hierarchy is expressed, negotiated, and renegotiated. For although there are regulations about which clans are able to be entitled, and which individuals are entitled to which rank, nonetheless, given the flexibility of the system, titles can be won by achievement as well as by ascription, and the result is a lively competition for promotions. Only the chiefs can grant promotions, but through their good graces, commoners can rise to titled rank, and people can advance more quickly through the title system then would normally occur. People from non-royal backgrounds can on occasion become chiefs:

> Promotions come about in part through bringing to feasts for presentation to chiefs larger and better and more frequent food offerings than other men, thus demonstrating industry, ability, loyalty and affection toward the chiefs. But more important than presentations at feasts are the direct offerings of first fruits. . . . All these types of presentation are known as Service. Perhaps even more important than regular offerings is the bringing of a valuable article, such as the fermented contents of a large breadfruit pit, to the Nahmwarki on a special occasion. . . . All of these acts are stored in the memories of the chiefs . . . and duly rewarded when vacant titles arise. (Riesenberg 1968, 76)

Although Riesenberg does not mention the fact in this particular passage, giant yams also figure as highly prized prestige items along with pit breadfruit.

Service is divided into the Great Work and the Little Work. The so-called Great Work consists of all the duties performed by his subjects for the Nahmwarki, who actually owns all valuable property as the representative of and on behalf of his state or section. These duties constitute effort devoted to benefiting the community and include communal labor, observance of correct etiquette, deference, and "Service" proper, that is, offerings of food and prestige gifts at feasts and on other occasions. The Little Work is service to the chief in war. The Great Work involves actions that are routine but time-consuming; the Little Work is of short duration but strenuous and of course risky. Outstanding performance in either of these may be the basis for a chief's granting a promotion to a higher rank or title than that to which a person would normally be entitled by the regular system. Thus, the stage is set for competition among all men to distinguish themselves in the Great and/or Little Work, in order to move up the status hierarchy. And one very important way to succeed in this competition is to bring the biggest prize yam to a feast.

So we can say that the huge yam is indeed an indicator of something, and not only of whether or not one is a good farmer, though of course it does indicate that too. Rather it is ipso facto evidence of one's commitment of effort to the Great Work on behalf of the chief and of the community which he represents. Getting up at two in the morning to line the pits in which giant yams are grown with alternating layers of soil and decaying vegetable matter as fertilizer is a demonstration to one's chief that one is sacrificing oneself, putting in the extra effort, for the good of the community rather than for the benefit of oneself and one's own immediate kin.[4] By the same token, one can now see why a household denying itself food on the table in order to ensure that the prize yams are competitive enough to win would be a very reasonable thing to do insofar as one accepts that the principal aim of life on Pohnpei for men is to advance as far as possible up the prestige ranking system, as it empirically appears very obviously to be.

Raising prize yams, then, along with voluntarily undertaking the other forms of service in the Great Work or the Little Work, is a way of competing with others for praise from the chiefs and with it enhanced prestige in concrete (symbolic) terms as movement up the ranked status ladder. Bascom stresses the endemic competition on Pohnpei for status in the first and last sentences of his article, comparing it to the more overtly competitive potlatch of the Kwakiutl:

The inhabitants of Ponape . . . have a system of prestige competition reminiscent in some ways of the potlatch of the American Indians of the Northwest Coast, but with a distinctive character deriving from Ponapean patterns of modesty. Instead of the distribution and destruction of property that marks the potlatch, contributions of certain foods to community feasts are the traditional means of achieving status. (211)

The retiring modesty of the Ponapeans, even in comparison with the peoples of neighboring islands of Micronesia, could have been more fully documented than it is in this article. Yet it is coupled with a system of winning prestige and status which is almost as competitive as that of the "Dionysian" Kwakiutl. (221)

The comparison with the potlatch of the Northwest Coast alerts us to the fact that underneath the harmonious surface of the chiefly feasts, at which kava is served by the chiefs to insure peaceful relations among all who partake of it, there is a symbolic war going on, conducted with food as a marker not only of one's expertise as a farmer but also of one's generosity and industriousness on behalf of the chief and of the whole community as a citizen of that community.

Riesenberg devotes a long section of his detailed monograph to the category of feasts that entails even more rivalry than ordinary feasts; indeed, he refers to these as "competitive feasts." Of course he recognizes that all feasts on Pohnpei are competitive in one sense:

The competition is between every man and all his peers simultaneously, each striving to outdo the others by showing that he is the most industrious, the most skillful in growing large yams, the most generous in making offerings, the most devoted to the chiefs. Thus he gains merit and earns titles. (1968, 90)

Feasts in which this ordinary sort of competition occurs are called "Serving Together."

But there is another sort of feast, called "Displaying Together," which

is not a general struggle among all men but a direct contention between two individuals, two groups of relatives, or two groups of coresidents. It is not directly concerned with personal ambition for personal advancement but with besting another man or group in order to avenge a slight, or upon order of a chief, or simply for ostentation and ego gratification. The feasting is done not to make offerings to chiefs but, as the means of expression of rivalry, to exchange goods with rivals. (90)

There are three kinds of such feasts. One of them is a friendly feast, in which the rivalry is muted and the main purpose is for the participants to demonstrate in an amicable way their admirable prosocial qualities by outdoing each other in generosity. The other Displaying Together feasts, which may be compared directly with the potlatch, "usually result from a disparaging remark to the effect that someone has no yams or pigs or kava; more rarely when someone has been seen to violate the secrecy of a yam plantation by sneaking in to observe what possessions the owner has" (91).

At these feasts, there are different and more stringent rules regarding the criteria for winning, since the point of the feast is to settle a dispute. Riesenberg continues:

> In the potlatch, the side that has been insulted or spied upon calls on all its resources for the preparation of the feast, to which it invites the challengers. . . . The challenged side kills all its pigs and brings yams and kava and whatever valuables it may have. The articles are displayed, then given ostentatiously to the other side, which accepts them, if it feels it can outdo the challenged parties at the return feast, held the next day or so; otherwise it rejects the gifts and acknowledges defeat. (92)

At Serving Together feasts, by contrast, where the aim is not to defeat a specific rival but simply to deliver the most prestigious offering, the customary modesty prevents the taunting and active striving for success, cheered on and abetted by the relatives of the hosts, that occurs at the unfriendly Displaying Together feasts. Instead, the man who has brought the largest yam modestly praises another's yams and disparages his own—though everyone knows that he has actually "won."

The Pohnpeian prestige system as I have described it certainly appears to be the culturally constructed equivalent of a dominance hierarchy. Much of this hierarchy is already determined and institutionalized in symbolically encoded form: it is shared public information that certain clans are noble, some royal, and others common, some rank above others within each category, and so on. Thus to a degree one's place in the hierarchy is ascribed by one's birth. But the system is also flexible enough, and allows enough room for movement up or down the prestige ladder, that the rank orders require to be constantly renewed, displayed at feasts, and renegotiated or even radically altered, for example by war, coup, or fission.[5]

A key problem that faces evolutionist thinkers in a case such as this remains, however: even if it is shown that raising big yams is highly "sensible," to use Boyd and Richerson's rather problematic concept (sensible to whom?), in that it is one of the main ways to advance in the prestige hierarchy, why is the quest for prestige important at all? In this perspective, such status competition distracts one from potentially more adaptive activities, thus representing a wasteful expenditure of valuable effort on an activity that does not serve the only value evolution actually rewards, namely, reproductive fitness. I will offer two different answers to this question in the case of Pohnpei.

The first concerns the ethnographic facts of the case. Boyd and Richerson assume that since the payoff for growing big yams is largely or only prestige, and since prestige is not in itself of evident value as a contribution to fitness, then we must assume that something that was once of value, such as working hard in one's garden and growing robust crops, has acquired a derived meaning as an exaggerated indicator of value in a process that has broken free from evolutionary constraints thanks to the independent nature of the cultural channel of inheritance. Since the recognized Darwinian process of sexual selection can be invoked as an analogy from the genetic realm, the existence of the yams is explained within the parameters of evolutionary theory, provided one recognizes that culture, which was designed to enhance fitness, is capable of going off on some occasions to produce maladaptive results in a runaway process.

It must be added in fairness that it is often assumed in dual inheritance theory that prestige does confer reproductive benefits, in that it is often correlated with the attraction of desirable mates; but Boyd and Richerson do not use this argument when discussing the growing of giant yams. However, they might well have done so, because it appears to be the case that high prestige does actually lead to genetic fitness on Pohnpei. We have already seen that the social hierarchy, while it mainly concerns issues of prestige, also entitles at least the highest ranks to great and, at the very top, absolute power. It also entails considerable freedom for high-ranking individuals from the ritual and customary constraints that guide the actions of ordinary people further down the chain. This is true in the realm of sex and matrimony as in other spheres of activity.

Sex is regarded on Pohnpei as an irresistible and pleasurable need that must be met regularly; premarital relations seem quite completely free, and even after marriage there is considerable leeway. Martha Ward (2005) observes:

The public line about extramarital sex is that married couples are straight, modest, fruitful, and relatively faithful. By all accounts, however, adultery is as frequent as it is disruptive or attractive. . . . Again and again, listening to gossip, I was aware of attitudes about sexuality that permeated Pohnpeian customs. When two people of the opposite sex are left alone, the inevitable will happen. (30–31)

In public, some insist that things have changed and that Pohnpeians are now abiding by the rules imposed by foreign influences. In private, however, they laugh, joke, and gossip about sex. In private, I think they are having ribald, exuberant, enthusiastic, playful sex lives. (28)

297

CONCLUSION

Beyond this general atmosphere of sexual tolerance, moreover, there are the prerogatives of chiefs:

> While a Nahmwarki or Nahnken would often have 10 or more wives, few lesser tribal chiefs had more than two. An informant past 90 years of age could not remember any polygynous commoners, and some informants deny that a commoner could have plural wives. Keimw Sapwasap, a Nahmwarki of Solehs, is said to have had some 30 women in his harem. . . . In addition, a chief would have sexual access to the female servants. (Riesenberg 1968, 72–73)

Furthermore, the chief could in principle demand any woman he desired from within his realm, whether she was married or unmarried. A husband who demurred would be punished; if he cooperated, he received a present from the chief. A high chief could also forbid a pretty young (prepubescent) girl to marry; he would bring her into his own house where she would be raised by his principal wife until the age of thirteen or fourteen, at which time she would be made a secondary wife.

On the basis of this evidence, then, it seems that there were very real reproductive advantages to having high status, in the form of the right not only to multiple wives but of unimpeded access to any other woman. This being so, it renders somewhat moot the objection to the quest for movement up the prestige hierarchy and, if possible, into a chiefly rank, that it does not serve genetic fitness: on the contrary, as is the case in many such graded hierarchies in other species, one of the perquisites of ruling status on Pohnpei is the right to multiple consorts, whether marital or not, a right denied to commoners.[6] This consideration does not, however, explain the efforts of a married commoner to advance in the chief's eyes and in the hierarchy of titles by working in his secret giant yam garden, since he is exceedingly unlikely to thereby acquire noble or royal status and thus the sexual prerogatives that go with it (although it might be possible that he may, through his prestige, attract extramarital mates who might then enhance his reproductive success).

However, my second response to the problem of the role of prestige and prestigious activities in human social life is a more general one. While in the case of Pohnpei there may in fact be demonstrable direct fitness advantages to very high prestige, in many cases the link, if there is one, is far less palpable; and even in the case of Pohnpei, it is unlikely that the extent to which the struggle for prestige, with all the time and effort it entails, dominates people's lives can entirely or even mainly be explained by the reproductive advantages that go along with high status. There seems to be something about prestige itself as a positive goal in human social life that is not best accounted for either by calculations of presumed genetic interest or as runaway processes.

Prestige

The reasons for the pursuit of prestige seem to me to have to do with the fact that human social life inherently depends, as I have shown, on there being a public arena in which symbols can be made available to perception and shared by many people. It is in this arena that the cultural channel of information transmission and reproduction takes place, which guarantees it a central role in human social life. I have argued throughout this book that the fact that the people in a group share in the perception of these symbols, and incorporate them into their own thinking, feeling, and identity, means that they experience their consociates as "kin," by which I mean as related to them in a real way through having the same shared substance. They recognize members of the same group whom they have not met before by virtue of the fact that they speak the same language, wear similar dress, observe similar customs, subscribe to the same sacred symbolic lore, and so on. This public arena thus serves as a sine qua non for social life and its reproduction over time. It is the realm of the tribal instincts that Richerson and Boyd identify, in contradistinction to the usually more cloistered realm of sexual reproduction.[7]

This wider arena also serves as the locus of social control, in that it is in the public eye that the people's actions are evaluated as good and right or incorrect and antisocial. Honor, shame, pride, admiration, envy, prestige—all these are qualities that emerge in the public drama of social life as it is observed by others as well as by oneself. In the public arena one's actions are evaluated according to their correspondence, or lack thereof, to the prevailing norms and values of the group, in gossip and other dimensions of group communication; reputations rise and fall on this basis. In this respect too, then, prestige, which is a measure of one's relative value in whatever symbolically constructed prestige system, is a necessary good sustained by the requirement of social life at the tribal level that people should obey the explicit or implicit regulations, strive to be thought well of by others, or pay the consequences of failing to do so. Only the existence of a system of symbols enables this to happen, at least to the degree achieved among humans.

Thus for the sake of the existence of the group and its social system, upon the resources of which one must rely for the necessities of existence, most individuals are motivated—most of the time—to fall in with this prescription.

Furthermore, the social compact in the arena of social life must set as values those actions that lead to harmony, cooperation, and unity if the tribal society is to sustain itself. Any other choice would obviously be self-destructive. Competition among humans certainly exists, and must be contained, managed, and tolerated; but unless it is turned away from violent and lethal forms—at least most of the time, and within the group—it threatens the group with disruption and in extreme cases annihilation. Therefore the symbol system, as Donald Campbell rightly noted, must set as values actions that are generous, altruistic, devoted to the good of the community, and equable. Actions that are too obviously marked by individual rivalry, ambition, selfishness, or violence, at least insofar as they occur within the group, are judged negatively in the court of public opinion, and produce negative social consequences, as well as bringing shame and dishonor to the perpetrator.

In addition, sexuality, which is essential in the genetic reproductive system, is largely eliminated from the ordinary discourse and practice of social life in the public arena. In this realm, reproduction of information occurs through the transmission in perceptual space of instructions in the form of symbols and practices that can be observed; reproductive sex has no place there. On the contrary, unless it is well-regulated and contained within institutional boundaries, opportunities for copulation would be a prime source of rivalry, disruption, fighting, feuding, and general disharmony (which of course they are often enough despite the precautions taken).

Yet the tribal instincts, and the public realm in which they carry on their existence, are not all there is to human social life: genetic reproduction happens, and must happen if the society is to go one existing (unless it resorts to extreme tactics like those of the Marind or Mbaya). Furthermore, the kinds of actions associated with genetic reproduction, with fitness maximization at their heart, continue to play a role in life no matter what the official public stance toward them might be. Individual actions continue to be motivated at least in part by competition, rivalry, and inclusive self-interest, including in the sexual realm, rather than by generalized prosocial aims alone. Some parts of the symbolic system itself, too, pertain to, justify, and valorize just this sort of behavior, at least some of the time.

Therefore one of the tasks that a prestige system accomplishes, more or less successfully, is to integrate the different imperatives implicit in the two kinds of society and the two kinds of information transmission that generate and maintain them. Here is how the trick is often done: *a prestige system, such as the one that exists on Pohnpei, allows and encourages people to compete, yes; but it directs them to compete not to further their own interests, but on the contrary to compete in being generous, altruistic, and otherwise loyal servants of the chief and of the community, and thus to demonstrate that they put the group's interests ahead of their own.* This is just what happens in the feasts on Pohnpei. The competition is every bit as real and as ruthless as it is anywhere else, but since it is not waged with violence or in an explicitly self-assertive format (except in certain Displaying Feasts), and is not directly concerned with sexual matters, but rather rewards those who demonstrate generosity and industrious work for the chief, and through him the group as a whole, it promotes the general welfare rather than setting each man against his neighbor. The man who brings the largest yam, in accord with the socially prescribed virtue of modesty that mutes overt expressions of self-assertion and self-promotion, does not act proudly or boast openly. He lets the yam speak for itself:

When others discuss the merits of his yam, he pretends not to listen. When they come up to tell him that his yam is the largest, he protests that it really isn't. He points to the next largest yam, claiming that it is better than his own. . . . A man who shows his pride openly is talked about and laughed at, and his prestige is turned to shame. (Bascom 1948, 215)

By expending farming energy on prize yams, which are expressions of commitment to the chief and to the polity at large, and not on his own family food garden, a man is thus showing that he is willing to sacrifice the needs of his near kin and himself, that is, the welfare of himself and his genetic reproductive unit, in favor of the tribal society, whether it is in the form of the subclan, the section, or the state. Thus, precisely in order to help maintain the social symbolic system that sustains them all and makes their social life possible and harmonious, the man growing prize yams is devoting energy to a task critically necessary to the maintenance of his own interest insofar as he is reliant on the group and its chief to provide peace and protection. (The gloves do come off, however, in "competitive feasting," in which the aim of a man, or a kin or local group, is explicitly to defeat the other in conspicuous and excessive generosity; in these symbolic bouts, promoting oneself and cheering for one's kin or allies to win is the norm and reveals another very different face of ritual feasting.)

Therefore we may conclude that while it is true that growing prize yams is not adaptive insofar as it is not for the satisfaction of the fundamental needs of nutrition for the inclusive kin group, it is highly adaptive given the requirement of human society that the interests of the group must be supported in the face of the constant pull of individual need and interest. This is the essential social work that prestige systems accomplish:

> Just as Ponapeans respect a host who serves his guests left-over food, so they praise a man for having kept a pit breadfruit for many years without having eaten it. Both show the abundance of food he has provided. (Bascom 1948, 213)

Serving leftover food would not necessarily be the mark of a good host in our own cultural system, but on Pohnpei it demonstrates that the person has not taken all the food he has produced for himself and his family but is willing to share what he has produced with others. Likewise he demonstrates his commitment to the common good by producing a very old pit breadfruit: he and his family could have eaten the year's crop of breadfruit themselves at some point, perhaps even in an hour of need, but instead he has saved it to present to a chief at a communal feast, where it will be consumed by everyone. The pit breadfruit is thus the sign of his self-denial, and his self-denial is the affirmation of his real membership in the social group.

A central task that must be performed by the symbolic system in the public arena, furthermore, is not just to manage competitiveness in general but to keep the men from actively destructive rivalry. This may be accomplished in a variety of ways. In societies with men's houses, as I have shown, the symbolic system may redirect their competitive aggression against enemies, or against women, or project it onto fierce ancestors; in other societies it may send them out to sea with fishing gear and then allow them to drown their sorrows in alcohol-induced self-pity. Many societies also provide some form of symbolic equivalent for fighting, whether it involves hitting rocks with a stick to see which one breaks first, or seeing who can sing the better song. In many societies, competitive "games" of wrestling, club fights, ball games, archery contests, log races, or pounding on another man's chest provide outlets for aggressive competition. On Pohnpei, this essential work is accomplished by feasting and all that goes with it. Once again, this function is supremely adaptive insofar as one accepts that society itself is necessary for the life and reproduction of any individual member, and therefore that sustaining it is a matter of both group interest and self-interest.

Now the entire Pohnpeian social system and the feasting and the prestige associated with it depends upon the widespread shared acceptance of the supreme authority of the chiefs. It is they who set the stakes for the competition, act as arbiters of it, and have the power to name the players by distributing titles not only according to customary precedence but also according to their own personal choice. They also have the power to punish and thus to enforce the rules. The chiefs with their near absolute authority are thus the keystone of the whole elaborate arch of the prestige system that enables social life on Pohnpei to operate with relative harmony. It might then be deduced that in the absence of a chief, the result might be chaos. And this is in fact the case. When a chief died in former times, those around him made every effort to keep his death secret until the other chief in the dual chieftainship had chosen his replacement, for fear of a civil war or attempted coup. Otherwise, the death of a chief was a society-wide crisis, since it was the chief who by his absolute power symbolized, authorized, and maintained the existing social order with its elaborate crosscutting hierarchies. With him gone, anything could happen:

All [the] early reports ... refer to lawlessness and to general destruction of coconut trees, yams, and dogs at the place where the dead chief lived and to a lesser degree elsewhere. His lands and other property would be taken and divided up among the other chiefs, sometimes in an orderly way, but often a rush was made by all to seize what he could. (Riesenberg 1968, 42)

Any dead man's possessions were sometimes taken by a mob and his widow turned out of his ransacked house. Following a chiefly death, there ensued a more general breakdown of social order:

For a time afterward, people would feign madness, running about wildly besmearing themselves with mud and throwing dung at each other, lose their normal inhibitions in dress and speech, and sometimes engage in sexual license. It was to avoid these excesses that the attempt at secrecy was made and the new [chief] was immediately installed in office. Ideally the general populace was not to know anything until they were summoned to a feast in the community house and saw the new incumbent in the place of honor on the main platform. (42)

So it seems that the threat of disorder, including unregulated sexual indulgence, was very much a real possibility and a danger that the institutions of the social organization, represented in the person of the chief, were in place to regulate and keep under control.

The Importance of Being Seen

The image of the new chief being presented as a surprise before the assembled populace leads to yet another important consideration in our analysis of human socio-cultural systems: that of the centrality of display as a strategy in creating prestige effects. We have already discussed the fact that public life is lived to some considerable extent as a drama performed in the social arena before an audience of the society by whom it is evaluated. In a similar way, the prestige system depends on the display of symbols of prestige for all to see and admire. It is only on public display that they do their work, by virtue of the admiration (and envy) they inspire. The feasts hosted by the chiefs of Pohnpei are partly about food as nutrition and partly about the redistribution of goods, true enough. But also, and perhaps more so, they are about food as a show, in the literal sense: feasts are the occasion for displaying the values of abundance, fecundity, generosity, and sociability in concrete form by the society to itself. The competition to provide offerings leads to more food being present than can be eaten, and so it can then be redistributed in a gesture of altruistic concern for the common people. At the feast, there is a prosocial enactment of social well-being, at the same time as there is relentless competition, always undertaken in a self-abnegating and prosocial guise, for individual advantage.

This emphasis on display is hardly limited to the Pohnpeians. Peter Wilson (1988) sums up much anthropological wisdom concerning this phenomenon, arguing that to try to understand nonmodern exchange and distribution systems in terms of contemporary economic theories is highly misleading: "The striking thing about these displays, at least to observers from a capitalist society, is the huge expenditure of labor, time, and resources, bearing in mind that such expenditures appear to have little or no material return. Anthropologists, however, have stressed the returns in terms of prestige, sociability and redistribution" (86). The use of yams for display purposes rather than as food is hardly limited to Pohnpei. Boyd and Richerson stress that, since symbols are (or can be) "arbitrary," size is not the only possible way in which a yam can be turned into a prestige item that will impress the audience: "in a different population starting under almost identical conditions the [runaway] process might result in a different indicator of prestige. It might be the color or shape of the yam that would be important, or another aspect of farming altogether" (1985, 277–78). One can illustrate this by reference to one of the best-known instances of yams used for display, that observed by Malinowski on the Trobriand Islands. There, unlike on Pohnpei, the yam is actually the main staple crop:

> No doubt yams are appreciated as food. But as much if not more value is attached to the size and appearance of yams because they are critical items in the creation of displays, which in turn govern much of Trobriand interaction and social evaluation. When yams are sent to the sister's husband as *Urigubu* payments they are arranged in conical heaps for display; when guests are entertained the yams are first exhibited, then cooked. And when they have been harvested and transferred to the village to be stored they are carefully arranged in what Malinowski calls storehouses but which might more accurately be called galleries. These galleries are the only village structures other than the chief's house that are decorated. (Wilson 1988, 83)

Here, then, it is the aesthetic arrangement of regular yams, rather than the size of one particular specimen, that is the symbolic expression of prestige.

One might argue that this way of using yams, unlike raising giant ones, does not require the expenditure of energy that appears so costly in the Pohnpeian system: all one has to do to convert yams from useful food items to generators of prestige is to arrange them in a certain artful way. Therefore it is not maladaptive and not a "runaway" procedure. That could of course be true. But if making that argument, one should also recall that these same Trobrianders are willing to hop in their canoes and paddle for hundreds of miles to engage in the *kula* trade for prestige items that have barely any intrinsic use value whatsoever, in an expenditure of energy in pursuit of prestige that makes the Pohnpeian giant yams look like small potatoes. If one is not to regard this as senseless behavior, one must, I think, recognize the crucial role that prestige plays in the maintenance of harmony in a society composed of men otherwise inclined to be violent rivals.

The reason display and performance are central to social life from a dual inheritance perspective is that what they address themselves to is precisely the shared public realm observable by the senses. Just as dance unites people by shared kinesic and proprioceptive experience, and music similarly unites people via the sensory channel of sound, just so the seeing of a display or performance by a crowd of people creates an eminently social reality: the fact that everyone is experiencing the same sensory perception of something impressive at the same time and is aware that this is so. This is, as I have argued, a wider social enhancement of the "shared intention" or "we intention" originating in the mother-child dyad now employed symbolically to bind individuals into a real social group.

This dissemination of a visual effect simultaneously throughout a group reinforces their sense of unity, of identification with each other, and a sense of belonging to a surrounding and containing group that is larger and more powerful than oneself but that has one's best interests at heart in the form of its prosocial activities. It is, as I have argued throughout this book, a key mechanism whereby the cultural channel creates real kinship in a way parallel to but partly independent of the genetic channel. True enough, not all symbols unite people: some may in fact set them at odds. And the duration of the sense of unity created may be short or long, requiring repetition of the rituals that create it. But the creation and maintenance of human society depends on the fact that at least on some important occasions, often marked as sacred or ritually set apart, the felt unity of the group is re-created through shared participation in and observance of performance and display.

Male Symbolic Reproduction

Yet one more aspect of social process into which the prestige system of Pohnpei and its giant yams provide insight is the frequent association of the competition for prestige and display with men. While there is no formal men's society on Pohnpei, in former times the feast houses served as a center for unmarried men, and men were the more prominent social actors in the public arena. The hierarchy of the prestige system was largely a matter among the men: only men could be chiefs or title-holders. Normally women would be given titles only insofar as they were married to men who themselves held titles (though there seem to have been some exceptions to this rule). This indicates that, as I have argued, the problem that needs to be solved and is addressed by the prestige ranking is the ordered regulation of men and the muting or redirecting of their potential for rivalry and competition within a manageable system.

The association of men with prestigious display highlights a phenomenon to which, as I have shown, dual inheritance often gives rise, namely, ideas about male reproduction. In the case of Pohnpei, there is no overt identification or linking of the farming and displaying of giant yams with human reproduction, but at another level it is obvious that what is on display is the ability of a man to make something grow, hidden away in a secret cavity out of sight, to fertilize it and bring it to term, and then to take it out from where it was hidden and display it to the community. Whether or not one is willing to see this as a metaphoric representation of fertilization, gestation, and birth, there is no doubt that it is a demonstration of the man's ability, through his cultural "labor," to produce a new and awe-inspiring living thing.

A much clearer identification of yams with male "childbirth" elsewhere in Oceania is to be found among the Northern Abelam, who live in the Sepik region of Papua New Guinea. There is a dual division in that society that creates ceremonial exchange partners who initiate each other and engage in competition by trying to outdo the other in giving away pigs and yams. They too have a yam cult that is linked with prestige and political control. The Abelam yam cult culminates in the ceremonial display and subsequent exchange of yams between partners from the different sides of the dual divisions. These exchanges are one of the main interests in the life of the men, who grow the yams used for exchange in plots separate from their ordinary gardens:

There is a close identification between a man and his finest yam; it is a symbol of his manhood and industry. Many of the longest yams (five to ten feet in length) are not eaten: they are displayed at harvest, stored, distributed, stored again and eventually planted. . . . The finest are normally reserved for presentation to a ceremonial partner once or twice a year. (Kaberry 1966, 340)

At the exchange ceremonies the yams given are carefully measured so that, in principle, a return of exactly equal proportions can be made; in reality, men on opposite sides of the dual division try to outdo those on the other side. The idea is to assert the superiority of the donor and his group and belittle the yams of the recipient and his group.

Women are taboo to the yams. Not only are women not allowed to participate in any of the activities associated with the yams, but the men are not allowed to have sexual intercourse during the period of the cultivation of ceremonial yams. Thus there appears to be both a parallel and a conceptual opposition between the realm of sexual reproduction and the realm of the cultural reproduction of the yams. The yams seem to be a form of male—hence necessarily "cultural" in opposition to "sexual"—reproduction. Abraham Rosman and Paula

CONCLUSION

Rubel (1978), in their admirable comparative study of New Guinea exchange systems, make it clear that this is indeed the case:

> Groups of men exchange women in order to reproduce those groups. Reproduction is achieved through sexual intercourse with and impregnation of the women exchanged. In the structure of exchange of long yams, men reproduce long yams, which are their symbolic equivalent, by means that demand the exclusion of women. . . . The yams are painted, given masks, and decorated as though they were men, for the ceremonial exchange. What is produced by male fertility is an emblem of maleness which is dressed up like a man. The yam cult is a statement that men are able to reproduce a metaphoric male society, without the aid of, and indeed through the exclusion of, women. The structure of yam exchange represents an inversion of the structure of exchange of women, both structures in different ways relating to sexuality and reproductiveness. (65–66)

While it is of course true that one cannot simply transpose the meanings of a symbol from one cultural setting to another, the North Abelam treatment of yams, which shows many similarities with the competitive displays of large yams in Pohnpei, lends support to the idea that the production of prize yams there too might carry the significance of male fertility and reproduction. Thus, insofar as this is so, while the ceremonial yam is a prime symbol of the tribal society and the asexuality and prosociality that characterize it, at the same time the symbol metaphorically draws its power precisely from the facts of sexual reproduction and childbirth. The scandal of birth from a woman's body and of infantile dependence on a woman is simultaneously denied while at the same time the positive affects that accompany it are employed to energize the symbolism without actually bringing the sexual into the public arena except in sublimated symbolic form.

Back to Dual Inheritance Theory Again

My analysis of the symbolism of the giant yams of Pohnpei seems at odds with Boyd and Richerson's analysis of it as a runaway process. But it is not a violation of the spirit of their formulation of dual inheritance theory. What I have done in fact is to take seriously their own theoretical propositions that culture evolved to do things the genes are unable to do, and that in the course of accomplishing this humans evolved a second form of social "instinct" that allows for the creation of social groups over and above nuclear mating units. From the genetic point of view, any organism needs to attend to its own maintenance by nutrition and protection from predation and to the reproduction of the genes that informed it. But humans have achieved the feat of going beyond what the genes can accomplish, creating a wider social arena that encompasses the mating units operating on the basis of genetic priorities. This has entailed the creation of a shared public arena in which externally represented information could be communicated directly, to many people, without sexuality and the competition it generates or the restrictions on widely shared sociality it imposes. This externally encoded information—what I have been calling the cultural symbol systems intrinsic to any human socio-cultural system—allows humans to overcome the barriers the genetic system places in the way of wide participation and contains and transmutes the disruptive passions associated with the genetic agenda, putting it into the service of the maintenance of the social system.

Because the socio-cultural system itself is of crucial survival value to each individual within it, it is in the every person's self-interest to uphold it, at the same time as each is also, on the basis of the genetic agenda, pursuing reproductive fitness. An additional task of the symbolic system is therefore the reconciliation of the two kinds of sociality resulting from the two kinds of information transmission. Far from being an add-on or a decorative frill, the symbolic system is as central and major a building block of human life and society as are the genes.

It follows, then, from the basic premises that flow from Boyd and Richerson's theory of dual inheritance, that, in addition to providing for nutrition and biological reproduction, any viable human socio-cultural system must ensure the maintenance of its public arena and of the symbols that are displayed, exchanged, and inherited within it. This public arena is where the distinctively human phenomenon of "prestige" does its work. The giant yams of Pohnpei, to allude only to this one instance, are not an example of a frivolous or maladaptive runaway process, but rather key necessary factors in creating and sustaining the prestige system that makes possible an organized harmonious social life on the island. While it may not, from the biological point of view, be sensible to work on giant yams to the neglect of one's own garden (a luxury made possible by the benign environment), from the socio-cultural perspective it is indeed an eminently sensible thing to do.

The implication for anthropology more generally, I submit, is that evolutionary and biological theorists should take into account more than they typically do the critical work the cultural symbol system

CONCLUSION

does in human social life, while social and cultural anthropologists ought to consider the need facing any human socio-cultural system to sustain biological reproduction, and forge an effective compromise. The prospects of this happening are rather dim; but, ironically, only because the dynamics I have identified in this book are also operative in the society of anthropologists: for is it merely an accident that we have biological theorists who tend to dismiss the importance of the socio-cultural sphere competing with socio-cultural anthropologists who generally ignore the role of biology and genetics in human social life? Or is it rather the genetic program and the cultural program duking it out once again, in our own public arena, and preventing us from seeing that a complete account of human social life cannot be "either/or," but must be "both/and" to be true to our subject matter—the human social animal?

Notes

INTRODUCTION

1. The best overall exposition of dual inheritance theory is *Not By Genes Alone* (Richerson and Boyd 2006). For those interested in these authors' use of mathematical modeling, the primary source would be *Culture and the Evolutionary Process* (Boyd and Richerson 1985).

CHAPTER 1

1. The concept of binary opposition, which is operative in both DNA and language, is due to Roman Jakobson (Jakobson and Halle 1956). It was on this commonality that Lévi-Strauss's structural analysis might have built a synthetic paradigm, but did not.
2. The assumption that the natural environment, except perhaps for some higher animals, is without intention is distinctive of our own current cultural climate; it is more usual in other societies to believe that non-human agents, including ones we judge to be inanimate or ethereal, do indeed have human-like intentions and can and do influence human life on the basis of them. Whether this is our problem or theirs is a question that divides much anthropological thinking. Recent defenses of "animist" thought include those of Nurit Bird-David (1999), Philippe Descola (2013), Eduardo Viveiros-de Castro (1998), and Rane Willemslev (2011).
3. David L. Hull (1990) considers species as emergent phenomena in evolutionary theory and discusses the contributions of Ghiselin, Verba, Jablonka, and others who concur. Most

NOTES TO CHAPTER ONE

evolutionists do not grant to a species its own internal structure and organization, and its own ability to reproduce itself, other than via the summed reproduction of the individuals that compose it.

4. Discoveries in the emerging field of epigenetics as well as the recognition of such phenomena as regulator genes, gene repair, jumping genes, and so on of course complicate this overly simple picture.
5. The term "Darwinian" is often used by contemporary evolutionists to refer to a formulation that actually postdated Darwin himself by quite a bit. Darwin himself did not of course know about the mechanism of the genetic transmission of information, which had not yet been discovered and theorized; and his approach to human evolution, and especially toward the importance of moral "communities" (as opposed to populations) in creating selective advantage at the group level, as expounded in his *Descent of Man* ([1871] 2007), is much more compatible with typical socio-cultural approaches than is a good deal of contemporary "neo-Darwinism."
6. Whether such replication has to be exact to count as Darwinian has been a matter of dispute among evolutionists; see for example Henrich, Boyd, and Richerson (2008).
7. This does not apply to several evolutionist thinkers who do see adaptive value in religion; one such is David Sloan Wilson, whom I discuss later in this chapter.
8. The idea of group selection has begun to win more adherents in the last few years; see for example Samuel Bowles and Herbert Gintis's *A Cooperative Species: Human Reciprocity and Its Evolution* (2011). Dual inheritance approaches to cultural group selection may be found in Henrich (2004) and Boyd and Richerson (2010).
9. Wilson's equation of sin with individual selection and virtue with group selection is certainly a vast oversimplification. The ethical implications of multi-level selection, though interesting, are beyond the scope of the present work.
10. Durham explains his choice of the word "meme" rather than "symbol" for the unit of cultural inheritance and selection giving two main reasons: "First, meme is not a term with a priori connotations from common usage as is 'symbols.' . . . Second, meme has appropriate, if sparse, antecedent usage" (1991, 189). Anthropological objections to the term "meme" may be found in Maurice Bloch (2005a, 2005b) and Terrence Deacon (1999).
11. Other dual inheritance theorists too maintain that the variations upon which selection operates need not be seen as blind or random to allow a Darwinian process to occur (Henrich, Boyd, and Richerson 2007).
12. The literature on the relations between culture and evolution is by now quite extensive. I list here only a selection of (mainly) recent book-length works directly relevant to the issue (in addition to the ones discussed in the main text): Blute (2011), Boehm (1999, 2012), Bowles and Gintis (2013),

Cronk (1999), Distin (2011), Laland and Brown (2002), Linquist (2010), Mesoudi (2011), Prager (2012), Prinz (2012), Pulliam and Dunford (1980), Richerson and Christenson (2013), Ridley (2004), Slingerland (2008), Slingerland and Collard (2012), Sterelny (2012), Tattersall (2012), and Tomasello (1999, 2008). Barnard (2011) examines human evolution using concepts from the perspective of British social anthropology. Hallpike (1986) and Ingold (1986), also both British social anthropologists of very different theoretical hues, have contributed critiques of the use of evolution in anthropological theory; but these were written before the real flourishing of Darwinian theory in the study of culture; they therefore are more concerned with theories about the supposed evolution of social forms than with a more strictly Darwinian approach to socio-cultural phenomena. Recent attempts to integrate socio-cultural and Darwinian evolutionary ideas can be found in articles by Bloch and Sperber (2002) and Bloch (2013).

CHAPTER 2

1. See for example Carsten (2004), Collier and Yanagisako (1987), Schneider (1984), and Sahlins (2011, 2013) among many others. This position can itself be faulted for sometimes minimizing or even negating the role of actual reproduction and descent in the organization of society. Maurice Bloch (2013) has recently offered a critique of this position, arguing, as I do here, for a position that gives proper weight to both genetic and cultural factors in human social life.
2. I reject the idea that any cultural fact, like any natural fact, can be "anomalous," since by existing it demonstrates that it is as possible a way of doing things as any other. The duck-billed platypus, a mammal that lays eggs, is not in itself "anomalous" just because we can't fit it neatly into our categorization of living things; from its perspective it has as much right to live in its own way on God's green earth as any other creature. The quality of being an "anomaly" lies in our system of categories, not in the platypus.
3. On the Comanche and their empire, see for example Collier (1988), Fehrenbach (1974), Gwynne (2011), Hamalainen (2008), Kavanagh (1996), and Wallace and Hoebel (1952).

CHAPTER 3

1. Some noteworthy recent works addressing the evolution of the human capacity for culture and symbolization are Barnard (2012), Deacon (1997), Donald (1991), R. Klein (2009), Lewis-Williams (2002), Mithen (1998), Tattersall (2012), and Tomasello (2008).
2. Richerson and Boyd write: "the human cultural system arose as an adaptation, because it can evolve fancy adaptations to changing environments rather more swiftly than is possible by genes alone" (2006, 7).

NOTES TO CHAPTER THREE

3. By "public" I mean visible or otherwise perceptible *en plein air* and not inside anyone's head. The members of the "public" can range from two to millions, and many aspects of cultural information are particular only to certain groups, or even closely guarded secrets conveyed only to select others. They nonetheless remain "public" in the sense intended here, that is, "out there" and in principle subject to being perceived rather than only in someone's mind and hence inaccessible to others.
4. This is Sperber's position in his book *Rethinking Symbolism* (1975). Descola deconstructs this position at length in his recent work *Beyond Nature and Culture* (2013).
5. The study of symbols, signs, or representations—what they are, how they work, what they mean—is such a vast enterprise that it would take a book much longer than this one to do justice to it. Suffice it to say that it has been central in such scholarly fields as semiotics, whether of the Saussurian, Peircean, or Prague varieties; symbolic interactionism in sociology; multiple subfields of linguistics; symbolic anthropology, cognitive anthropology, psychoanalytic anthropology, and structuralism; cultural studies; and of course in philosophy, psychology, literary theory, art history, and many more besides. My too hasty treatment here will have to suffice for present purposes.
6. I have previously used this analogy in my article *"Civilization and Its Discontents* from an Anthropological Perspective, Eighty Years On" (2012).
7. See Solnit (2010) on the phenomenon of positive communal reactions arising in the wake of natural disasters.
8. Shared intention and "we consciousness" as basic to human sociality are central to the work of scholars such as Michael Tomasello (Tomasello and Carpenter 2007, Tomasello 2009); see also Axel Seemann, ed., *Joint Attention* (2011).
9. On the mystery of how and why humans cooperate, see such works as Bowles and Gintis (2011), Cronk and Leech (2013), Henrich and Henrich (2007), and Tomasello (2009), among others. Cooperation is of course only a mystery from a neo-Darwinian perspective; social and cultural anthropologists—as well as Darwin himself—have long taken it for granted that humans cooperate, the question being how such cooperation is organized, and how it is related to competition and conflict. Margaret Mead (1937) edited a useful and prescient volume of comparative ethnography bearing on the issue of competition and cooperation in different societies.
10. James Watson (1990) has proposed a distinction between "Lamarckian" and "Mendelian" ways of constructing identity, depending on whether shared being is thought of as acquired during one's lifetime by certain acts or procedures, or whether it is presumed to inhere in biogenetic descent, however it is understood culturally.
11. See Raymond Kelly (2000) and Otterbein (2004), for example.

12. The assumption among many evolutionists was that elementary bands of foragers could cooperate because their members are closely related genetically. This was rendered problematic by the work of Hill et al. (2011), which showed on the basis of the empirical study of the composition of contemporary foraging groups that the assumption of close genetic relatedness within such groups is incorrect.

CHAPTER 4

1. A recent book by Jesse Prinz, *Beyond Human Nature: How Culture and Experience Shape the Human Mind* (2012), argues forcefully in favor of the primacy of cultural over genetic influence on human thought and action.
2. A taste of this debate can be gotten from Ardener (1975), the various articles in MacCormack and Strathern (1980), and Valeri (1990).
3. The question of what "marriage" is has taken on new and highly charged political meanings in our contemporary cultural climate. In the anthropological literature it has traditionally been used to describe the institution involving opposite-sex couples whose sexual relations and resulting offspring are socially recognized and sanctioned by rite or custom. I use the term here in this sense as a nominal, not a substantive, definition, to employ a distinction made by Spiro (1966) in reference to the term "religion." It is an example of a term drawn from our own conceptual system that does not always map well onto the totality of ethnographic reality but that must be used for want of a better term. It is well-known, for example, that same-sex marriage between women is a traditional recognized practice in much of Eastern Africa (Njambi and O'Brien 2000).
4. Normative avoidance and/or prohibition of sexual relations, much less marriage, between full first-degree relatives is nearly universal but not quite. Besides the instances of sibling marriage by divine rulers such as those in the Inca Empire or Hawaii, the best-documented case of sanctioned sibling marriage throughout wider sectors of society is that of Roman Egypt (Hopkins 1980).

CHAPTER 5

1. It is often asserted that birdsong is a form of culture; for a recent discussion of birdsong from a dual inheritance perspective, see Darren E. Irwin (2012). But while it is true that birdsongs are key external systems of meaningful communication, they are not generative and do not contribute to the creation of diverse socio-cultural systems within a species like the ones humans have. But it makes no difference to my argument whether other creatures do or don't have culture, and if they do, more power to them. On animal culture, see Bonner (1983) and Laland et al. (2009).

NOTES TO CHAPTER FIVE

2. Silk et al. (2005); de Waal (2010).
3. I certainly do not mean to suggest that human societies are never hierarchical—most of human history argues against that proposition. If, however, social relations among contemporary nomadic foraging societies have anything to tell us about our prehistoric past, it is likely that elementary human societies were and are egalitarian, as both Christopher Boehm (1999, 2012) and Bruce Knauft (1993) have argued. Hierarchy arises with such factors as food production, settled communities, and greater population densities.
4. Flinn et al. (2012), in a study of testosterone levels among Dominican men, found that while testosterone was high in the presence of available women and of potential rival men, it was reduced in the presence of women already claimed by another man. Whether this is because humans have an evolved biological mechanism like the hamadryas one that underlies our cultural prohibitions, or whether it reflects the reaction of testosterone to symbolically recognized boundaries, is not a question to be decided here.
5. On the role of monogamous marriage in constraining male-male mating competition, see Henrich, Boyd, and Richerson (2012).
6. Actually, though Ulysses was indeed a crafty fellow, the scheme to hear the Sirens' song and live to tell about it was not his idea but was suggested to him by the sorceress Circe. Nor is it clear why he had to instruct his crew not to pay attention to his shouts if they had beeswax in their ears and could not hear him.
7. On the employment by centralized states of nonreproducing actors in key political roles to avoid the threat of nepotism, see Balch (1985).
8. I illustrate the historical supersession of genetic by symbolic inheritance in the transition from Israelite to Christian religion in my article "The Genealogy of Civilization" (1998).
9. On the wide variations among different contemporary hunter-gatherer societies, see Robert F. Kelly (1995).
10. There is a debate in the kinship literature about whether the Dravidian system of kin terminology differs enough from the Iroquois system to justify being considered as a separate category. This matter may safely be ignored for present purposes.
11. Morris is here reacting to a suggestion by Peter Gardner (1966) that among another South Indian group of foragers, the Paliyan, relationships are superficial and brittle.
12. I regard the question of the origin and inculcation of incest avoidance in humans as a problem not yet fully understood or explained. I have attempted a dual inheritance interpretation of it in my article "Incest Avoidance: Oedipal and Pre-Oedipal, Natural and Cultural" (2010). I currently tend to believe that genetic relationship too close for breeding is probably signaled by chemical information of some sort, as argued by Ingham and

Spain (2005); solving this problem is not central to the present argument.

CHAPTER 6

1. On adoption as exchange, see Lallemand (1993).
2. See note 12 to chapter 5.

CHAPTER 7

1. True unanimity is of course a rarity in human experience. But as Roy Rappaport (1979) argued in an important essay, whatever dissident or unique thoughts a person participating a ritual may harbor, the outward actions of all participants, unified by their common enactment of what Rappaport calls the "liturgy" of the ritual—an external script that they themselves have not created—creates a public appearance and shared perception of unanimity that transcends these difference and can create a group level reality. As I discuss further in chapter 10, there are also motivational rewards and punishments sanctioning and regulating conformity to group norms and practices.
2. Actually this question is no more valid than to ask how DNA, a tiny invisible molecule, could create a whole person. The power of encoded information, whether genetic or symbolic, lies not in its physical force but in its ability to structure, through its own formal organization, the action of living matter.
3. I have now made this claim several times, but of course it is no longer true in recent times thanks to new reproductive technologies. This means that the account I have given in this book applies to an evolutionary era in human history that is now over. The social, legal, and political ramifications of the de-coupling of genetic reproduction from copulation have given rise to a burgeoning subfield of anthropological research. Unfortunately, space precludes my pursuing the topic further here.
4. On ritualized homosexuality in men's societies in Melanesia, see Herdt (1984).
5. Envy may be an individual emotion, but I think it is one essential to the constitution of society. This is because it is the negative response of the collectivity to individuals who "cheat" and thus stand out by violating the social norm of equality. It is thus a prime aspect of social control in smaller and more egalitarian societies. Even in hierarchical societies, this dynamic is often operative within any given stratum of the hierarchy. As a Sherpa informant told me, a member of the goldsmith caste is not envious of higher caste Brahmins and Chhetris but only of other goldsmiths who are doing better than he is.
6. For a recent discussion of jealousy and envy, see Wurmser and Jarass (2007).

CHAPTER 8

1. It is an irony of patriliny that in order to reproduce, men have to appropriate a whole unrelated woman and deal with her kinsfolk, much as they might regret having to do so; in matrilineal societies, at least in principle, this is not the case: the Na matriline only requires a few minutes of a man's time and a few drops of his semen to reproduce itself. Given this fact, it becomes a good question why more societies haven't capitalized on this asymmetry. The Amazons were supposed to reproduce in this way, but they suffer from the problem of not having existed.
2. It should be noted that until very recently such groups in our own society were exclusively or predominantly male in membership.
3. As Melford Spiro (1971) has shown in the case of Burmese Buddhists, and as is true elsewhere in the Buddhist world, this "kammatic" (related to karma, or in Pali, *kamma*) goal is much more widely sought than is the more abstract goal of complete libration in *nibbana* (Pali for "nirvana").
4. On Sherpa monasticism, see von Fürer-Haimendorf (1964), Ortner (1989), and Paul (1990, 1996). For a monographic examination of Tibetan cultural symbolism in which these issues are explored further, see my book *The Tibetan Symbolic World* (1982).
5. It was on the basis of the ethnography of the Yaghan (Yamana) Indians and their near neighbors in Tierra del Fuego, the Selk'nam (Ona), that Joan Bamberger (1974) developed her foundational study of the widespread "myth of the matriarchy," that is, the myth that the secrets of the men's society were first owned by women and stolen from them by men, who must be on guard lest women steal them back. For critiques of Bamberger's thesis, see Deborah Gewertz (1988).
6. For the Hain ceremony, which is no longer performed since the Selk'nam as a society have become extinct, I have relied on the reconstruction by Anne Chapman (1982) based on the reports of a few surviving informants. For a full ethnography of the Selk'nam, see Gusinde (1931); an English translation of the latter work is available online via the Human Relations Area Files.
7. There is an intriguing parallel here to the well-known Rangda and Barong drama performed by the Balinese. Geertz (1973b), one of a long series of distinguished anthropologists to discuss this ritual, writes: "To ask, as I once did, a man who has *been* Rangda whether he thinks she is real is to leave oneself open to the suspicion of idiocy" (118). In the Balinese case, the terrifying witch Rangda has disciples who play with the corpse of a baby, while a pregnant woman is tormented by grave diggers; meanwhile the benign figure of the dragon Barong restores "life" to the participants who have gone into trance under Rangda's malignant influence.

CHAPTER 9

1. Wm. Blake Tyrell (1984) interprets the myth of the society of Amazons as a systematic set of reversals of the real gender relations in ancient Greece: "Classical Athens was a patriarchy, a social system organized along the lines of sexual asymmetry of male privilege. The cultural ideal, the adult male warrior, depended upon the imperative that boys become warriors and fathers, and girls become wives and mothers of sons. The genesis of the Amazon myth is the reversal of that imperative: Amazons go to war and refuse to become mothers of sons" (xiii–xiv). On the historical and archaeological foundations informing the myth, see Lyn Webster Wilde (2000).
2. The idea that children are only prevented from being serious antisocial menaces by their small size goes back to St. Augustine, as Sahlins (2008, 99) notes. Sahlins means his remark as a critique of what he sees as Western society's aberrant and wrong-headed view that children are little animals who require constraint by society. As the argument and evidence presented in this book show, however, this view is in fact neither exclusively Western nor entirely wrong-headed.
3. Has there really ever been a society in which women dominated in an asymmetrical way parallel to the way men dominate in "patriarchal" societies? This would require that women rule not only in fact but in ideological theory, holding the keys to public discourse, to justice, to access to the divine and so on. Janet Montgomery McGovern (1922), who studied the aboriginal people of Formosa (now Taiwan) early in the twentieth century, before either Japanese or Chinese acculturation had penetrated the interior, writes that prior to her fieldwork she had been skeptical about the existence of a true matriarchate. "But a land which is, as regards aboriginal inhabitants . . . sufficiently matripotestal to justify its being called a matriarchate, I have found" (28). Many of these aboriginal groups have not only an unusually strong matrilineal descent system but also are ruled by a "chief-priestess" or "queen," who one sees "borne upon the shoulders of her subjects, as she goes about the village, so that her feet may not touch the ground" (120–21). Furthermore, disputes in the community "are almost always settled either by the queen, or chief-priestess alone, or by a 'palaver' or meeting of remonstrance on the part of all the elderly women of the group" (126–27). Unfortunately, her book does not expand on these tantalizing observations.
4. The equation of "nature" with the cosmic order, surely one of its wide array of meanings in Western usage, is of course at odds with the idea of "nature" as "the wild" or "wilderness" as opposed to "the domesticated" or "the settlement." Both these opposite meanings of "nature" appear side by side in the Minangkabau symbolic system, as my further exposition

CHAPTER 10

1. Recent experimental research shows that small infants appear to have innate, and therefore probably evolved, tendencies toward prosociality and against antisocial action (Wynn 2007, 2009, Hamlin and Wynn 2011).
2. As therapeutic work in the tradition of Fairbairn (1996) and Winnicott (2005) has strongly suggested, disruptions of the early mother-child bond tend to result in later antisocial disorders of varying degrees of severity, thus indicating that the innate predisposition to prosociality must be bolstered and given shape in subsequent development beginning with the mother-child dyad. The evolutionary roots of human norm-generating and norm-obeying proclivities are explored in Chudek and Henrich (2011).
3. James Chisholm (1999) has offered a case for the integration of attachment theory into the study of human evolution. Weingarten and Chisholm (2009) argue for the role of attachment theory in the formation of cooperative groups via the institution of religion. The concept of "cooperative breeding" (Hrdy 2006) expands on ideas from attachment theory, emphasizing the importance of allo-mothering in the evolution of human sociality. The theories of both Bowlby and Goldschmidt were anticipated by Freud, who proposed, as a biological factor leading to the distinctively human tendency to experience inner conflict, "the long period of time during which the young of the human species is in a condition of helplessness and dependence. Its intra-uterine existence seems short in comparison with that of most animals, and it is sent into the world in a less finished state. As a result, the influence of the real external world upon it is intensified and an early differentiation between the ego and the id is promoted. Moreover, the dangers of the external world have a greater importance for it, so that the value of the object [the mother] which can alone protect it from them and take the place of the former intra-uterine life is enormously enhanced. The biological factor, then, establishes the early situations of danger and creates the need to be loved which will accompany the child through the rest of its life" (Freud 1926, 153–54). I have suggested (Paul 2012) that Freud's structural theory, as well as his drive theory, could be profitably reimagined from the vantage point of dual inheritance theory.
4. These considerations offer an understanding of the way in which a community can come to feel responsibility in a communal way for its children, who would otherwise represent the competitive pawns in the struggle for individual fitness by their progenitors: the community is en-

abled to experience collectively by means of public symbolism displayed in rituals the reciprocal nurturant feelings that originate, probably both phylogenetically and ontologically, in the mother-infant dyad. On the importance of allo-mothering in human societies, see Sarah Blaffer Hrdy's *Mothers and Others: The Evolutionary Origins of Mutual Understanding* (2009).
5. Classical experiments in ethology by Niko Tinbergen (1953) demonstrated that individual animals shown "supernormal" sign stimuli, or releasers, prefer these to the real evolved releasers provided by nature. Thus a baby gull with the fixed action pattern of pecking at the red spot on the mother's bill to obtain food will preferentially peck at a popsicle stick with a painted red spot that is brighter and bigger than the actual spots on real gulls' bills. I assume that cultural symbols redirect innate drives in very much the same way. This idea has been developed by Deirdre Barrett in her book *Supernormal Stimuli: How Primal Urges Overran Their Evolutionary Purpose* (2010).

CONCLUSION

1. The island was previously known in the literature as "Ponape"; Boyd and Richerson spell it as "Ponapae."
2. Boyd and Richerson's account of how a sensible product turns into a prestige item that does not seem to make sense from an adaptive point of view ties in well with the idea I proposed in note 5 of chapter 10 about symbols employing the overriding effect of supernormal releasers.
3. John Fischer (1957, 174) makes this comparison of the Pohnpeian Nahnken to the "talking chiefs" of Samoa.
4. It is thus quite different from a bribe, which is how some Western observers might see it. It does of course advance one's own private interest in rising in the hierarchy, but the gift itself is public and done on behalf of the community, rather than a covert payment for the return favor of advancement.
5. A good detailed ethnographic description of a case of the fission of a subsection on Pohnpei is Glenn Peterson's book *One Man Cannot Rule a Thousand: Fission in a Ponapean Chiefdom* (1982).
6. Laura Betzig (1986), who did ethnographic work in Micronesia, shows a correlation between "despotism" and multiple mating opportunities for those of high rank, such as those provided for in harems and similar institutions.
7. An admirable ethnographic evocation of social life as a drama enacted on the stage of the public arena is to be found in Thomas Gregor's book *Mehinaku: The Drama of Daily Life in a Brazilian Indian Village* (1977).

References

Abegglen, J.-J. 1984. *On Socialization in Hamadryas Baboons.* Cranbury, N.J.: Associated University Presses.

Alexander, Richard D. 1979. *Darwinism and Human Affairs.* Seattle: University of Washington Press.

Alpern, Stanley B. 1998. *Amazons of Black Sparta: The Warrior Women of Dahomey.* London: Hurst; New York: New York University Press.

Anderson, Benedict. 1991. *Imagined Communities: Reflections on the Origin and Spread of Nationalism.* London, New York: Verso.

Aoki, Masahiko. 2001. *Toward a Comparative Institutional Analysis.* Cambridge, Mass.: MIT Press.

Ardener, Edwin. 1975. "Belief and the Problem of Women" and "The Problem Revisited." In *Perceiving Women,* edited by Shirley Ardener, 1–28. London: Malaby.

Atkinson, Jane. 1990. "How Gender Makes a Difference in Wana Society." In *Power and Difference: Gender in Island Southeast Asia,* edited by Jane Monnig Atkinson and Shelly Errington, 59–93. Stanford, Calif.: Stanford University Press.

Atran, Scott. 2002. *In Gods We Trust: The Evolutionary Landscape of Religion.* Oxford and New York: Oxford University Press.

Ayalon, David. 1977. *Studies on the Mamluks of Egypt (1250–1517).* London: Variorum Press.

Ayalon, David. 1979. *The Mamluk Military Society: Collected Studies.* London: Variorum Press.

Baal, Jan van. 1966. *Dema: Description and Analysis of Marind-anim Culture (South New Guinea).* The Hague: Martinus Nijhoff.

Baal, Jan van. 1984. "The Dialectics of Sex in Marind-anim Culture." In *Ritualized Homosexuality in Melanesia,* edited by

REFERENCES

Gilbert H. Herdt, 128–66. Berkeley and Los Angeles: University of California Press.

Balch, Stephen H. 1985. "The Neutered Civil Servant: Eunuchs, Celibates, Abductees, and the Maintenance of Organizational Loyalty." *Journal of Social and Biological Structures* 8:322.

Bamberger, Joan. 1974. "The Myth of Matriarchy: Why Men Rule in Primitive Society." In *Woman, Culture, and Society,* edited by Michelle Zimbalist Rosaldo and Louise Lamphere, 263–80. Stanford, Calif.: Stanford University Press.

Bamford, Sandra. 2007. *Biology Unmoored: Melanesian Reflections on Life and Biotechnology.* Berkeley and Los Angeles: University of California Press.

Barnard, Alan. 2011. *Social Anthropology and Human Origins.* Cambridge: Cambridge University Press.

Barnard, Alan. 2012. *Genesis of Symbolic Thought.* Cambridge: Cambridge University Press.

Barrett, Deirdre. 2010. *Supernormal Stimuli: How Primal Urges Overran Their Evolutionary Purpose.* New York: W. W. Norton.

Bascom, William R. 1948. "Ponapean Prestige Economy." *Southwestern Journal of Anthropology* 4:211–21.

Bay, Edna. 1998. *Wives of the Leopard: Gender, Politics, and Culture in the Kingdom of Dahomey.* Charlottesville: University of Virginia Press.

Beckerman, Stephen, and Paul Valentine, eds. 2002. *Cultures of Multiple Fathers: The Theory and Practice of Partible Paternity in Lowland South America.* Gainesville: University of Florida Press.

Beidelman, T. O. 1966. "The Ox and Nuer Sacrifice: Some Freudian Hypotheses About Nuer Symbolism." *Man* 1:453–67.

Bennett, Lynn. 1983. *Dangerous Wives and Sacred Sisters: Social and Symbolic Roles of High Caste Women in Nepal.* New York: Columbia University Press.

Bettelheim, Bruno. 1954. *Symbolic Wounds: Puberty Rites and the Envious Male.* Glencoe, Ill.: Free Press.

Betzig, Laura. 1986. *Despotism and Differential Reproduction: A Darwinian View of History.* New York: Aldine.

Bird-David, Nurit. 1999. "'Animism' Revisited: Personhood, Environment, and Relational Epistemology." *Current Anthropology* 40:S67-S91.

Bloch, Maurice. 1986. *From Blessing to Violence: History and Ideology in the Circumcision Ritual of the Merina of Madagascar.* Cambridge: Cambridge University Press.

Bloch, Maurice. 1989. "Descent and Sources of Contradiction in Representations of Woman and Kinship." In Maurice Bloch, *Ritual, History and Power,* 152–65, London and Atlantic Highlands, N.J.: Athlone.

Bloch, Maurice. 2005a. "Where Did Anthropology Go? Or the Need for 'Human Nature.'" In *Essays on Cultural Transmission,* by Maurice Bloch, 1–19. Oxford and New York: Berg.

Bloch, Maurice. 2005b. "A Well-disposed Social Anthropologist's Problem with Memes." In *Essays on Cultural Transmission*, by Maurice Bloch, 87–101. Oxford and New York: Berg.

Bloch, Maurice. 2013. "Durkheimian Anthropology and Religion: Going In and Out of Each Other's Bodies." In *In and Out of Each Other's Bodies*, by Maurice Bloch, 1–21. Boulder, Colo., and London: Paradigm.

Bloch, Maurice, and S. Guggenheim. 1981. "Compadrazgo: Baptism and the Symbolism of a Second Birth." *Man* 16:376–86.

Bloch, Maurice, and Jonathan Parry. "Introduction: Death and the Regeneration of Life." In *Death and the Regeneration of Life*, edited by Maurice Bloch and Jonathan Perry, 1–44. Cambridge: Cambridge University Press.

Bloch, Maurice, and Dan Sperber. 2002. "Kinship and Evolved Psychological Dispositions." *Current Anthropology* 43:723–48.

Blute, Marion. 2011. *Darwinian Sociocultural Evolution: Solutions to Dilemmas in Cultural and Social Theory*. Cambridge: Cambridge University Press.

Boehm, Christopher. 1978. "Rational Pre-Selection from Hamadryas to *Homo Sapiens*: The Place of Decisions in Adaptive Process." *American Anthropologist* 80:265–96.

Boehm, Christopher. 1993. "Egalitarian Behavior and Reverse Dominance Hierarchy." *Current Anthropology* 34:227–54.

Boehm, Christopher. 1999. *Hierarchy in the Forest: The Evolution of Egalitarian Behavior*. Cambridge, Mass.: Harvard University Press.

Boehm, Christopher. 2012. *Moral Origins: The Evolution of Virtue, Altruism, and Shame*. New York: Basic Books.

Boggiani, Guido. n.d. *Os Caduveo*, with a preface and a historical and ethnographic study by G. A. Colini. Translated by Amadeu Amaral Jr. Sao Paulo: Livraria Martins Editora.

Bonner, John T. 1983. *The Evolution of Culture in Animals*. Princeton: Princeton University Press.

Bowlby, John. 1983. *Attachment and Loss*, vol. 1: *Attachment*. New York: Basic Books.

Bowles, Samuel, and Herbert Gintis. 2003. "Origins of Human Cooperation." In *Genetic and Cultural Evolution of Cooperation*, edited by Peter Hammerstein, 429–43. Cambridge, Mass.: MIT Press.

Bowles, Samuel, and Herbert Gintis. 2006. "The Evolutionary Basis of Collective Action." In *The Oxford Handbook of Political Economy*, edited by Barry G. Weingast and Donald A. Wittman, 951–70. Oxford: Oxford University Press.

Bowles, Samuel, and Herbert Gintis. 2013. *A Cooperative Species: Human Reciprocity and Its Evolution*. Princeton: Princeton University Press.

Boyd, Robert, and Peter J. Richerson. 1985. *Culture and the Evolutionary Process*. Chicago: University of Chicago Press.

Boyd, Robert, and Peter J. Richerson. 2005. *The Origin and Evolution of Cultures*. New York: Oxford University Press.

Boyer, Pascal. 2001. *Religion Explained: The Evolutionary Origins of Religious Thought*. New York: Basic Books.
Brady, Ivan, ed. 1976. *Transactions in Kinship: Adoption and Fosterage in Oceania*. Honolulu: University of Hawaii Press.
Brøgger, Jan. 1989. *Pre-Bureaucratic Europeans: A Study of a Portuguese Fishing Village*. Oslo: Norwegian University Press.
Brown, Donald E. 1991. *Human Universals*. Philadelphia: Temple University Press.
Cai Hua. 2001. *A Society without Fathers or Husbands: The Na of China*. Translated by Asti Hustvedt. New York: Zone Press.
Campbell, Donald T. 1975. "On the Conflicts between Biological and Social Evolution and Between Psychology and Moral Tradition." *American Psychologist* 30:1103–26.
Carroll, Vern, ed. 1970. *Adoption in Eastern Oceania*. Honolulu: University of Hawaii Press.
Carsten, Janet. 2004. *After Kinship*. Cambridge: Cambridge University Press.
Cavalli-Sforza, L. L., and M. Feldman. 1973. *Cultural Transmission and Evolution: A Quantitative Approach*. Princeton: Princeton University Press.
Chance, M.R.A. 1961. "The Nature and Special Features of the Instinctual Social Bond of Primates." In *The Social Life of Early Man*, edited by Sherwood L. Washburn, 17–33. London: Routledge.
Chance, M.R.A. 1962."Social Behavior and Primate Evolution." In *Culture and the Evolution of Man*, edited by M. F. Ashley Montague, 84–130. New York: Oxford University Press.
Chapais, Bernard. 2008. *Primeval Kinship: How Pair-Bonding Gave Birth to Human Society*. Cambridge, Mass.: Harvard University Press.
Chapman, Anne. 1982. *Drama and Power in a Hunting Society: The Selk'nam of Tierra del Fuego*. Cambridge: Cambridge University Press.
Chisholm, James S. 1999. *Death, Hope and Sex: Steps to an Evolutionary Ecology of Mind and Morality*. Cambridge: Cambridge University Press.
Chudek, Maciej, and Joseph Henrich. 2011. "Culture-gene Coevolution, Norm-Psychology, and the Emergence of Human Prosociality." *Trends in Cognitive Sciences* 20:1–9.
Clastres, Pierre. 1998. *Chronicle of the Guayaki Indians*. Translated by Paul Auster. New York: Zone Books.
Cloak, F. T. Jr. 1975. "Is a Human Ethology Possible?" *Human Ecology* 3:161–82.
Collier, Jane Fishburne. 1988. *Marriage and Inequality in Classless Societies*. Stanford, Calif.: Stanford University Press.
Collier, Jane F., and Michelle Z. Rosaldo. 1981. "Politics and Gender in Simple Societies." In *Sexual Meanings: The Cultural Construction of Gender and Sexuality*, 275–329. Cambridge: Cambridge University Press.
Collier, Jane Fishburne, and Sylvia Yanagisako, eds. 1987. *Gender and Kinship: Essays Toward a Unified Analysis*. Stanford, Calif.: Stanford University Press.

REFERENCES

Cosmides, Leda, John Tooby, and Jerome H. Barkow. 1992. "Introduction: Evolutionary Psychology and Conceptual Integration." In *The Adapted Mind: Evolutionary Psychology and the Generation of Culture*, edited by Jerome H. Barkow, Leda Cosmides, and John Tooby, 3–18. Oxford: Oxford University Press.

Crocker, Jon Christopher. 1985. *Vital Souls: Bororo Cosmology, Natural Symbolism, and Shamanism*. Tucson: University of Arizona Press.

Cronk, Lee. 1999. *That Complex Whole: Culture and the Evolution of Human Behavior*. Boulder, Colo.: Westview Press.

Cronk, Lee, and Beth L. Leech. 2013. *Meeting at Grand Central: Understanding the Social and Evolutionary Roots of Cooperation*. Princeton: Princeton University Press.

Daniélou, Jean. 1956. *The Bible and the Liturgy*. Notre Dame, Ind.: University of Notre Dame Press.

Darwin, Charles. 2007 [1871]. *The Descent of Man, and Selection in Relation to Sex: The Concise Edition*. New York: PLUME Books.

Dawkins, Richard. 1976. *The Selfish Gene*. Oxford: Oxford University Press.

Dawkins, Richard. 1982. *The Extended Phenotype: The Gene as the Unit of Selection*. Oxford: Oxford University Press.

Deacon, Terrence W. 1997. *The Symbolic Species: The Co-evolution of Language and the Brain*. New York: W. W. Norton.

Deacon, Terrence W. 1999. "The Trouble with Memes (And What to Do About It)." *The Semiotic Review of Books* 10, no. 3:1–3.

Deacon, Terrence W. 2012. *Incomplete Nature: How Mind Emerged from Matter*. New York: W. W. Norton.

Delaney, Carol. 1991. *The Seed and the Soil: Gender and Cosmology in Turkish Village Society*. Berkeley and Los Angeles: University of California Press.

Descola, Philippe. 1994. *In the Society of Nature: A Native Ecology in Amazonia*. Translated by Nora Scott. Cambridge: Cambridge University Press.

Descola, Philippe. 2013. *Beyond Nature and Culture*. Chicago: University of Chicago Press.

Descola, Philippe, and Gisli Palsson, eds. 1996. *Nature and Society: Anthropological Perspectives*. London and New York: Routledge.

de Waal, Frans. 1989. *Peacemaking among Primates*. Cambridge, Mass.: Harvard University Press.

de Waal, Frans. 1996. *Good Natured: The Origins of Right and Wrong in Humans and Other Animals*. Cambridge, Mass.: Harvard University Press.

de Waal, Frans. 1998. *Chimpanzee Politics: Power and Sex among Apes*. Baltimore: Johns Hopkins University Press.

de Waal, Frans. 2009. *The Age of Empathy: Nature's Plan for a Kinder Society*. New York: Harmony Books.

de Waal, Frans, and Frans Lanting. 1997. *Bonobo: The Forgotten Ape*. Berkeley and Los Angeles: University of California Press.

Distin, Kate. 2011. *Cultural Evolution*. Cambridge: Cambridge University Press.

REFERENCES

Donald, Malcolm. 1991. *Origins of the Modern Mind: Three Stages in the Evolution of Culture and Cognition.* Cambridge, Mass.: Harvard University Press.

Douglas, Mary. 1873. *Rules and Meanings: The Anthropology of Everyday Life.* London: Routledge.

Durkheim, Emile. 1964 [1895]. *The Rules of Sociological Method.* Translated by Sir George Edward Gordon Catlin. New York: Free Press.

Durkheim, Emile. 1965 [1915]. *The Elementary Forms of the Religious Life.* Translated by Joseph Ward Swain. New York: Free Press.

Durkheim, Emile. 1973 [1914]. "The Dualism of Human Nature and Its Social Conditions." In *Emile Durkheim: On Morality and Society*, edited by Robert N. Bellah, 149–63. Chicago: University of Chicago Press.

Durham, William H. 1991. *Coevolution: Genes, Culture, and Human Diversity.* Stanford, Calif.: Stanford University Press.

Edgerton, Robert B. 2000. *Warrior Women: The Amazons of Dahomey and the Nature of War.* Boulder, Colo.: Westview Press.

Elster, Jon. 1978. *Ulysses and the Sirens: Studies in Rationality and Irrationality.* Cambridge: Cambridge University Press.

Evans-Pritchard, E. E. 1940. *The Nuer.* Oxford: Clarendon Press.

Fairbairn, W.R.D. 1990. *Psychoanalytic Studies of the Personality.* London and New York: Routledge.

Fajans, Jane. 1997. *They Make Themselves: Work and Play among the Baining of Papua New Guinea.* Chicago: University of Chicago Press.

Fathauer, George. 1954. "The Structure and Causation of Mohave Warfare." *Southwestern Journal of Anthropology* 10:97–118.

Fehrenbach, T. R. 1974. *The Comanches: Destruction of a People.* New York: Alfred A. Knopf.

Fischer, John L. 1957. *The Eastern Carolines.* New Haven: Human Relations Area Files, Behavior Science Monographs.

Flinn, Mark V., David Ponzi, and Michael P. Muchlenbein. 2012. "Hormonal Mechanisms for Regulation of Aggression in Human Coalitions." *Human Nature* 23:68–88.

Forge, Anthony. 1966. "Art and Environment in the Sepik." *Proceedings of the Royal Anthropological Institute,* 23–31.

Forge, Anthony. 1973. "Style and Meaning in Sepik Art." In *Primitive Art and Society*, edited by Anthony Forge, 169–92. London: Oxford University Press.

Fortes, Meyer. 1969. *Kinship and the Social Order: The Legacy of Lewis Henry Morgan.* London: Routledge and Kegan Paul.

Fox, Richard G. 1969. "'Professional Primitives': Hunters and Gatherers in Nuclear South Asia." *Man in India* 49:139–60.

Fox, Robin. 1980. *The Red Lamp of Incest.* New York: E. P. Dutton.

Fox, Robin. 1991. "Comment" on "The Human Community as a Primate Society" by Lars Rodseth, Richard W. Wrangham, Alisa M. Harrigan, and Barbara Smuts. *Current Anthropology* 32:242–43.

Freud, Sigmund. 1955 [1926]. *Inhibitions, Symptoms, and Anxiety.* In *The Standard Edition of the Complete Psychological Works of Sigmund Freud,* translated and edited by James Strachey, Alix Strachey, and Alan Tyson, vol. 20:87–174. London: The Hogarth Press.

Fried, Morton. 1967. *The Evolution of Political Society.* New York: Random House.

Fürer-Haimendorf, Christoph von. 1964. *The Sherpas of Nepal: Buddhist Highlanders.* Berkeley and Los Angeles: University of California Press.

Gardner, Peter M. 1966. "Symmetric Respect and Memorate Knowledge: The Structure and Ecology of Individualistic Culture." *Southwestern Journal of Anthropology* 22:389–415.

Gardner, Peter M. 1972. "The Paliyans." In *Hunters and Gatherers Today,* edited by M. G. Bichieri, 404–47. New York: Holt, Rinehart and Winston.

Geertz, Clifford. 1973a. "The Growth of Culture and the Evolution of Mind." In *The Interpretation of Cultures,* by Clifford Geertz, 55–83. New York: Basic Books.

Geertz, Clifford. 1973b. "Religion as a Cultural System." In *The Interpretation of Cultures,* by Clifford Geertz, 87–125. New York: Basic Books.

Gennep, Arnold van. 1960 [1909]. *The Rites of Passage.* Translated by Monika B. Vizedom and Gabrielle L. Caffe. Chicago: University of Chicago Press.

Gewertz, Deborah, ed. 1988. *Myths of Matriarchy Reconsidered.* Sydney: Oceania Monographs.

Gillison, Gillian. 1980. "Images of Nature in Gimi Thought." In *Nature, Culture and Gender,* edited by Carol P. MacCormack and Marilyn Strathern, 143–73. Cambridge: Cambridge University Press.

Gilmore, David D. 2001. *Misogyny: The Male Malady.* Philadelphia: University of Pennsylvania Press.

Goldschmidt, Walter. 2006. *The Bridge to Humanity: How Affect Hunger Trumps the Selfish Gene.* New York and Oxford: Oxford University Press.

Goodale, Jane C. 1980. "Gender, Sexuality and Marriage: A Kaulong Model of Nature and Culture." In *Nature, Culture and Gender,* edited by Carol P. MacCormack and Marilyn Strathern, 119–42. Cambridge: Cambridge University Press.

Goodale, Jane C. 1983. "Siblings as Spouses: The Reproduction and Replacement of Kaulong Society." In *Siblingship in Oceania: Studies on the Meaning of Kin Relations,* edited by Mac Marshall, 275–305. Lanham, Md.: University Press of America.

Goody, Jack, ed. 1968. *Literature in Traditional Societies.* Cambridge: Cambridge University Press.

Goody, Jack. 1977. *The Domestication of the Savage Mind.* Cambridge: Cambridge University Press.

Goody, Jack. 1987. *The Interface between the Written and the Oral.* Cambridge: Cambridge University Press.

Goody, Jack. 2000. *The Power of the Written Tradition.* Washington, D.C.: Smithsonian Institution Press.

REFERENCES

Gregor, Thomas. 1977. *Mehinaku: The Drama of Everyday Life in a Brazilian Indian Village*. Chicago: University of Chicago Press.

Gregor, Thomas A., and Donald Tuzin. 1985. "The Anguish of Gender: Men's Cults and Moral Contradictions in Amazonia and Melanesia." In *Gender in Amazonia and Melanesia: An Exploration in the Comparative Method,* edited by Thomas A. Gregor and Donald Tuzin, 309–36. Berkeley and Los Angeles: University of California Press.

Greif, Avner. 2006. *Institutions and the Path to the Modern Economy: Lessons from Medieval Trade*. Cambridge: Cambridge University Press.

Gudeman, Stephen. 1972. "The Compadrazgo as a Reflection of the Spiritual and Natural Person." *Proceedings of the Royal Anthropological Institute*, 1972: 45–71.

Gusinde, Martin. 1931. *Die Feuerland-Indianer,* vol. 1: *Die Selk'nam*. Moedling bei Wien: Anthropos Verlag.

Gwynne, S. C. 2011. *Empire of the Summer Moon: Quanah Parker and the Rise and Fall of the Comanches, the Most Powerful Indian Tribe in American History*. New York: Scribner.

Hallpike, C. R. 1986. *The Principles of Social Evolution*. Oxford: Clarendon Press.

Hamalainen, Pekka. 2008. *The Comanche Empire*. New Haven: Yale University Press.

Hamilton, William D. 1964. "The Genetical Evolution of Social Behavior, I and II." *Journal of Theoretical Biology* 7:1–52.

Hamlin, J. Kiley, and Karen Wynn. 2011. "Young Infants Prefer Prosocial to Antisocial Others." *Cognitive Development* 26:30–39.

Hardin, Garrett. 1968. "The Tragedy of the Commons." *Science* 162:1243–48.

Harrison, Simon. 1992. *The Mask of War: Violence, Ritual, and the Self in Melanesia*. Manchester: Manchester University Press.

Henrich, Joseph. 2004. "Cultural Group Selection, Coevolutionary Processes, and Large-Scale Cooperation." *Journal of Economic Behavior and Organization* 53:3–35.

Henrich, Joseph, Robert Boyd, and Peter J. Richerson. 2008. "Five Misunderstandings about Cultural Evolution." *Human Nature* 19:119–37.

Henrich, Natalie, and Joseph Henrich. 2007. *Why Humans Cooperate: A Cultural and Evolutionary Explanation*. Oxford and New York: Oxford University Press.

Henry, Jules. 1941. *Jungle People: A Kaingang Tribe of the Highlands of Brazil*. New York: J. J. Augustin.

Herdt, Gilbert H. 1981. *Guardians of the Flutes: Idioms of Masculinity*. New York: McGraw-Hill.

Herdt, Gilbert H., ed. 1984. *Ritualized Homosexuality in Melanesia*. Berkeley and Los Angeles: University of California Press.

Hiatt, L. R. 1971. "Secret Pseudo-Procreation Rites among the Australian Aborigines." In *Anthropology in Oceania: Essays Presented to Ian Hogbin,* 77–88. San Francisco: Chandler Publishing Company.

Hill, Kim, and A. Magdalena Hurtado. 1996. *Ache Life History: The Ecology and Demography of a Foraging People*. New York: Aldine de Gruyter.
Hill, Kim, Robert S. Walker, Miran Bozecevic, James Eder, Thomas Headland, Barry Hewlett, A. Magdalena Hurtado, Frank Marlowe, Polly Weissner, and Brian Wood. 2011. "Co-residence Patterns in Hunter-Gatherer Societies Show Unique Human Social Structure." *Science* 331:1286–89.
Hopkins, Keith. 1980. "Brother-Sister Marriage in Roman Egypt." *Comparative Studies in Society and History* 22:305–54.
Horney, Karen. 1926. "The Flight from Womanhood." *International Journal of Psycho-Analysis* 7:324–39.
Hrdy, Sarah Blaffer. 1981. *The Woman That Never Evolved*. Cambridge, Mass.: Harvard University Press.
Hrdy, Sarah Blaffer. 2006. "The Evolutionary Context of Human Development." In *Attachment and Bonding: A New Synthesis*, edited by C. S Carter, L. Ahnert, K. E. Grossmann, S. B. Hrdy, M. E. Lamb, Stephen Porges, and N. Sachser, 9–32. Cambridge, Mass.: MIT Press.
Hrdy, Sarah Blaffer. 2009. *Mothers and Others: Female and Male Strategies of Reproduction*. Cambridge, Mass.: Belknap Press of Harvard University Press.
Hugh-Jones, Stephen. 1979. *The Palm and the Pleiades: Initiation and Cosmology in Northwest Amazonia*. Cambridge: Cambridge University Press.
Hull, David L. 1990. *Science as a Process: An Evolutionary Account of the Social and Conceptual Development of Science*. Chicago: University of Chicago Press.
Huntington, Richard, and Peter Metcalf. 1979. *Celebrations of Death: The Anthropology of Mortuary Ritual*. Cambridge: Cambridge University Press.
Ingham, John M. and David H. Spain. 2005. "Sensual Attachment and Incest Avoidance in Human Evolution and Child Development." *Journal of the Royal Anthropological Institute* 11:677–701.
Ingold, Tim. 1986. *Evolution and Social Life*. Cambridge: Cambridge University Press.
Jablonka, Eva, and Marion J. Lamb. 2005. *Evolution in Four Dimensions: Genetic, Epigenetic, Behavioral, and Symbolic Variation in the History of Life*. Cambridge, Mass.: MIT Press.
Jackson, Jean E. 1996. "Coping with the Dilemmas of Affinity and Female Sexuality: Male Rebirth in the Central Northwest Amazon." In *Denying Biology: Essays in Gender and Pseudo-Procreation*, edited by Warren Shapiro and Uli Linke, 89–127. Lanham, Md.: University Press of America.
Jakobson, Roman, and Morris Halle. 1956. *Fundamentals of Language*. The Hague: Mouton.
Jones, David E. 1997. *Woman Warriors: A History*. Washington, D.C., and London: Brassey's.
Kaberry, Phyllis M. 1940–41. "The Abelam Tribe, Sepik District, New Guinea: A Preliminary Report." *Oceania* 11:233–58, 345–67.
Kavanagh, Thomas W. 1996. *The Comanches: A History, 1706–1875*. Lincoln: University of Nebraska Press.

REFERENCES

Keesing, Roger M. 1982. *Kwaio Religion: The Living and the Dead in a Solomon Island Society.* New York: Columbia University Press.

Keller, Evelyn Fox. 2010. *The Mirage of a Space between Nature and Nurture.* Durham, N.C.: Duke University Press.

Kelly, Raymond C. 2000. *Warless Societies and the Origin of War.* Ann Arbor: University of Michigan Press.

Kelly, Robert L. 1995. *The Foraging Spectrum: Diversity in Hunter-Gatherer Lifeways.* Washington, D.C.: Smithsonian Institution Press.

Kitch, Sally. 1989. *Chaste Liberation: Celibacy and Female Cultural Status.* Urbana: University of Illinois Press.

Klein, Melanie. 1975. *Envy and Gratitude, and Other Works, 1946–63.* New York: Free Press.

Klein, Richard G. 2009. *The Human Career: Human Biological and Cultural Origins.* Chicago: University of Chicago Press.

Knauft, Bruce. 1991. "Violence and Sociality in Human Evolution." *Current Anthropology* 32:391–428.

Knauft, Bruce. 1993. *South Coast New Guinea Cultures: History, Comparison, Dialectic.* Cambridge: Cambridge University Press.

Kojève, Alexandre. 1969. *Introduction to the Reading of Hegel: Lectures on the Phenomenology of Spirit Assembled by Raymond Queneau.* Edited by Allen Bloom, translated from the French by James H. Nichols Jr. New York: Basic Books.

Kulick, Don. 1997. *Language Shift and Cultural Reproduction: Socialization, Self, and Syncretism in a Papua New Guinea Village.* Cambridge: Cambridge University Press.

Kummer, Hans. 1971. *Primate Societies: Group Techniques of Ecological Adaptation.* Chicago: Aldine.

La Barre, Weston. 1984. *Muelos: A Stone Age Superstition about Sexuality.* New York: Columbia University Press.

La Fontaine, J. S. 1981. "The Domestication of the Savage Male." *Man* 16:333–49.

La Fontaine, J. S. 1986. *Initiation.* Manchester: Manchester University Press.

Laland, Kevin N., and Gillian R. Brown. 2002. *Sense and Nonsense: Evolutionary Perspectives on Human Behavior.* Oxford: Oxford University Press.

Laland, Kevin N., and Bennett G. Galef, eds. 2009. *The Question of Animal Cultures.* Cambridge, Mass.: Harvard University Press.

Lallemand, Suzanne. 1993. *La circulation des enfants en société traditionnelle: prêt, don, échange.* Paris: Harmattan.

Langer, Susanne. 1957. *Philosophy in a New Key: A Study in the Symbolism of Reason, Rite and Art.* Cambridge, Mass.: Harvard University Press.

Leitao, David D. 1912. *The Pregnant Male as Myth and Metaphor in Classical Greek Literature.* Cambridge: Cambridge University Press.

Lévi-Strauss, Claude. 1967a. "Structural Analysis in Linguistics and in Anthropology." In *Structural Anthropology,* by Claude Lévi-Strauss, translated by

Claire Jacobson and Brooke Grundfest Schoepf, 29–53. Garden City, N.Y.: Anchor Books.
Lévi-Strauss, Claude. 1967b. "The Structural Study of Myth." In *Structural Anthropology*, by Claude Lévi-Strauss, translated by Claire Jacobson and Brooke Grundfest Schoepf, 202–28. Garden City, N.Y.: Anchor Books.
Lévi-Strauss, Claude. 1969. *The Elementary Structures of Kinship*. Translated by James Harle Bell, John Richard von Sturmer, and Rodney Needham, ed. Boston: Beacon Press.
Lévi-Strauss, Claude. 1973. *Tristes Tropiques*. Translated by John and Doreen Weightman. New York: Atheneum.
Lévi-Strauss, Claude. 1983. *The Raw and the Cooked*. Translated by John and Doreen Weightman. Chicago: University of Chicago Press.
Lewis-Williams, David. 2002. *The Mind in the Cave: Consciousness and the Origins of Art*. London: Thames and Hudson.
Linquist, Stefan Paul, ed. 2010. *The Evolution of Culture*. Farnham, England, and Burlington, Vt.: Ashgate Publishers.
Linton, Ralph. 1939. "Marquesan Culture." In *The Individual and His Society: The Psychodynamics of Primitive Social Organization*, by Abram Kardiner, with a foreword and two ethnological reports by Ralph Linton, 137–96. New York: Columbia University Press.
Lowie, Robert H. 1920. *Primitive Society*. New York: Horace Liveright.
Lumsden, Charles J., and Edward O. Wilson. 1981. *Genes, Mind and Culture: The Co-evolutionary Process*. Cambridge, Mass.: Harvard University Press.
MacCormack, Carol P. 1980. "Proto-adult to Adult: A Sherbro Transformation." In *Nature, Culture and Gender*, edited by Carol P. MacCormack and Marilyn Strathern, 95–118. Cambridge: Cambridge University Press.
MacCormack, Carol P., and Marilyn Strathern, eds. 1980. *Nature, Culture and Gender*. Cambridge: Cambridge University Press.
Malinowski, Bronislaw. 1929. *The Sexual Life of Savages: In North-western Melanesia*. New York: Halcyon House.
Marlowe, Frank W. 2010. *The Hadza: Hunter-Gatherers of Tanzania*. Berkeley and Los Angeles: University of California Press.
McElvaine, Robert S. 2001. *Eve's Seed: Biology, the Sexes, and the Course of History*. New York: McGraw-Hill.
McGovern, Janet B. Montgomery. 1922. *Among The Head-Hunters of Formosa*. London: T. Fisher Unwin Ltd.
McNeill, William H. 1995. *Keeping Together in Time: Dance and Drill in Human History*. Cambridge, Mass.: Harvard University Press.
Mead, Margaret, ed. 1937. *Cooperation and Competition among Primitive Peoples*. New York: McGraw-Hill.
Mead, Margaret. 1949. "Human Reproductivity." In *Male and Female: A Study of the Sexes in a Changing World*, by Margaret Mead, 223–41. New York: William Morrow and Company.

Mech, David L. 1970. *The Wolf: The Ecology and Behavior of an Endangered Species*. Garden City, N.Y.: The Natural History Press.

Meggitt, Mervyn. 1962. *Desert People: A Study of the Walbiri Aborigines of Central Australia*. Sydney: Angus and Robertson.

Meigs, Anna S. 1984. *Food, Sex, and Pollution: A New Guinea Religion*. New Brunswick, N.J.: Rutgers University Press.

Merlan, Francesca, and Alan Rumsey. 1991. *Ku Waru: Language and Segmentary Politics in the Nebilyer Valley, Papua New Guinea*. Cambridge: Cambridge University Press.

Mesoudi, Alex. 2011. *Cultural Evolution: How Darwinian Theory Can Explain Human Culture and Synthesize the Social Sciences*. Chicago: University of Chicago Press.

Michod, Richard E. 1999. "Individuality, Immortality, and Sex." In *Levels of Selection in Evolution*, edited by Laurent Keller, 53–74. Princeton: Princeton University Press.

Mithen, Steven. 1996. *The Prehistory of the Mind: A Search for the Origins of Art, Religion and Science*. London: Phoenix.

Mithen, Steven. 2005. *The Singing Neanderthals: The Origins of Music, Language, Mind and Body*. London: Weidenfeld and Nicolson.

Morris, Brian. 1982a. "Economy, Affinity and Inter-Cultural Pressure: Notes around Hill Pandaram Group Structure." *Man* 17:452–61.

Morris, Brian. 1982b. *Forest Traders: A Socio-Economic Study of the Hill Pandaram*. Atlantic Highlands, N.J.: The Athlone Press.

Murphy, Robert F. 1957. "Intergroup Hostility and Social Cohesion." *American Anthropologist* 59:1018–35.

Murphy, Robert F. 1960. *Headhunter's Heritage: Social and Economic Change among the Mundurucu Indians*. Berkeley and Los Angeles: University of California Press.

Murphy, Yolanda, and Robert F. Murphy. 1974. *Women of the Forest*. New York: Columbia University Press.

Myers, Fred R. 1986. *Pintupi Self, Pintupi Country: Sentiment, Place, and Politics among Western Australian Aborigines*. Washington, D.C.: Smithsonian Institution Press.

Nadelson, Leslee. 1981. "Pigs, Women, and the Men's House in Amazonia: An Analysis of Six Mundurucu Myths." In *Sexual Meanings: The Cultural Construction of Gender and Sexuality*, edited by Sherry B. Ortner and Harriet Whitehead, 240–72. Cambridge: Cambridge University Press.

Njambi, Wairimu Ngaruiya, and William E. O'Brien. 2000. "Revisiting 'Woman-Woman Marriage': Notes on Gikuyu Women." *NSWA Journal* 12:1–23.

Nordhoff, Charles. 1993 [1875]. *American Utopias*. Stockbridge, Mass.: Berkshire House Publishers.

Oberg, Kalervo. 1949. *The Terena and the Caduveo of Southern Mato Grosso, Brazil*. Washington, D.C.: Smithsonian Institution Publication 9.

Oberg, Kalervo. 1955. "Types of Social Structure among the Lowland Tribes of South and Central America." *American Anthropologist* 57:472–87.
O'Flaherty, Wendy Doniger. 1973. *Asceticism and Eroticism in the Mythology of Siva*. Oxford: Oxford University Press.
Onians, Richard Broxton. 1951. *The Origins of European Thought: About the Body, the Mind, the Soul, the World, Time, and Fate*. Cambridge: Cambridge University Press.
Ortner, Sherry B. 1974. "Is Female to Male as Nature is to Culture?" In *Woman, Culture, and Society*, edited by Michelle Zimbalist Rosaldo and Louise Lamphere, 67–87. Stanford, Calif.: Stanford University Press.
Ortner, Sherry B. 1989. *High Religion: A Cultural and Political History of Sherpa Monasticism*. Princeton: Princeton University Press.
Ottenberg, Simon. 1988. *Boyhood Rituals in an African Society: An Interpretation*. Seattle: University of Washington Press.
Otterbein, Keith. 2004. *How War Began*. College Station: Texas A & M Press.
Paige, Karen Ericksen, and Jeffery M. Paige. 1981. *The Politics of Reproductive Ritual*. Berkeley and Los Angeles: University of California Press.
Paul, Robert A. 1982. *The Tibetan Symbolic World: Psychoanalytic Explorations*. Chicago: University of Chicago Press.
Paul, Robert A. 1990. "Recruitment to Monasticism among the Sherpas." In *Personality and the Cultural Construction of Society*, edited by David K. Jordan and Marc J. Swartz, 254–74. Tuscaloosa: University of Alabama Press.
Paul, Robert A. 1996. "Symbolic Reproduction and Sherpa Monasticism." In *Denying Biology: Essays on Gender and Pseudo-Procreation*, edited by Warren Shapiro and Uli Linke, 52–73. Lanham, Md.: University Press of America.
Paul, Robert A. 1998. "The Genealogy of Civilization." *American Anthropologist* 100:387–96.
Paul, Robert A. 2010. "Incest Avoidance: Oedipal and Preoedipal, Natural and Cultural." *Journal of the American Psychoanalytic Association* 58:1087–112.
Paul, Robert A. 2012. "*Civilization and Its Discontents* in Anthropological Perspective: Eight Decades On." *Psychoanalytic Inquiry* 32:582–95.
Peletz, Michael G. 1996. *Reason and Passion: Representations of Gender in a Malay Society*. Berkeley and Los Angeles: University of California Press.
Petersen, Glenn. 1982. *One Man Cannot Rule a Thousand: Fission in a Ponapean Chiefdom*. Ann Arbor: University of Michigan Press.
Pollock, Donald K. 1985. "Looking for a Sister: Culina Relationship and Affinity." In *The Sibling Relationship in Lowland South America*, edited by Kenneth M. Kensinger, 8–15. Bennington, Vt.: Bennington College Press.
Prager, Mark. 2012. *Wired for Culture: Origins of the Human Social Mind*. New York: W. W. Norton.
Prinz, Jesse J. 2012. *Beyond Human Nature: How Culture and Experience Shape Our Lives*. London and New York: Allen Lane.
Pulliam, H. Ronald, and Christopher Dunford. 1980. *Programmed to Learn: An Essay on the Evolution of Culture*. New York: Columbia University Press.

REFERENCES

Radcliffe-Brown, A. R. 1933. *The Andaman Islanders.* Cambridge: Cambridge University Press.

Rank, Otto. 1952. *The Myth of the Birth of the Hero: A Psychological Interpretation of Mythology.* Translated by F. Robbins and Smith Ely Jelliffe. New York: R. Brunner.

Rappaport, Roy A. 1979. "The Obvious Aspects of Ritual." In *Ecology, Meaning, and Religion,* by Roy A. Rappaport, 173–221. Berkeley, Calif.: North Atlantic Books.

Rappaport, Roy. A. 1999. *Ritual and Religion in the Making of Humanity.* Cambridge: Cambridge University Press.

Richerson, Peter J., and Robert Boyd. 1978. "A Dual Inheritance Model of the Human Evolutionary Process: Basic Postulates and a Simple Model." *Journal of Social and Behavioral Structures* 1:127–54.

Richerson, Peter J., and Robert Boyd. 2006. *Not By Genes Alone: How Culture Transformed Human Evolution.* Chicago: University of Chicago Press.

Richerson, Peter J., and Morten H. Christensen, eds. 2013. *Cultural Evolution: Society, Technology, Language, and Religion.* Cambridge, Mass.: MIT Press.

Ridley, Matt. 2004. *Nature via Nurture: Genes, Experience, and What Makes Us Human.* New York and London: HarperCollins.

Riesenberg, Saul H. 1968. *The Native Polity of Ponape.* Washington, D.C.: Smithsonian Institution Press.

Rodseth, Lars, Richard W. Wrangham, Alisa M. Harrigan, and Barbara B. Smuts. 1991. "The Human Community as a Primate Society." *Current Anthropology* 32:221–54.

Róheim, Géza. 1925. *Australian Totemism.* London: Allen and Unwin.

Rubel, Paula G., and Abraham Rosman. 1978. *Your Own Pigs You May Not Eat: A Comparative Study of New Guinea Societies.* Chicago: University of Chicago Press.

Sahlins, Marshall. 1976a. *Culture and Practical Reason.* Chicago: University of Chicago Press.

Sahlins, Marshall. 1976b. *The Uses and Abuses of Biology: An Anthropological Critique of Sociobiology.* Ann Arbor: University of Michigan Press.

Sahlins, Marshall. 2008. *The Western Illusion of Human Nature.* Chicago: Prickly Paradigm Press.

Sahlins, Marshall. 2011. "What Kinship Is (Parts 1 and 2)." *Journal of the Royal Anthropological Institute* 17:2–19, 227–42.

Sahlins, Marshall. 2013. *What Kinship Is . . . And Is Not.* Chicago: University of Chicago Press.

Sanday, Peggy Reeves. 1981. *Female Power and Male Dominance: On the Origins of Sexual Inequality.* Cambridge: Cambridge University Press.

Sanday, Peggy Reeves. 2002. *Women at the Center: Life in a Modern Matriarchy.* Ithaca, N.Y.: Cornell University Press.

Saperstein, Marc. 1980. *Decoding the Rabbis: A Thirteenth-Century Commentary on the Aggadah.* Cambridge, Mass.: Harvard University Press.

Schneider, David M. 1955. "Abortion and Depopulation on a Pacific Island." In *Health, Culture and Community: Case Studies of Public Reactions to Health Programs*, edited by Benjamin D. Paul, 211–35. New York: Russell Sage Foundation.

Schneider, David M. 1980. *American Kinship: A Cultural Account*. Chicago: University of Chicago Press.

Schneider, David M. 1984. *A Critique of the Study of Kinship*. Ann Arbor: University of Michigan Press.

Seemann, Axel, ed. 2011. *Joint Attention: New Developments in Psychology, Philosophy of Mind, and Social Neuroscience*. Cambridge, Mass.: MIT Press.

Sered, Susan Starr. 1994. *Priestess, Mother, Sacred Sister: Religions Dominated by Women*. New York and Oxford: Oxford University Press.

Sered, Susan Starr. 1999. *Women of the Sacred Groves: Divine Priestesses of Okinawa*. New York and Oxford: Oxford University Press.

Shore, Bradd. 1981. "Sexuality and Gender in Samoa: Conceptions and Missed Conceptions." In *Sexual Meanings: The Cultural Construction of Gender and Sexuality*, edited by Sherry B. Ortner and Harriet Whitehead, 192–215. Cambridge: Cambridge University Press.

Silk, Joan B., Sarah F. Brosnan, Jennifer Vonk, Joseph Henrich, Daniel J. Povinelli, Amanda S. Richardson, Susan P. Lambeth, Jenny Mascaro, and Steven J. Schapiro. 2005. "Chimpanzees Are Indifferent to the Welfare of Unrelated Group Members." *Nature* 437:1357–59.

Slingerland, Edward. 2008. *What Science Offers the Humanities: Integrating Body and Culture*. Cambridge: Cambridge University Press.

Slingerland, Edward, and Mark Collard, eds. 2012. *Creating Consilience: Integrating the Sciences and the Humanities*. Oxford and New York: Oxford University Press.

Smuts, Barbara B., Dorothy L. Cheney, Robert M. Seyfarth, Richard W. Wrangham, and Thomas T. Struhsaker, eds. 1987. *Primate Societies*. Chicago: University of Chicago Press.

Sober, Elliott, and David Sloan Wilson. 1998. *Unto Others: The Evolution and Psychology of Unselfish Behavior*. Cambridge, Mass.: Harvard University Press.

Solnit, Rebecca. 2009. *A Paradise Built in Hell: The Extraordinary Communities that Arise in Disasters*. New York: Viking Press.

Sperber, Dan. 1974. *Rethinking Symbolism*. Translated by Alice L. Morton. Cambridge: Cambridge University Press.

Sperber, Dan. 1985a. "Anthropology and Psychology: Towards an Epidemiology of Representations." *Man* 20:73–89.

Sperber, Dan. 1985b. *On Anthropological Knowledge: Three Essays*. Cambridge: Cambridge University Press.

Sperber, Dan. 1996. *Explaining Culture: A Naturalistic Approach*. Oxford: Blackwell.

Spiro, Melford E. 1966. "Religion: Problems of Definition and Explanation." In *Anthropological Approaches to the Study of Religion*, edited by Michael Banton, 85–126. London: Tavistock.

REFERENCES

Spiro, Melford E. 1970. *Buddhism and Society: A Great Tradition and its Burmese Vicissitudes.* New York: Harper and Row.

Stanner, W.E.H. 1964. *On Aboriginal Religion.* Sydney: Oceania Monographs 11.

Sterelny, Kim. 2012. *The Evolved Apprentice: How Evolution Made Humans Unique.* Cambridge, Mass.: MIT Press.

Strathern, Marilyn. 1980. "No Nature, No Culture: The Hagen Case." In *Nature, Culture and Gender,* edited by Carol P. MacCormack and Marilyn Strathern, 174–222. Cambridge: Cambridge University Press.

Strathern, Marilyn. 1988. *The Gender of the Gift: Problems with Women and Problems with Society in Melanesia.* Berkeley and Los Angeles: University of California Press.

Struhsaker, Thomas T. 1975. *The Red Colobus Monkey.* Chicago: University of Chicago Press.

Swedell, Larissa. 2005. *Strategies of Sex and Survival in Hamadryas Baboons: Through a Female Lens.* Upper Saddle River, N.J.: Pearson.

Tanner, Nancy. 1974. "Matrifocality in Indonesia and Africa and among Black Americans." In *Woman, Culture, and Society,* edited by Michelle Zimbalist Rosaldo and Louise Lamphere, 129–56. Stanford, Calif.: Stanford University Press.

Tattersall, Ian. 2012. *Masters of the Planet: The Search for Our Human Origins.* New York: Palgrave Macmillan.

Thoden van Velsen, H.U.E., and W. Wetering. 1960. "Residence, Power Groups and Inter-Societal Aggression: An Enquiry into the Conditions Leading to Peacefulness within Non-Stratified Societies." *International Archives of Ethnography* 49:169–200.

Tinbergen, Niko. 1953. *The Herring Gull's World.* London: Collins.

Tomasello, Michael. 1999. *The Cultural Origins of Human Cognition.* Cambridge, Mass.: Harvard University Press.

Tomasello, Michael. 2008. *Origins of Human Communication.* Cambridge, Mass.: MIT Press.

Tomasello. Michael. 2009. *Why We Cooperate.* Cambridge, Mass.: MIT Press.

Tomasello, Michael, and M. Carpenter. 2007. "Shared Intentionality." *Developmental Science* 10:121–25.

Trivers, Robert L. 1971. "The Evolution of Reciprocal Altruism." *The Quarterly Review of Biology* 46:35–57.

Turner, Victor. 1967. *The Forest of Symbols: Aspects of Ndembu Ritual.* Ithaca, N.Y.: Cornell University Press.

Tuzin, Donald. 1995. "Art and Procreative Illusion in the Sepik: Comparing the Abelam and the Arapesh." *Oceania* 65:289–303.

Tuzin, Donald. 1997. *The Cassowary's Revenge: The Life and Death of Masculinity in a New Guinea Society.* Chicago: University of Chicago Press.

Tyrrell, Wm. Blake. 1984. *Amazons: A Study in Athenian Mythmaking.* Baltimore: Johns Hopkins University Press.

Valeri, Valerio. 1990. "Both Nature and Culture: Reflections on Menstrual and Parturitional Taboos in Huaulu [Seram]." In *Power and Difference: Gender in Island Southeast Asia,* edited by Jane Monnin Atkinson and Shelly Errington, 235–72. Stanford, Calif.: Stanford University Press.

Viveiros de Castro, Eduardo. 1998. "Cosmological Deixis and Amerindian Perspectivism." *Journal of the Royal Anthropological Institute* 4: 469–88.

Wallace, Ernest, and E. Adamson Hoebel. 1952. *The Comanches: Lords of the South Plains.* Norman: University of Oklahoma Press.

Ward, Martha. 2005. *Nest in the Wind: Adventures in Anthropology on a Tropical Island.* Long Grove, Ill.: Waveland Press.

Watson, James B. 1990. "Other People Do Other Things: Lamarckian Identities in Kainantu Subdistrict, Papua New Guinea." In *Cultural Identity and Ethnicity in the Pacific,* edited by Jocelyn Linnekin and Lin Poyer, 17–42. Honolulu: University of Hawaii Press.

Weingarten, Carol Popp, and James S. Chishom. 2009. "Attachment and Cooperation in Religious Groups." *Current Anthropology:* 50:759–85.

Wilde, Lyn Webster. 2000. *On the Trail of the Women Warriors: The Amazons in Myth and History.* New York: Thomas Dunne Publishers.

Willemslev, I. Rane. 2011. "Frazer Strikes Back from the Armchair: A New Search for the Animist Soul." *Journal of the Royal Anthropological Institute* 17:504–26.

Wilson, David Sloan. 2002. *Darwin's Cathedral: Evolution, Religion, and the Nature of Society.* Chicago: University of Chicago Press.

Wilson, David Sloan, and Edward O. Wilson. 2007. "Rethinking the Theoretical Foundation of Sociobiology." *The Quarterly Review of Biology* 82:327–48.

Wilson, Edward O. 1975. *Sociobiology: The New Synthesis.* Cambridge, Mass.: Harvard University Press.

Wilson, Edward O. 1979. *On Human Nature.* Cambridge, Mass.: Harvard University Press.

Wilson, Edward O. 2012. *The Social Conquest of Earth.* New York: Liveright.

Wilson, Peter. 1988. *The Domestication of the Human Species.* New Haven: Yale University Press.

Winnicott, Donald. 2005. *Playing and Reality.* London and New York: Routledge.

Woodburn, James. 1982. "Egalitarian Societies." *Man* 17:431–51.

Wurmser, Leon, and Heidrun Jarass, eds. 2008. *Jealousy and Envy: New Views about Two Powerful Feelings.* New York and London: The Analytic Press.

Wynn, Karen. 2007. "Some Innate Foundations of Social and Moral Cognition." In *The Innate Mind: Foundations and the Future,* edited by P. Carruthers, S. Laurence and S. Stich, 3:330–47. Oxford: Oxford University Press.

Wynn, Karen. 2009. Constraints on Natural Altruism. *British Journal of Psychology* 100:481–85.

Wynne-Edwards, V. C. 1962. *Animal Dispersion in Relation to Social Behaviour.* Edinburgh: Oliver and Boyd.

Index

Abegglen, J.-J., 122
Abelam, 197
Ache (Guayaki) people: child theft, 265; *kybairu* festival, 266–67; male rage, assuaging of, 265–66; physical touch, avoidance of, 264–66; *proaa* game, 267; tickling ceremony, as deregulated eroticism, 266–68; violence among, 265
affect hunger: attachment, idea of, 262; mother-infant dyad, 260–62; selfish gene, trumping of, 260–61
Africa, 119, 121, 173, 192
Agta people, 240
Alaska, 65
Alexander, Richard D., 34
alleles, 63–64, 77, 257
allomeme, 33, 35–36
al-Ma'mun, Caliph, 133
al-Mansur, Caliph, 225
al-Mulk, Nizam, 226
altruism, 25–26, 28, 84, 117; as reciprocal, 27
Amazon, 197
Amazonia, 198, 200
Amazons, 242, 318n1, 319n1
American Psychological Association, 23
Andaman Islanders, 80
Anderson, Benedict: imagined communities concept, 78
animal species, 127–28; cheaters and free riders in, 178–79
anthropology, 21, 27, 93–94; and recruitment, 210

Aquinas, Thomas, 204–5
artificial selection, 19–20; natural selection, 18
Asante, 241
asexual reproduction, 148, 168–70; sexual reproduction, conflict between, 163
Athena, 222
Athens (Greece), 319n1
Atkinson, Jane, 221
Atran, Scott, 22, 24
attachment theory: cooperative breeding, 320n3; mother-infant bond, 262; non-relatives, affiliation to, 279–80
Augustine, 319n2
Australia, 231; aboriginal groups of, 195–97; men's societies, 196; Pintupi of, 97–98
Avatip (Papua New Guinea), 102–4, 176–77, 186, 191, 210; collective identity in, 179; and fish, 180; male initiations in, 181; moiety system in, 180; reciprocity of men in, 180; ritual symbolism system of, 181–84; uterine kinship, 180; war-magic groups, 181; yam festival, 181; yams, significance of, 179–80
Ayalon, David, 225, 228–29
Azara, Felix de, 52

Bacchus, 222
Baining people, 101, 157, 163, 172–73; adoption, importance of to, 153–56; *aios* (bush

341

INDEX

Baining people (*continued*)
creatures) in, 100; children, attitude toward, 154–55; communal ritual in, and core values, 156; egalitarian ethos of, 153; equality between sexes in, 155; marriage, attitude toward, 154; nature and social, distinction between in, 155; sex, shameful attitude toward by, 154; social units in, 153; songs among, 81; as swidden cultivators, 153; work, as social activities, 154
Balinese, 318n7
Bamberger, Joan, 318n5
baptism, 203, 268; baptismal font, and womb, as symbolic replica of, 204; gender ideology in, 206; genetic and cultural relatedness, reflection of, 206; godparenthood, 205–6
Bara people, 218, 220; death, ritual handling of, 219
Barasana people, 197, 200
Barkow, Jerome H., 92
Barrett, Deirdre, 321n5
Bascom, William R., 290–91, 294, 301–2
Basque people, 207
Bateson, Gregory, 81
Battlestar Galactica (television series), 132
Baybars, 226
Beauvoir, Simone de, 93
behavioral learning, 15. *See also* social learning
Beidelman, T. O., 276–79, 281
Bemba, 207
Benin, 241. *See also* Dahomey
Bennett, Lynn, 165
ben Yedaiah, Isaac, 274
Betzig, Laura, 321n6
Beyond Human Nature (Prinz), 315n1
Beyond Nature and Culture (Descola), 314n4
binarism: binary opposition, 311n1 (chap. 1); of nature and nurture, 14
biology, 3–5, 9, 21, 39; life history theory, 106; and phenotypes, 73–74
Bird-David, Nurit, 311n2
birdsongs, 315n1 (chap. 5)
Bloch, Maurice, 169, 171, 206, 312n10, 313n1 (chap. 2)
Boas, Franz, 2, 91
Boehm, Christopher, 19, 33, 123, 138, 140, 316n3
Boggiani, Guido, 52

bonobos, 90, 120, 137, 281; harmonious group life of, 121. *See also* chimpanzees; gibbons; gorillas; great apes; hamadryas baboons; monkeys; orangutans; primates
Bororo people: *aroe* and *bope*, antithetical dyad of, 98–99; and *raka*, 99–100
Bowlby, John, 262–63, 279, 320n3
Bowles, Samuel, 83
Boyd, Robert, 5, 14, 35–39. *See also* Boyd and Richerson
Boyd and Richerson, 11, 13, 37–38, 39, 73, 87, 89, 93, 95, 104, 109, 256, 285, 296–97, 304, 309, 313n2; on culture, 64–65; runaway process, in cultural adaptation, 284, 286–90, 308; social instincts, 88; tribal instincts, 88, 90, 114, 120, 255, 299. *See also* Boyd, Robert; Richerson, Peter
Boyer, Pascal, 22, 24
Brady, Ivan, 155
Brahman and Chetri (*bahun-chhetri*) castes, 317n5; agnatic patriline, 163–66; ancestors, worship of, 163–65; asceticism and eroticism, reconciling of, 164–65; ascetic practice, and virility, 167–68; asexuality, 168–69; *bartaman* (male initiation ritual), 167; and celibacy, 164, 167–69; and *dharma*, 164; festival of Devali, 166; *gotra* (descent group), 165; householder path, 163–64; immorality, two paths of, 163–64; marriage ceremony in, 167; *rishis* (sages), 165; ritual purification, 164; *samsara* (material existence), 164; *sanyasin* (ascetic) in, 164; wedding ceremony, 168–69; women, role in, 165–69
Brazil, 42, 50–51, 98, 110, 243, 263
Bridge to Humanity, The (Goldschmidt), 260
Brøgger, Jan, 236–39
Brown, Donald, 92
Buddhism, 243; Tibetan Buddhism, 212

Cadmus, 180
Caduveo Indians, 50–52. *See also* Mbaya people
Cairo (Egypt), 226–27
Calvin, John: and Calvinism, 31
Campbell, Donald T., 14, 23–24, 27, 29, 35–36, 88–89, 97, 183, 256, 300
Candomble (Brazil), 243

canids: dominance hierarchies among, 118–19
Carroll, Vern, 155
Cassirer, Ernst, 69
Catholicism, 243
Caucasus, 259
Cavalli-Sforza, Luca, 14, 17–18, 20–21, 36, 284
celibacy, 164, 167–69, 211–12; monasticism, 214–15; women warriors, 242
Central Asia, 225
Chance, Michael, 127
Chapais, Bernard, 137–38, 146–47
Chapman, Anne, 318n6
childbirth symbolism, 268; childhood symbolism, and affective ties, 271
child purchase, 224. *See also* recruitment
chimpanzees, 15, 90–91, 281; cooperation among, 120; fraternal band in, 146; peacekeeping functions among, 120–21; scramble competition among, 120, 137. *See also* bonobos; gibbons; gorillas; great apes; hamadryas baboons; monkeys; orangutans; primates
China, 190
Chisholm, James, 320n3
Christianity, 243
Chudek, Maciej, 320n2
Circassians, 133–34, 226–28, 259
Clastres, Pierre, 265, 267
Cloak, F. T., Jr., 21
coevolutionary theory, 27
Coevolution (Durham), 31, 41
cognitive anthropology, 4
Collier, Jane, 152
Colombia, 195–96
Comanche Indians, 57; Comanche Empire, 54; infertility, and horseback riding, 55; women and children, capture and adoption of, 54–55
Comancheria, 54
communitas, 79
compadrazgo (godparenthood), 268; genetic and cultural relatedness, reflection of, 20; "hot hand," ambivalent meaning of, 205–6; men, association with, 206
competition, 82, 255, 300; cultural managing of, 128; deflection of, 185; lethal weapons, possession of, 137–38, 216; male rivalry, 302, 306; among males, 137, 216; mating opportunities, 118;

social harmony, 139; symbolic system, 302
Confucianism, 243
cooperation, 26, 255, 300, 314n9; as adaptive benefit, 118; in animals, 117–20; cheaters, 178; among chimpanzees and bonobos, 90–91; cultural evolution, 88; cultural kinship, 86; cultural symbol system, 279; and culture, 62, 78–79, 82–83; free riders, 178; human genetics, mutating, in direction of, 88; among proto-humans, 87–88, 90
Cosmides, Leda, 3, 92
Council of Trent, 205
couvade, 222
creation myths, 222
Crocker, Jon Christopher, 99–100
Cronk, Lee, 92
Culina Indians, 115, 128, 135, 148, 162, 190, 258; genetic kinship, 179; *madiha* (groups), organization of into, 110–14; siblingship system in, 111–14, 179; as *wemekute* (real relatives), 110–13
cultural anthropology, 32, 49–50
cultural channel: and genetic channel, 62, 74, 86, 93, 97, 105, 107–9, 115, 126–27, 129–30, 193; and kinship, 306; social space, transmission of in, 86–87, 89
cultural evolution: and cooperation, 88; genetic fitness, independent of, 20; as Lamarckian, 18–19, 39; population thinking, 36; runaway process in, traits of, 288
cultural genomes, 31
cultural information, 1–2, 18, 64; artifacts, information storage in, 65–66; cultural kinship, and cooperation, 86; genetic information, 7; horizontal transmission of, 20; as phenotype, 73–74; shared identity, 78; transmission of, into public world, 68, 86
cultural inheritance, 18, 65; genetic fitness, in opposition to, 59; symbols, and sexual reproduction, tension between, 45, 59; tribal instincts, 89
cultural knowledge: information transmission of, 66–68; public space, transmitting of into, 66–67; and symbols, 68, 70; transmission of, 70
cultural opposition: and maladaptation, 45, 47

343

INDEX

cultural program: genetic fitness, 55, 59; genetic program, 109, 134, 156; genetic reproduction, 105–6
cultural relativism, 91
cultural reproduction: cultural practices, 59; genetic fitness, opposition to, 53; genetic reproduction, tension between, 11–12, 59; and women, 12
cultural selection, 33, 35
cultural symbolism, 16–17, 31, 38, 129; genetic reproduction, 175; sexual reproduction, channeling of, 192–93
cultural symbol systems, 7, 33, 40, 199, 215–16, 279, 309–10; cooperative social groupings, 11; emotions, evoking of, 280; genetic information, 6–7; male competition, control of, 199; and sociality, 279; Texas Hold'em, analogy of, 75–76, 85
culture, 1, 14–15, 61, 87–88, 187, 283; adaptation, enhancement of, 20; adaptive advantages of, 62–63, 78–79, 91–92; assertions of, 32; biology, 27; competition, 82; cooperation, 62, 78–79, 82–83; cultural identity, cheaters and free riders, 179; cultural innovation, and social learning, 19; cultural norms, copulation and reproduction, restrictions of, 107–8; cultural system, containment and devaluation, 151; cultural variation, genetic one, as parallel to, 5–6; as defined, 64–65; evolution of, 62–63; genes, 41, 50; genes, as different from, 38–39, 308–9; genetic change, 32–33; genetic fitness, 62; genetic fitness, in opposition to, 45; genetics, 45; group solidarity, 81; and human nature, 91; as identical, 77–78; and imitation, 36; information storage, 64–65; kinship, 82; men, association with, 189, 199, 253–54; natural selection, 20; v. nature, 93–96, 99, 101, 189; phenotypes, 73–74; prosociality, and genetic changes, 62–63; as publicly shared, 77–78; and runaway process, 297; sensory perception, 11; sexual impulses, containing of, 150; shared identity, 78–79; shared symbolic markers, 89; social learning, 36, 64, 66; as symbol systems, 10–11, 38, 68–74, 77; Texas Hold'em, analogy of, 74–77;

variation in, and jukeboxes, analogy of, 92–93, 132; virus, comparison to, 22

Dahomey: women warriors in, 241–43. *See also* Benin
Daniélou, Jean, 204
Darwin, Charles, 17–18, 26, 37, 80, 83, 88, 312n5, 314n9; Darwinian evolution, 19, 21, 23, 283, 290
Dawkins, Richard, 3, 14; meme, concept of, 20–22; memetic replication, and genetic reproduction, 22–23
Deacon, Terrence, 141, 149, 312n10
Delaney, Carol, 208–9
Dema ceremonies, 45–46, 202
Descent of Man (Darwin), 312n5
Descola, Philippe, 311n2
DNA, 2, 6, 32–33, 67–68, 74, 85, 133, 136, 272, 285, 317n2 (chap. 7); copulation, transmission of, 11; genetic instructions, 7, 65; random mutation, 18; and RNA, 132
Dobzhansky, Theodosius, 3
dominance hierarchy, 121, 128; among canids, 118–19; in monkeys, 119; in Pohnpei, 296; weapons, acquiring of, 138
Dominican Republic, 316n4
Donald, Merlin: memetic culture theory, 80–81
Douglas, Mary, 98
Dravidian kinship model, 143–44, 148, 316n10
dual inheritance theory, 8–10, 12–14, 35–36, 39, 50, 59, 63, 73, 84, 86, 100, 182, 199, 203, 215, 230, 278–79, 283, 308, 320n3; binary opposition, 15; biological evolution, and social evolution, 24–25; cultural phenomena, and genetic fitness, 22–23; and culture, 15, 66; development of, 22–25, 31–32; display and performance, 305–6; external symbolic system, 272; male reproduction, 306–7; moiety system, 180; nature/culture concept, 95; origins of, 17; premises of, 6; and prestige, 297; "shared intention," 306; social consequences of, 109, 115; social learning, 66; symbols, role of in, 285–87; as term, 5, 14; as two channels, independent of each other, 61–62, 102

344

Durham, William, 14, 31–32, 45, 55–57, 59, 284, 312n10; allomemes, 33, 35–36; comparative modes, 34; interactive mode, 34; on opposition, 41–44; primary values, 33, 50; secondary values, 33–34, 50
Durkheim, Emile, 2, 27, 82, 97, 183, 267; on double consciousness, 29–30, 79–80; ritual, power of, 80; theory of religion, 30–31

East Africa, 243
Eastern Africa, 315n3
Eastern Carolines, 287
Edgerton, Robert B., 242
Egypt, 133–34, 224, 226–27, 242, 315n4
elementary societies, 192; asexual grouping principle in, 148; envy, 179; equality, 179; fraternal bands, as building block of, 146; hamadryas baboons, difference between, 148; kinship, 142; proto-humans, 87–88, 90
Elementary Structures of Kinship, The (Lévi-Strauss), 93
Elster, Jon, 131
envy, 179, 317n5
epigenesis studies, 15
Erer-Gota population, 125
Ethiopia, 121
ethnography, 8–10, 262–63
ethnoscience, 4
ethology, 262, 321n5 (chap. 10)
Eurasia, 231
Europe, 51, 235–36
Evans-Pritchard, E. E., 7–8, 275–76
evolution, 35; and genotype, 74; and natural selection, 5, 63; and speciation, 17
evolutionary biology, 283
evolutionary process, 255; intention, elimination of, 19; self-interest, transcending of, 260
evolutionary theory, 3–4, 6, 17, 286, 297; group selection, 24; inclusive fitness, 178; reciprocal altruism, 178
Evolution in Four Dimensions (Jablonka and Lamb), 15
evolutionism, 14
exogamous moieties, 180

Fairbairn, W. R. D., 320n2
Fajans, Jane, 81, 100, 153–55
Fathauer, George, 55
Federated States of Micronesia, 287. *See also* Micronesia
Fehrenbach, T. R., 55
Feldman, Marcus, 14, 17–18, 20–21, 284
female secret societies, 231, 243, 281
feminism, 93–94
Fischer, John, 321n3
Flinn, Mark V., 316n4
foraging groups, 315n12
Formosa, 319n3
Fortes, Meyer, 103
Fox, Richard, 147
Fox, Robin, 140, 142, 148
Freud, Sigmund, 69, 320n3
Fried, Morton, 140

Gapun (Papua New Guinea): *hed/kokir* (self), 96; *save* (self), 97
Gardner, Peter, 147, 316n11
Geertz, Clifford, 2, 4, 32–33, 71, 318n7
gender: male superiority, 192; men, and death, association with, 193, 216–18, 220; order, association with men, 218; vitality, association with women, 218–19; women, and birth, association with, 216–18
gender antagonism, 198–201
gender asymmetry, 231, 235, 253
Genes, Mind and Culture (Lumsden and Wilson), 27
genetic channel: cultural channel, 40, 93, 234; cultural modes of reproduction, conflict between, 173, 254; heterosexual intercourse, transmission of, 74, 86–87, 89–90
genetic evolution, 15; runaway processes in, 37
genetic fitness, 106; cultural beliefs, 47; cultural evolution, 20; cultural inheritance, 59; cultural program, 55, 59; cultural reproduction, 53; and culture, 45; dual inheritance theory, 22–23; genetic program, 130–32; social learning, 37; symbolic identity, 179
genetic information, 3, 285; and phenotypes, 272
genetic inheritance, 89, 104; cultural learning, 284; fitness, optimizing of, 106
genetic kinship, 179, 285

INDEX

genetic material: vertical transmission of, 20
genetic procreation: symbolic system, conflict between, 199
genetic program, 40, 190, 256; copulation, shame of, 105; cultural program, 310; genetic fitness, 130–32; selfish urges, 109, 183, 256–57
genetic reproduction, 187, 208, 300; cultural reproduction, tension between, 11–12; and culture, 20; mutations, 17–18
genetics, 5; and culture, 45; genetic code, 14–15
genetic variation, 6; and mutations, 63
genital surgery, 273; circumcision rituals, 274–75; social cohesion, 280
Germany, 287
Ghiselin, Michael, 311–12n3
gibbons, 119–21. *See also* bonobos; chimpanzees; gorillas; great apes; hamadryas baboons; monkeys; orangutans; primates
Gillson, Gillian, 94
Gilmore, David D., 216
Gimi people, 220
Gintis, Herbert, 83
Goldschmidt, Walter, 260–62, 279, 320n3
Goodale, Jane, 157, 160
Goody, Jack, 66
gorillas, 120–21. *See also* bonobos; chimpanzees; gibbons; great apes; hamadryas baboons; monkeys; orangutans; primates
great apes, 119. *See also* bonobos; chimpanzees; gibbons; gorillas; hamadryas baboons; monkeys; orangutans; primates
Greece, 180, 319n1; "one-seed theory," 222
Gregor, Thomas A., 198
group evolution, 84
Group Psychology and the Analysis of the Ego (Freud), 278
group selection, 24–25, 28, 85; natural selection, 30; religion, 30–31; selective advantage of, 29; shared symbolic markers, 89; social and tribal instincts, 88; strong reciprocity in, 83–84
Guana people, 51
Guaycuru people, 51–52
Gudeman, Stephen, 205–6
Guggenheim, S., 206

Hadza people, 80, 138–39, 142
hamadryas baboons, 11, 130, 135, 137, 142, 149, 316n4; adult males, ferocity of, 122–23; band, social unit of, 122–23, 128; clan, social unit of, 122–23; competition for reproductive sexuality among, 125–26; cooperation among, 123; as diurnal, 122; elementary human groups, difference between, 148; inhibition of, 128; kin-group exogamy, 140–41; male competition, 124–25; mate-guarding strategy of, 122; one male unit (OMU) of, 122–24, 128; pair-bonds, 148; social organization of, 121–27; troop, social unit of, 122–24, 128. *See also* bonobos; chimpanzees; gibbons; gorillas; great apes; monkeys; orangutans; primates
Hamilton, William, 26–27, 29
Hardin, Garrett, 28
Harlow, Harry, 262
Harrison, Simon, 102–4, 186
Hawaii, 315n4
Hegel, Georg Wilhelm Friedrich: Master/Slave dialectic of, 134–35
Henrich, Joseph, 320n2
Henry, Jules, 263, 264
Henza (Okinawa), 252; equality between sexes, 248; female priesthood in, 244–46; men's rituals, 246; *noro*, 246–47; *obon* (annual festival), 244
Herdt, Gilbert H., 180
Hiatt, L. R., 195–97, 202
Hill, Kim, 106–7, 265
Hill Pandaram society, 142, 162; camp life in, 146–47; children in, 145–46; Dravidian kinship system in, 148; dyadic relations in, 146; as endogamous, 143; equality among sexes in, 143; honey-gathering, key role in, 145, 147; incest avoidance, cooptation of in, 148–49; male cooperation in, 146–47; marriage in, 143–46; as nomadic, 144; pair-bonds, 147–48
Hindu ideology, 163–64, 172; ascetic practice, 167–68; women in, 165–66
hominids: impulse control, 127
Homo erectus: and hominid transition, 142
Homo sapiens, 6, 62–63, 116, 129; lasting mating pairs, evolutionary leap by, 121
Hopi Indians, 269–70

Horney, Karen, 194
Hrdy, Sarah Blaffer, 127
Hua people, 220–21
Hull, David L., 311–12n3
"Human Community as a Primate Society, The" (Rodseth et al.), 140
Human Relations Area Files, 8
human society, 1–2, 8–10, 102, 106; alliance model of, 140–41; antireproductive measures in, 106; collective social life, advantages of, 129; collectivity, as emergent organization, 85; constraints, effect of on, and social organizations, 149–50; cultures of, as symbolic worlds, 16; and death, 161; genetic program, 95; genetic reproduction and cultural reproduction, tension between, 90; groups, rivalry among males, 177; in-breeding, aversion to, 162; incest taboo, 93, 162; institutions, importance of to, 73; kin-group alliance, 140–41; lethal weapons, crucial role in, 239–40; male dominance, 93; male solidarity, 177; monogamy, 106, 138; prestige, 161–62; sexuality, shame of, 105, 154, 161, 259; shared human universals, 92; social instincts of, 88; social systems, and biological process, opposition to, 153; status, loss of, 161–62; tribal instincts, 88–89; as unified entity, 85
hunter-gatherers, 65, 84, 95, 135, 152, 191–92, 222, 264; contemporary examples, 136–40, 143, 147; dancing and singing activities, 80; hunting, as male activity, 240; weapons, acquiring of, 137–38; women, role of, 240
Hurtado, A. Magdalena, 106–7, 265

Iatmul people, 102
Ibn Khaldun, 226–27
Ibn Taghri Birdi, 228–29
Igbo people: of Akifpo Village-Group, 173; circumcision, 174; genital excision, 174; reproductive fertility among, 173–74; sexes, separation of, 175; sexual abstinence, 175; supernatural spirits, 174; women, control of, by men, 175
ihram, 209
Ilahita Arapesh people, 198–99; men's cult of, 194

Inca Empire, 315n4
inclusive fitness, 11, 26–28, 87, 116, 255, 285–86; altruism, 84; evolutionary theory, 178
India: Hill Pandaram society of, 142–48
Indonesia, 45, 221, 250, 253
initiation, 271, 280; Chinese Triad society, 269; formalized rituals, 268; of Hopi Indian Tribal males, 269–70; and males, 167, 181; in Nyoro society, 269; Poro people, 270; rituals, 232, 270; and secret societies, 269
intention, 33
Inuit people, 136
Ireland, 191
Islam, 226–27, 243, 250, 253; Mamluk regiments, 133; military slavery, 133, 225

Jablonka, Eva, 15, 311–12n3
Jackson, Jean, 195
Jakobson, Roman, 311n1 (chap. 1)
James, William, 183
Japan, 243, 246, 287
Judaism, 275
Ju/'hoansi (!Kung) people, 139

Kaingang people, 263, 267–68; solidarity in, 264
Kalahari, 65–66
Kamea people, 258
Kaulong people, 168, 170, 173, 175, 192; adoption in, 158; brother-sister pairs, 158–59, 162; celibacy among, 157–58, 168; forest plants, as ideal model for human life, 158, 160; marriage, as fatal condition, 157, 160, 168; marriage customs, 157–60; networking and trade, importance of to, 158; sexual relations, as shameful, 158–60; sexual reproduction among, 159–60, 162–63; women, attitude toward, 160
Keeping Together in Time (McNeill), 81
Keimw Sapwasap, 298
Keller, Evelyn Fox, 14–15
Kelly, Robert, 240
Keraki people, 220–21
Kinshasa (Democratic Republic of the Congo), 92–93
kinship, 9–10, 49, 50, 299; amity, 82; cultural channel, 88; and culture, 82;

kinship (*continued*)
elementary societies, 142; kin selection, 11, 78; mutuality of being, 82; shared identity, 79, 82; siblingship, metaphor of, 111; and symbols, 285; unity, 83
Kitch, Sally, 212
Klein, Melanie, 194
Knauft, Bruce, 316n3
Kojève, Alexandre, 134
Kolonia (Pohnpei), 287
Korea, 243
Kpelle secret society, 231
Kulick, Don, 96–97
Kummer, Hans, 122–26, 137
Kwaio people, 216, 220; *abu* (taboos), 217; *adalo*, 217; menstrual hut, 217–18
Kwakiutl people, 136, 294–95

La Fontaine, J. S., 206–7, 268–70, 274
Lamb, Marion, 15
Langer, Susanne: significant form, 16, 67–68, 70
language, 15–16, 257; singing and dancing, 80; as system of symbols, 7, 141
Latin America, 205
League of Nations, 287
Lévi-Strauss, Claude, 4, 51, 53, 93, 140, 180, 259, 311n1 (chap. 1)
Levy-Bruhl, Lucien, 82
linguistics, 21, 65
Linton, Ralph, 190
literacy, 66; literary studies, 21
Lowie, Robert H., 197
Lumsden, Charles J., 21, 27, 32

Macumba (Brazil), 243
Madagascar, 218. *See also* Malagasy Republic
Malagasy Republic, 169. *See also* Madagascar
Malaita (Solomon Islands), 216
Malaysia, 140, 222, 250
male aggression, 182; containing of, 200
male parturition. *See* male reproduction
male reproduction, 220–24, 306; procreative rituals, and "pseudo-procreation," 195; uterine rites, 196–97; yams, identification with, 307–8
male womb envy, 193–98, 221
Malinowski, Bronislaw, 259, 305

Mamluks, 150–51, 224, 229, 242–43; child purchase, 226; master-slave relationships, 225–28; Sultanate, 133–34
Manambu people, 102–3; male solidarity, 177; men's cult among, 176–77; natural fertility, 177; and war, 177
Marind-Anim people, 50–51, 53, 55, 57, 105–6, 150, 224, 300; ceremonial life of, 45; Dema ceremonies of, 45–46, 202; as headhunters, 45; kidnapping practices of, 47–48, 62; maladaptive practices of, 47–49; *otiv bombari*, practice of, 46, 49, 62, 74, 85; population decline, 46; ritual life of, 46; venereal disease epidemic, 46
Marlowe, Frank, 80
Marquesas Islands, 190
marriage, 167, 257, 282, 315n3; advantages of, 138–39; competitive males, effect on, 139; cultural inhibitions in, 130; in elementary social systems, 135–42; and monogamy, 138; sexual procreation, 281; sibling model, 258; and socialization, 281
matriarchy, 248–50, 318n5
matriliny, 271
Matrix, The (film), 132
Mbaya people, 50, 55, 57, 105–6, 150, 190, 300; abortion and infanticide, as common practice, 51–52; child capture, 53, 62, 224; genetic reproduction, as hostile to, 51; hierarchy of, 53; procreation, negative reaction toward, 51–53; recruitment, 224. *See also* Caduveo Indians
Mbuti people, 139
McElvaine, Robert S., 194
McGovern, Janet Montgomery, 319n3
McNeill, William, 81
Mead, Margaret, 106–7, 194, 314n9
Meeker, Michael, 208
Meggitt, Mervyn, 196
Mehinaku (Gregor), 321n7
Meigs, Anna, 220–21
Melanesia, 103, 156, 163, 177, 192, 291
memes, 20, 22, 33; as term, 21
memetics, 21
Mende secret society, 231
men's cults, 176, 178, 194; all-male groups, 12; as alternative kin group, 201; of Avatip, 179–81; collective identity of, 179; competition, deflection of,

184–85; competition, and rivalry, for women, 199–200; femaleness, evoking of, 180–81, 185; gender antagonism, 198–200; male solidarity, 177; moiety system, 180; natural fertility, 177; outsiders, as dangerous, 179; reciprocity, 180; ritual cycles, 197; rituals, control over, 181; ritual system symbolism, 184–85, 201; "self" in, as enlarged, 182; sexual reproduction, conflict over, 188–89, 192; shared identification, 182; socialization, mechanism of, 182; supernatural, role of, 184–86, 193; and war, 177; re: male re: cause of competition = rivalry for sexual access to women, 199–200; and yams, 179–80
men's secret societies, 231, 281
Merina people, 175, 190, 248; ancestral cult, 169–70; birth rituals, 171–72; circumcision ritual, 171, 180–81; descent groups, 169–70, 172, 203; funeral rituals, 170; gender differentiation and gender equality in, 234; women in, 170–72
Merlan, Francesca, 82
Mexico, 54
Micronesia, 287, 291, 295, 321n6. *See also* Federated States of Micronesia
Middle East, 192
Minangkabau people, 281, 319–20n4; *adat* (customary law or tradition), 250–53; *adat ibu* and *adat limbago*, distinction between, 251–52; core concept of, and motherhood, 250; evil forces in nature, control of, 251; fishing in, 249; Islam, strong influence of, 250, 253; as matriarchy, 248–50; men, roles of in, 249; social life of, 250–51; women, as decision makers in, 249, 252–53
Mithen, Steven, 80
Mnemosyne (goddess), 21
Mohave Indians, 55
moiety system, 99; and dual inheritance, 180
monasticism: and recruitment, 212. *See also* Sherpa monasticism
monkeys: dominance hierarchies among, 119. *See also* bonobos; chimpanzees; gibbons; gorillas; great apes; hamadryas baboons; orangutans; primates
Montague, Ashley, 194

Morris, Brian, 142, 144–47, 316n11
Mundurucu Indians, 85, 150, 216; captive-taking, 57–58; *Dajeboiši*, status of, 42–44, 59, 84; headhunting raids, 34, 45, 55–58; war, attitude toward, 55–56, 58–59; women's role, 57–58
Murdock, G. P., 8
Murinbata people, 196
Murphy, Robert, 55–58, 84
Murphy, Yolanda, 57–58
Myanmar (Burma), 243

Nadelson, Leslee, 192, 199
Na people, 190, 235, 318n1
natural selection, 6, 22, 28, 105, 129; artificial selection, 18–19; cultural inheritance, 18; and evolution, 63; group selection, 30; infantile immaturity, 260; mother-infant dyad, 260–61; runaway effect, 37; sexual selection, 37, 284; social learning, 36–37, 63–64
nature: v. culture, 93–96, 99, 101; v. Nature, 28, 135; women, association with, 189
nat worship, 243
Nazaré (Portugal), 245, 281; fishing industry in, 235–39; machismo, absence of in, 238; marriage in, 236; men, role of in, 237–39; Okinawa, parallels with, 247–48; social structure in, 237; taverns in, 239; women, dominance of in, 236–39
Ndembu people, 271
Negeri Sembilan (Malaysia), 222
neo-Darwinian evolutionary theory, 3, 25–27, 29, 33–35, 39
Nepal, 163, 212; Nepalis, 170, 172–73, 175, 240
New Britain, 81, 100, 153, 157
New Guinea, 197–98, 231, 308; cannibalism in, 42; Fore people of, 42; kinship, conceptualization of, 82; *kuru* (disease), spread of, 42; Marind-Anim, 45
Ngoni people, 206–7
Nigeria, 173
Nordhoff, Charles, 211
Northern Abelam people: exchange ceremonies, 307; and women, 307; yam cult, 307–8
Nuer people, 275; ox symbolism, 276–79
Nyoro people, 269

INDEX

Oberg, Kalervo, 52
Oceania, 155, 307
Odyssey, The (Homer), 131
O'Flaherty, Wendy Doniger, 164
Okinawa (Japan), 246, 249, 253, 281; female priests in, 243–44; *kami-sama*, 244; as matriarchy, 250; men, and death, association with, 247; Nazaré, parallels with, 247–48; women, and birth, association with, 247
One Man Cannot Rule a Thousand (Peterson), 321n5 (conc.)
On Human Nature (Wilson), 27
"On the Conflicts Between Biological and Social Evolution and Between Psychology and Moral Traditions" (Campbell), 23
On the Origin of Species (Darwin), 17
orangutans: as solitaries, 119. *See also* bonobos; chimpanzees; gibbons; gorillas; great apes; hamadryas baboons; monkeys; primates
Ortner, Sherry, 93–94, 189
Ottenberg, Simon, 173
Ottomans, 134, 226

Paliyan foragers, 316n11
Papio, 121
Papua New Guinea, 220, 258, 307; Gapun of, 96–97; Gimi of, 94; Manambu men, 176–77; Manambu people of, 102–3; men's cults in, 180; *simbuk* (war magician), 176
Paraguay, 264, 265
Parker, Cynthia Ann, 55
Parker, Quanah, 55
patriarchy, 319n1, 319n3
patriliny, 206, 209–10, 236, 318n1; agnatic principle, 207–8
Peirce, C. S., 69
Peru, 110
Phenomenology of Spirit (Hegel), 134
Philippines, 240
philosophy, 21
Pilaga Indians, 50
Pintupi people: *walytja* (relatedness), concept of, 97–98
Plains Indians groups, 54
Pleistocene era, 84; environment of evolutionary adaptation (EEA), 92

Pohnpei (Micronesia), 13, 39, 321n3; common good, commitment to, 302; competition in, 294–95, 301, 306; Displaying Together feasts, 295–96; dominance hierarchy, 296; dual chieftainship of, 291–94, 298, 303–4; feasts, as central feature of, 287, 293, 295–96, 301–4; giant yams of, 37–38, 286–91, 293–98, 301–2, 304–6, 308–9; giant yams, and male reproduction, 307; Great Work and Little Work in, 293–94; kava, 287, 295; matrilineal clans of, 292; pit breadfruits, 287, 291, 293, 302; political system of, 291–93; potlatch, comparison with, 294–96; prestige and genetic fitness in, 297; prestige system of, 287–90, 292–99, 301–6; public display, significance of, 304–6; runaway process in, 287–88, 299, 305, 309; Serving Together feasts, 295–96; sex, attitude toward in, 297–98; tribal society, 301, 308
Pollock, Donald, 111–12, 114
polygyny, 276, 277
Polynesia, 291
Ponape. *See* Pohnpei (Micronesia)
population biology: cheaters and defectors, 83
population genetics, 64, 283
population thinking: evolutionary model based on, 76–77
Poro secret society, 231, 243; Gbende spirit, 270; and *hinga*, 270
Portugal, 235
presentational symbols, 16
prestige system: competition, within community, 301; genetic fitness, 297; harmony, role of in, 305–6; male rivalry, 306; public display, 304, 309; system of symbols, 299–300
primates, 117. *See also* bonobos; chimpanzees; gibbons; gorillas; great apes; hamadrayas baboons; monkeys; orangutans; primates
Primeval Kinship (Chapais), 137
procreative symbolism, 270
prosociality, 12, 91, 103, 308; competition, offset by, 13; and culture, 62–63; genetic changes, 62–63; self-interest, offset by, 13
Pschav people, 259

350

psychoanalytic anthropology, 4
psychology, 23

Qipchak people, 134, 226
Qipchak Turks, 133

Radcliffe-Brown, A. R., 27, 80
Rank, Otto, 204
Rappaport, Roy, 81, 317n1; logoi, concept of, 70–71
Raw and the Cooked, The (Lévi-Strauss), 93
rebirth: symbolism of, 268–69
recruitment, 224, 226, 242, 257; men's group, 210–11; moiety organization, 211; monasticism, 212; and Shakers, 211. *See also* child purchase
Red Lamp of Incest, The (Fox), 140
religion, 22–23, 315n3; as adaptation, 30–31; group selection, 30–31; religious rituals, 203; ritual symbolism, 183–84; as social fact, 183; unity, of identification, 183
"Religion as a Cultural System" (Geertz), 32
reproduction: cultural and genetic forms of, 40–41; human society, 10; reproductive fitness, 296; sexual v. mystical, 190
Rethinking Symbolism (Sperber), 314n4
Richards, Audrey, 206
Richerson and Boyd. *See* Boyd and Richerson
Richerson, Peter, 5, 14, 35–39. *See also* Boyd and Richerson
Riesenberg, Saul, 291–93, 295–96
rituals, 202; dancing and singing, as communal activity, 80; and death, 219; and male cults, 197; performance and display, 306; ritual purification, 164
ritual symbolism, 181–84, 206, 270–71, 280–81
Rodseth, Lars, 140–42
Róheim, Géza, 196
Rondonopolis (Brazil), 98
Rosaldo, Michelle, 152
Rosman, Abraham, 307–8
Rubel, Paula, 307–8
Rumsey, Alan, 82

Sabbathday Lake (Maine), 211
Sahlins, Marshall, 10, 38, 82, 91, 94, 96, 111, 319n2

Sambia people, 180
Samoa, 321n3; *aga* (social norms) in, 101–2; *amio* (self/motivations) in, 101–2
Sanchez Labrador, Jose, 50–52
Sanday, Peggy, 248–49, 252
Sande secret society, 231, 243
Saussure, Ferdinand de, 69
schismogenesis, 4
Schneider, David, 10, 49, 258
secret societies, 231–33; initiation ceremonies, 269
self, 96–100
Selk'nam (Ona) people, 233; *Klokoten/Hain* ritual, 222–24; K'terrnen, 224; Olum, 223; and Xalpen (deity), 223–24
semiotics, 4, 21, 69, 314n5. *See also* symbols
Sered, Susan, 243–48
sexual reproduction, 188, 257; contraception, 191; control of, 200; cultural symbolism, 192–93; and marriage, 258
sexual segregation, 191–92
sexual selection: genetic realm, 297; natural selection, 37; prestige value, 288
Shakers, 212; recruitment, 211
shamanism, 243
shared symbolism: as identity marker, 286
Sherbro people, 232–34, 248, 270; children, animal-like appetites of, 235
Sherpa monasticism, 212–13; reincarnate lamas, institution of, 214–15
Sierra Leone, 231
Shore, Bradd, 101–3
Sober, Elliott, 30
social animals, 116
Social Conquest of Earth, The (Wilson), 29
sociality, 254, 279, 281–82; cooperation, as basis for, 83; tribal instincts, 255
socialization, 1, 11; mother-child dyad, 260–61, 281, 320n2; shared intention, 306; sibling pair, 281
social learning, 75, 132; adaptive information, and modeling, 63–64; conformity bias in, 88; cultural innovation, 19; and culture, 36, 64, 66; direct bias, 36; direct imitation, 68; frequency-dependent bias, 36–37; genetic fitness, 37; and imitation, 15, 65; indirect bias, 36–37; natural selection, 36; runaway effect, 37; as symbolic act, 70. *See also* behavioral learning

351

social organization, 9; disruptive potential, of adult males, 127; exogamy, as key to, 140
sociobiology, 4, 116
Sociobiology (Wilson), 27
socio-cultural anthropology, 3–5, 7, 14, 31, 39
socio-cultural systems, 1–2, 8–10, 40, 45, 69, 77, 91, 310; cooperation, 85; cultural symbols, dependence on, 285; culture and DNA, organization of, 11; external system of, 76; of genetic systems, 3; genetic variation, 6; of information systems, 3; and institutions, 73; male competition, minimizing and containing of, 128; and prestige, 304; prestige and public arena, 309; social structure and social organization, 72–73; symbolic systems in, as quasi-autonomous, 41; symbols, constructed by means of, 85
sociolinguistics, 4
sociology, 27
songs: as publicly shared "gossip," 81; and social harmony, 81–82
South America, 44, 50–52, 93, 136, 197, 222, 231
Spain, 287
Spanish-American War, 287
Sperber, Dan, 22, 24
Spiro, Melford, 315n3, 318n3
Sterelny, Kim, 84
Strathern, Marilyn, 102
structuralism, 4, 69
Sudan, 275
Sulawesi, 221
supernatural, 192–93; symbol systems, 187–88
Supernormal Stimuli (Barrett), 321n5
symbolic anthropology, 4, 32
symbolic culture, 132, 137
symbolic kinship group: and tribal societies, 285–86
symbolic reproduction, 12, 202–3, 229; procreation, model of, 233
symbolism, 284, 288; prosocial bonds, 141–42
symbols, 6, 12, 16, 74, 257, 271–72, 314n5; and culture, 10–11, 68, 70–72; cultural inheritance, 68; cultural instructions in, 7; dual inheritance theory, 285; as external, 77–78, 83, 85; and kinship, 285; performance and display, 306; semiotics of, 69; sensory world, medium of, 285; shared identity, 78; signs, distinction between, 69–70; social interaction, 73; symbolic code, and culture, 6. *See also* semiotics
symbol systems, 6, 40–41, 84, 115, 172–73; central task of, 302; and culture, 38, 77, 83; as external, 85–86; genetic program, 272–73; as maladaptive, 38; men, equation with, 189; mother-infant bond, 262; phenotypes, 73–74; prestige, 299–300; rebirth, 268; runaway process, 289–90; social cooperation, as making possible, 85–86; social interaction, 73; society, as central to, 309; socio-cultural groups, 7; and supernatural, 187–88

Taiwan. *See* Formosa
Tambaran, 198–99
Tanzania, 80; Hadza of, 138–39
Tasmania, 273
Tauna Awa people, 220
Temne secret society, 231
Tereno people, 51
Texas, 54–55
Thebes, 180
theory of evolution. *See* evolutionary theory
Thoma secret society, 231–32, 270; Gbana Bom figure, 233
Tierra del Fuego, 222
Tinbergen, Niko, 321n5 (chap. 10)
Toba Indians, 50
Tooby, John, 3, 92
"Tragedy of the Commons" (Hardin), 28
Transoxania, 225
tribal instincts, 11, 88–89, 114
Tristes Tropiques (Lévi-Strauss), 51
Trivers, Robert, 3, 26–27
Trobriand Islands, 259–60, 305
Tukanoan people, 195
Tunde, 199
Turkey: *ocah* (hearth), 208; patrilineal system in, 208–10
Turner, Victor, 270, 280
Tuzin, Donald, 194, 198–99

Tylor, Edward Burnett, 140
Tyrell, Wm. Blake, 319n1

Uganda, 269
United States, 54–55, 287

van Baal, Jan, 47–48
van de Kolk, Bessel, 47–48
van Gennep, Arnold, 268, 270
Veraguas Province (Panama), 205–6
Verba, Sidney, 311–12n3
Viveiros-de Castro, Eduardo, 311n2

Walbiri people, 136, 196
Wana people, 221
Ward, Martha, 297
Watson, James, 314n10
Weingarten, Carol Popp, 320n3
West Africa, 231, 243
Western Iatmul people, 184
Western Sumatra, 248
White, Leslie, 20, 140
Willemslev, Rane, 311n2
Wilson, David Sloan, 14, 30–32, 35, 39, 183, 285, 312n7

Wilson, Edward O., 3, 14, 21, 27, 32, 83–84, 92, 97, 285; earlier beliefs, recanting of, 28–29
Wilson, Peter, 304
Winnicott, Donald, 320n2
wolves, 118
women, 152–53, 165–69, 229–30, 235; biological reproduction, 206–7; childbirth, 206; cultural reproduction, 12; female priesthood, 243–45; reproduction and genetic inheritance, equation with, 189; sexual reproduction, association with, 253–54; sexual reproduction, male control over, 200; as warriors, 240–43
Woodburn, James, 139–40, 142
World War I, 287
World War II, 287
Wynne-Edwards, V. C., 25–26

Yap (Micronesia), 190

zar possession cult, 243
Zeno (bishop of Verona), 204–5
Zeus, 222

Lightning Source UK Ltd.
Milton Keynes UK
UKHW022053141021
392206UK00007B/351